Endocrine Sex Differentiation in Fish

Series on

Sex Differentiation in Fish

Endocrine Sex Differentiation in Fish

T.J. Pandian
Visiting Professor
CAS in Marine Biology
Annamalai University
Parangipettai 608502
Tamilnadu, India

CRC Press
Taylor & Francis Group
Boca Raton London New York

CRC Press is an imprint of the
Taylor & Francis Group, an **informa** business
A SCIENCE PUBLISHERS BOOK

CRC Press
Taylor & Francis Group
6000 Broken Sound Parkway NW, Suite 300
Boca Raton, FL 33487-2742

Cover illustration: Fishes share with other vertebrates many central mechanisms underlying sex differentiation. However, sex change is unique to all hermaphroditic and a few secondary gonochoric fishes. This figure illustrates the transdifferentiation commencing from behavior to morphological and finally to gonadal sex change from female to male in a grouper (from Kline RJ et al., 2011. Behavior, colour change and time required for sex inversion in the protogynous *Epinephelus adscensionis*. PLoS ONE, 6: e19576 doi: 10.1371/journalpone00119576).

Printed in the United States of America on acid-free paper

International Standard Book Number: 978-1-4665-7560-8 (Hardback)

Visit the Taylor & Francis Web site at
http://www.taylorandfrancis.com

CRC Press Web site at
http://www.crcpress.com

Science Publishers Web site at
http://www.scipub.net

Preface to the Series

Fishes constitute more than 50% of the living vertebrate species. The level of sexual plasticity and flexibility displayed by them has no parallel among other vertebrates. With most divergent expression of sex including many morphotypes within a sex, sexuality in fishes ranges from gonochorism to unisexualism and to an array of unbelievable patterns of functional hermaphroditism; their mode of reproduction also ranges from sexual to gynogenesis and hybridogenesis, each of these modes involves external fertilization and oviparity or internal fertilization and viviparity, which itself ranges from zygoparity to embryoparity. In fact, no other vertebrate group offers as much diversity as fishes to study the processes of sex determination and sex differentiation.

This book series is a continuum of my books on 'Sexuality in Fishes' and 'Sex Determination in Fish'. Information available on Sex Differentiation in Fish is rich and diverse, academically and economically so important that the original plan of bringing out a single volume has to be changed to three parts: 1. Genetic Differentiation, 2. Endocrine Differentiation and 3. Environment Induced Differentiation. An overview of a large number of publications concerning 'Sex Differentiation in Fish' suggests a transition from an experimental to analytical approach during the last decade. With the advent of molecular biology and arrival of more sophisticated tools and techniques, publications in this area are covered by increasing numbers and kinds of journals. Consequently there is a wide gap between molecular biologists including molecular endocrinologists publishing in high profile journals and fisheries scientists, who continue to publish in conventional fishery biology journals. A major objective of this book series is to bridge the widening gap between them. These days single author book series have become a rarity but an attempt has been made to comprehensively synthesize the available and relevant publications in this area.

October, 2011 T.J. Pandian
Madurai 625 014

Preface

Ever since estrogen was synthesized and became available for research in 1936, consistent endeavors have been made to understand the endocrine mechanism of sex differentiation in fishes. Hence, endocrine sex differentiation is the earliest to receive scientific attention among the genetic, endocrine—behavioral and environmental mechanisms of sex differentiation. Not surprisingly, a voluminous body of literature and a large number of books are available. However, even the most recent book on this topic has not covered viviparity, a phenomenon that occurs in 577 (about 20%) fish species and unisexualism. This book represents a comprehensive account covering relevant and related areas of endocrine and behavioral mechanisms of differentiation and transdifferentiation of sex in fish.

Commencing from neuroendocrine regulation, the book explores ovarian and testicular differentiation in gonochores and unisexuals, differentiation of morphotypes and transdifferentiation in sex changers, especially hermaphrodites. The book is perhaps the first to focus on factors that regulate the quantum of yolk accumulation in eggs. Vitellogenesis accumulates widely varying quantum of yolk in the piscine oocyte, whose size ranges form 350 μm (alecithal) to 20 mm (lecithal). Interestingly, for every 0.17 mm increase in egg size, an additional vertebra is formed within North American 71 cyprinid species. The book also brings to focus the obligate need for estradiol, a female hormone in the male to renew Spermatogonial Stem Cells (SSCs). From an analysis of the publications reporting the transplantation of Premordial Germ Cells (PGCs), Oogonial Stem Cells (OSCs) and SSCs, it confirms that differentiation in primary and secondary gonochores means the loss of bisexual potency of germ cells supporting somatic cells (GCSSCs). But the hermaphrodites retain bisexual potency of both PGCs and GCSSCs.

Unisexualism is an enigmatic phenomenon among fishes, although it occurs in less than a dozen species of hybrid origin. To solve the problem of hybrid sterility, the female unisexuals produce unreduced eggs through premeiotic endomitosis. To avoid the onslaught of mutational meltdown, they have devised the following escape mechanisms: (i) paternal leakage, (ii) paternal genome replacement and/or (iii) paternal genome addition.

The book is perhaps one of the few, that projects the endocrine and behavioral role in sex differentiation among gonochores and transdifferentiation in sex changers. Classifying and describing the Alternate Mating Strategy (AMS) prevalent among fishes, it shows that the subordinates have minimized the disadvantages of their status by adopting AMS that makes the best of a bad situation. It conclusively shows that the changes in morphotypes are intrasexual in gonochores but intersexual in hermaphrodites.

This book recognizes the important role played by behavior in transdifferentiation among hermaphrodites and sex changing gonochores as well as in induction of social status among gonochores that displays AMS. Hitherto, the removal of harem master/mistress is known to induce dramatic sex change among one of the harem members. While collecting the widely scattered information, the book brings to focus that housing in isolation of either one or more individuals of the same sex also induces sex change rather slowly in one of the hermaphrodites. Similar isolation of subordinates or dominants results in ascendance of one of the two subordinates or descendance of one of the dominants in social status of gonochorics that display AMS. Hence social mediation plays a major role in induction of sex change in hermaphrodites and social status in gonochoric morphotypes.

Despite wide diversity in the structure of reproductive organs and system of hermaphrodites, the endocrine-behavior mechanism of sex differentiation remains common and thereby confirms the origin of hermaphrodites from fishes like *Macropodus opercularis* which changes sex more than once in either direction. Sex change and hermaphroditism are unique but an intriguing phenomenon, especially from the point of genetic basis of sex determination. Raveling through the maze of gonochore-hermaphrodite continuum, the book shows for the first time that all hermaphroditic fishes are potential hermaphrodites but functionally gonochores. Consequently, all of them are capable of changing sex once in a single direction or more than once in either direction. Only the frequency is minimal in simultaneous hermaphrodites and maximal in serial hermaphrodites.

To have a better insight into the endocrine mechanism of sex differentiation and to produce monosex progenies, steroids and other chemicals including from herba have been used to induce sex reversal in commercially important fishes. Considering the negative effects of these inducers on survival, growth and especially reproductive performance, the need for the production of YY and WW brooders is shown. But the steroid residuals in marketable fish or in the aquatic systems are too little to harm humans and waters.

The book is a comprehensive synthesis of over 1200 relevant publications collected from a widely scattered 248 journals and other 80 literature sources. Incisive analysis of the cited literature has culminated in proposing many new hypotheses.

September, 2012
Madurai 625014

T.J. Pandian

Acknowledgements

It is with great pleasure that I wish to thank Drs K P Joy and E Vivekanandan for critically reading parts of the manuscript of this book and offering valuable suggestions. Special thanks are due to Prof T Balasubramanian for encouragement and support. I am grateful to Drs R Jeyabaskaran, A. Gopalakrishnan R Koteeswaran, G Kumaresan, P Murugesan, AG Ponniah and M Rajasekhar for useful discussions. I gratefully appreciate my students Drs T George, S Kavumpurath, S Kirankumar (UK), R Koteeswaran, SG Sheela (USA) and S Varadaraj (USA), whose publications from my laboratory indeed have helped me to launch this book. The Central Marine Fisheries Research Institute (Kochi) and Central Institute for Fisheries Education (Mumbai) provide the best libraries for fishery science and excellent service to visitors. I wish to record my sincere thanks to the Department of Science and Technology, New Delhi and Annamalai University for their support. The manuscript of the book was prepared by Ms S Akila B.Sc., B.L. and I wish to thank her for patience, cooperation and excellent work.

To reproduce figures and tables from published domain, I need to thank many. Firstly, I wish to thank my guru Prof O Kinne for providing me copies of his journals. I wish to gratefully appreciate the open door policy and record my sincere thanks to The American Society of Ichthyologists and Herpetalogists (*Copeia*), The Fishery Society of the British Isle/John Wiley & Sons Ltd (*Journal of Fish Biology*) and The Ichthyological Society of Fishery Science (*Ichthyological Research*). I am equally thankful to the editors/publishers of Cambridge University Press, CRC Press and Science Publishers for issuing permissions to reproduce figures and tables. I welcome and gratefully appreciate the new policy of *BMG Genomics, PLoS ONE, Functional Ecology, Integrative Comparative Biology* and *Ohio Journal of Science*. Accordingly, I thank very sincerely my fellow scientists Drs E Bonnet, RM Dawley, AL Filby, O Kah, RJ Kline, RF Oliveira, BJA Pollux, DN Reznick and T Shibuno. I also record my sincere thanks to Drs T Ajithkumar, T Balasumbramanian and R Koteeswaran for their consent to include their unpublished data. For advancing our knowledge in this area through their rich contributions, I thank all my fellow scientists, whose publications are cited in this book.

T.J. Pandian

Contents

1

Introduction

Introduction

Constituting more than 50% of living vertebrates, fishes are known for sexual plasticity and genetic diversity. The root causes for genetic diversity have been traced to the following (Kah and Dufour, 2011): 1. Consequent to the third genome doubling after the emergence of teleost fishes (see Pandian, 2011), the availability of duplicate sequences for evolution of new functions (Steinke et al., 2006), 2. The relatively higher frequency of gene linkage distruption leading to faster evolution of protein-gene coding sequences (Ravi and Venkatesh, 2008), and rapid radiation and speciation (e.g., the African cichlids, see Pandian, 2012) and 3. Remarkable capacity to innovate unique reproductive strategies due to sexual plasticity (see Pandian, 2010) at both the brain (e.g., *cyp19a1b*) and gonad (*cyp19a1a*, see also Pandian, 2012). Despite the enormity of genetic diversity and reproductive strategy, the regulatory mechanisms of endocrine sex differentiation have remained highly conserved (Oliveira and Goncalves, 2008). With the advent of molecular biology, molecular endocrinology has provided considerable insights into the regulatory mechanisms of endocrine sex differentiation. Nevertheless, fish farming and monosex aquaculture encounter many problems. Captive rearing of fish in farms has led to reproductive dysfunction; males produce less and low quality milt (Rurangwa et al., 2004), and females fail to ovulate and/or spawn (Zohar and Mylonas, 2001). The objective of this book is to comprehensively synthesize relevant information on the role of endocrines and behavior on sex differentiation in fishes and explore the possibilities of providing a remedy to the problems encountered by aquaculturists.

1.1 Sex determination

In fishes, the cytological sex determining mechanisms of hereditary transmission by chromosomes as carriers of genes have received much attention. Yet, of the 30,000 and odd fish species, cytogenetic and genetic analyses so far made are limited to about 1,700 species. The presence of sex chromosomes and heterogamety has been recognized only in 264 fish species belonging to 160 families. Of them, 67% and 33% are male (XX/XY) and female (ZZ/ZW) heterogametics, respectively (Pandian, 2011). High androgen ratios in male heterogametic vertebrates may be the consequence of Y-linked gene, whose product is an inhibitor of *cyp19*, whereas the W chromosome may carry a gene inducer of *cyp19* in female-heterogametics (Bogart, 1987). Incidentally, 9% of the 264 fish species, in which sex chromosomes have been recognized, employ multiple sex chromosomes (Pandian, 2011). The presence of male and female heterogametic sex determination systems as well as simple and multiple sex chromosomes clearly indicates the absence of a common, universal sex determining chromosome system in fishes, unlike in mammals.

Le Page et al. (2010) have indicated that in mammals sex is decisively determined by a single *Sry* gene located on a morphologically distinguishable sex chromosome. Among vertebrates, the only other decisive sex determining single gene is *Dmy/Dmrt1bY* present in *Oryzias latipes* (Matsuda et al., 2002; Nanda et al., 2002) and *O. curvinotus* (Matsuda et al., 2003). However, the *Dmy* is not the sex determining gene in other fishes including those belonging to genus *Oryzias* (Kondo et al., 2003; Volf et al., 2003). Besides these two medaka fishes, sex determination remains indecisive in others, as sex is determined by many genes located on different, not readily distinguishable chromosomes. For instance, neither a sex chromosome (Wallace and Wallace, 2003; however see Kochakpour and Moens, 2008), nor sex linked molecular marker (see Pandian, 2011) is identified in zebrafish *Danio rerio*. Nevertheless, employing genome-wide linkage, Bradley et al. (2011) have recognized the presence of *Dmrt1* and *cyp21la2* as sex determining genes in zebrafish and these genes are harbored in chromosomes 5 and 16, respectively. Interestingly, leading scientists have concluded that many (sex determining) genes at the top of the genetic cascade are not conserved, while the downstream genes are homologous in wide spectrum of fishes (e.g., Herpin and Schartl, 2011).

Sex determination and differentiation are interlinked and are considered together. The following examples may show that sex is determined by more than one gene residing on different chromosomes in fishes and a host of sex differentiation genes follows the event of sex determination. The platyfish *Xiphophorus maculatus* Jp163A strain is male heterogametic and a cross between it and *X. helleri* Sarabia male produces homogametic

YY male progenies. Hence *X. maculatus* Jp163A strain is regarded as an ideal model to understand sex determination and differentiation processes. Employing the 454-FLX massively parallel DNA sequencing platform and Poisson-based enriched testing, Zhang et al. (2011) recognized 2,250 male-enriched and 2,304 female-enriched contigs with various degrees of difference. Among these contigs, 22 and 23 are female and male predominant transcripts, respectively. With these contigs, eight female predominant and nine male predominant contigs are recognized as differentially expressing sex genes. Santos et al. (2007) found that the gonadal transcriptomes differ substantially between sexes in zebrafish. Microarray analysis of individuals has revealed the consistent expression of about 8,769 genes in the gonad, of which 7,976 and 7,060 are consistently expressed in ovaries and testes, respectively. Of them, statistical analyses have identified 2,940 and 1,370 genes that are differently expressed in females and males, respectively. Among the genes differentially expressed between ovaries and testes in the array dataset, 53 genes have been further identified to show a 10-fold difference in expression between sexes. A magnitude of 3,000-fold difference in expression of some genes between the ovary and testis is likely to play a crucial role in the context of sex specific gonadal function. Gene ontological analysis has shown that *sox11b, sox21a* and *sox31* are overexpressed in the ovaries; in contrast the absence of elevated expression of *sox9a* and the relative higher expression of *amh* and its receptor induce not only the differentiation and development of the testis but also the maintenance of sexual dimorphism in the secondary gonochoric zebrafish. Hence it is the co-action of *sox9a* and *amh* that masculinizes 50% of the presumptive males that had earlier developed an ovary. A novel gene ontology sequence analysis of *Sparus aurata* by Yufera et al. (2012) has revealed that the FDR 8.90E-04 EST sequence is overexpressed during gamete generation.

The expression of one or another sex differentiation gene results in synthesis and release of a specific hormone; it is the chronological sequence of action of the endocrine cascade that completes and maintains the sexualization. Despite the enormity of diversity in differentiation, the most conserved factor in nearly all fishes is the control of ovarian differentiation by the gonadal aromatase gene *cyp19a1a.* This gene codes for the enzyme cytochrome p450 aromatase (p450 arom), a key rate-limiting enzyme in conversion of c_{19} steroids (e.g., testosteronet) to the primary estrogen (e.g., estradiol-17β [E_2]) of fishes. The proportion of androgens and estrogens is critical for sex differentiation in many teleost species and is limited by the quantum of aromatase availability. p450 arom regulates the amount, especially estrogen (E_2) and maintains its level within adequate range for ovarian differentiation (Trant et al., 2001; Fenske and Segner, 2004). *cyp19a1* (aromatase) is considered as the most conserved factor common to nearly all fishes in controlling ovarian differentiation. However, there have been several studies implicating the *fox12* in ovarian development, granulosa cell differentiation and hence maintenance of ovarian follicle (e.g., Govourn et al., 2004; Yao, 2005; Wang DS et al., 2007). *fox12* is the earliest known sexually dimorphic marker initially expressed in the somatic cells and subsequently in granulosa cells surrounding the oocytes (e.g., Govoroun et al., 2004). Using monosex fry of *Oreochromis niloticus*, Wang DS et al. (2007) showed a decisive role played by *fox12* in early sex differentiation through activation of *cyp19a1* transcription either directly or in conjunction with Ad4BP/SF1, the binding site in the promoter region of *cyp19a1* and increased estrogen production.

1.2 Ontogeny of gonads

In teleost fishes, early gonadal development consists of four major morphological and physiological events: 1. Germ cells enter into cysts and commence oogenesis or spermatogenesis (see Pandian, 2012), 2. A layer of germ cells supporting somatic cells surrounds each of the follicle/basal laminal (see Schulz et al., 2010) layer. Hence the gonads are composed of germinal (germ cells) and supporting somatic (structural) elements, 3. The germ cells supporting cells begin to elongate to form the ovarian cavity or efferent duct leading to the formation of gonadal ducts (see Pandian, 2012). 4. Steroid producing cells (SPCs) appear intertitially. Sex steroids produced by the SPCs are considered to play a role in both gonadal sex differentiation and development (see Suzuki et al., 2004). The physiological mechanism of sex differentiation involving sex steroids is considered in Chapters 2 and 3.

From his studies on seven fish species belonging to five families, Nakamura (1978) showed that the somatic elements of the gonads of teleosts are derived only from cells of peritoneal wall; hence they are unitary in origin. Contrastingly, the stromal cells in the developing gonads of tetrapods have dual embryological origin namely, the medulla from the peritoneal wall and the cortex from the mesonephric or inter-renal blastema. This dual origin is the basis for the process of sex differentiation in tetrapods, known as cortico-medullary antagonism, under which the predominance of cortex or medulla during development establishes the ovary or testis (Strussmann and Nakamura, 2002). The unitary origin of teleostean gonad from only the medulla-derived peritoneal wall is attributed as one of the causes for the observed plasticity in sex determination and differentiation.

Another distinguishing developmental feature of the teleost gonads is related to the development of the gonadal duct system to provide an exit for the ovulated eggs and spermatozoa released from the gonad during spawning and spermiation. In tetrapods, the Mullerian and Wolfian ducts play embryologically an important role in the formation of the gonoducts in females and males, respectively. Hence there are structural connections between the excretory and genital systems. In teleosts, however there are no structural connections between the genital and excretory systems. Strussmann and Nakamura (2002) described three different morphogenetic processes by which the ovarian cavity is formed in molted sculpin *Cotus bairdii*, goldfish *Carassius auratus* and medaka *Oryzias latipes.* Essentially, these processes involve extension from the posterior end of the gonad, forming an ovarian cavity and its extension as oviduct in the female, and acinus precursors, differentiating into efferent duct and its extension as sperm duct in the male.

The first steps towards the formation of functional ovaries and testes are the differentiation of the primordial germ cells (PGCs) from the totipotent blastomeres during early embryogenesis, the formation of genital ridge and the migration of PGCs into the ridges. Relevant information on this aspect has been summarized elsewhere (see Pandian, 2011). The early gonads of putative ovaries have more germ cells than those in testes (Fig. 1) of *Gasterosteus aculeatus* (Gasterosteidae, Shimizu and Takahashi, 1980; Lewis et al., 2008), *Oncorhynchus mykiss* (Salmonidae, Lebrun et al., 1982), *Odontesthes bonariensis* (Atherinidae, Strussmann and Patino, 1999), *Puntius conchonius* (Cyprinidae, Cek et al., 1998), *Oreochromis niloticus* (Cichlidae, Kobayashi et al., 2008b), *O. latipes* (Adrianichthyidae, Saito et al., 2007) and even in the self-fertilizing hermaphrodite *Kryptolebias marmoratus* (see Pandian, 2012). This sexual dimorphism in germ cell number is brought about by the rapid mitosis of germ cells in females forming characteristic clusters, and is generally followed by the onset of meiosis (Strussmann and Nakamura, 2002). Nevertheless, a contrasting picture has been reported for *Silurus meridionalis,* in which there are more germ cells in the male than female (Fig. 1, Siluridae, Liu et al., 2008). A delayed proliferation of germ cells in ovaries commencing at the same time as the testis of the golden rabbitfish *Siganus guttatus* has also been reported (Komatsu et al., 2006). Apart from this, the initial ovarian ontogeny varies with species and is first indicated by the difference in the number of germ cells or by the appearance of the ovarian lumen (Nakumara et al., 1998). For example, the entrance of oogonia into meiotic division and development of primary growth stage oocytes occur prior to the formation of the ovarian cavity in *Cichlasoma dimerus, Anguilla anguilla,* and *O. latipes,* but it is after the formation of the lumen in *Paralichthys olivaceus, P. lethostigma, O. niloticus, O. aureus* (see Luckenbach et al., 2003) and *S. meridionalis* (Liu et al., 2008). In *C. auratus, C. bairdii* and *O. mossambicus,* the oogonial entry into meiosis and formation of ovarian lumen occur simultaneously. Incidentally, a true ovarian cavity is not present in some eels (Colombo and Grandi, 1996) and salmonids (e.g., *O. masou,* Nakamura, 1981).

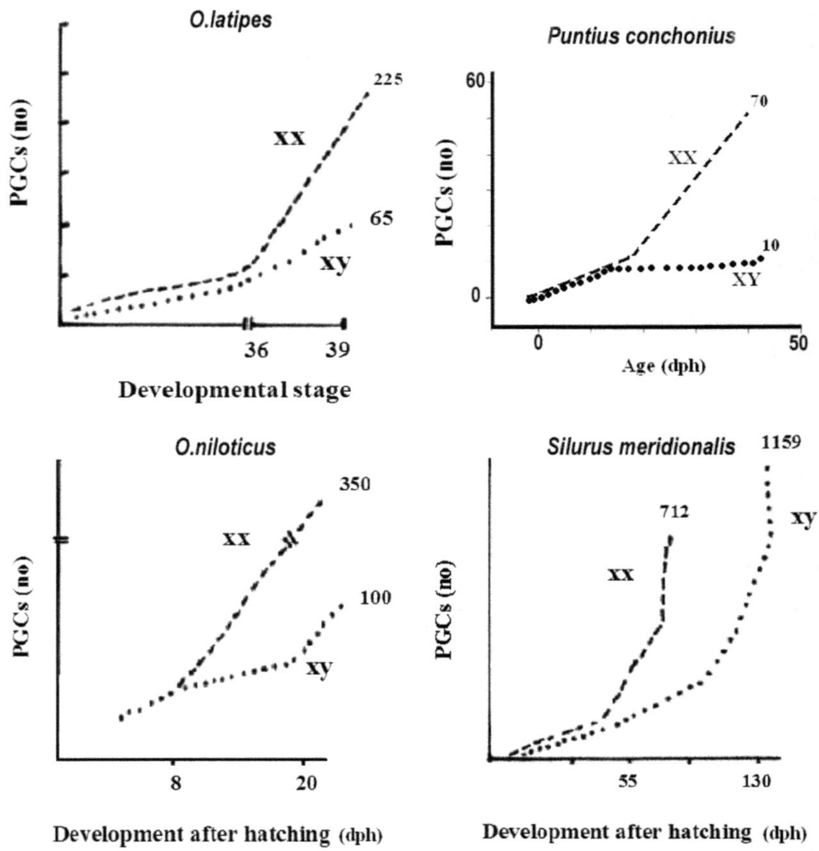

Fig. 1. Schematic representation of PGC proliferation as function of developmental stage/ days post hatching in selected model fishes: Upper panels: Left, *Oryzias latipes*, Right, *Puntius conchonius* Lower left panel: *Oreochromis niloticus* and Lower right panel: *Silurus meridionalis* (compiled from Saito et al., 2007; Kobayashi et al., 2008b; Cek et al., 1998 and Liu et al., 2008).

Subsequent to the migration and entry of the PGCs into the ridges, the rudimentary gonad is formed; the gonads then undergo a period of slow somatic growth with no apparent histological differentiation or germinal activity, which may last for days, as in *O. latipes*, months, as in *Corgonus larvaretus* and years, as in *Acipenser sinensis*. Consistent with proliferation and greater number of germ cells and possibly also of somatic cells, the cross sectional area of the gonads in presumptive females is larger than that of males in pejerrey *O. bonariensis* (Strussmann et al., 1996) and channel catfish *Ictalurus punctatus* (Patino et al., 1996). This may be one reason for the several times greater gonado-somatic index of females than that of males (see Table 6, Pandian, 2010). Despite the initial slow proliferation of germ cells in testis, males in many teleosts mature first; for example, male *A. sinensis* mature at the age of 14 year (yr), whereas females by 21 yr of age (Wei et al., 1997). In others like *C. larvaretus,* both male and female mature at the age of 2.7 yr (Bernet et al., 2009). In yet others the female matures first and the male later.

1.3 Regulatory mechanisms

Relevant information on the regulatory mechanisms has been compiled from Balment et al. (2006); Oliveira and Goncalves (2008); Dufour et al. (2010); Migaud et al. (2010); Unniappan (2010); Kah and Dufour (2011); Diotel et al. (2011); Zohar and Mylonas (2001); Zohar et al. (2010). Across vertebrates, the regulatory mechanisms of neuroendocrines and hormones are highly conserved. As in other vertebrates, the neuroendocrine system is structured in a hierarchial order of brain/hypothalamus → pituitary → adenophysial (or gonadal) hormones and thereby the so called brain/ hypothalamus-pituitary–gonad (BPG/HPG) axis is established (Fig. 2). The pituitary gland of fish consists of two types of tissue, the adenohypophysis and neurohypophysis (Oliveira and Goncalves, 2008). The adenohypophysis contains different cells secreting the pituitary hormones and thus represents the glandular part of the pituitary; the neuropypophysis consists of bundles of nerve fibers originating from different parts of the brain and secretes various peptides in the vicinity of the pituitary cells (Zohar et al., 2010). In general, the pituitary cells of a given type are regionalized in the adenohypophysis. Secretions of the anteriorily located lactotropes and corticotropes are prolactin (PRL) and corticotropin (ACTH), respectively; gonadotropes, which secrete FSH and LH and somatotropes, secreting growth hormone (GH) are located at the center. "However, unlike other vertebrates, teleosts lack the hypothalamic-hypophysial portal system. Therefore, in teleosts, the relevant releasing factors controlling hypophysial function are still produced in neurosecretory hypothalamic neurons that project to the pars distalis of the adenohypophysis, thus making the hypothalmo-hpophysial axis" (Oliveira and Goncalves, 2008). Interestingly, Guin et al. (1993) reported that addition of thyroid hormone, triiodothyronine (T_3) to oocytes of *Anabas testudineus in vitro* results in a dose-dependent increase in progesterone release. The strong correlation between thyroid (TH) and nutritional status suggests that TH plays a key role in energy partitioning during the reproductive cycle, shifting the mobilization of nutrient reserve for growth to reproductive maturation (Raine, 2011). Surprisingly, very few studies have been made on the role of thyroid in superfoetacious viviparous fishes, in which nutrient mobilization

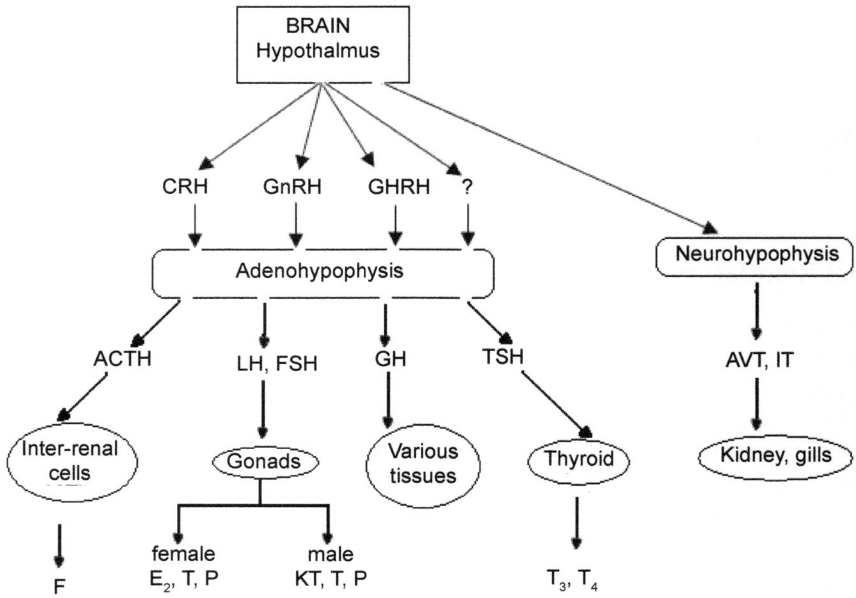

Fig. 2. Organization of the nueroendocrine cascade with major hypothalamic-hypophysial axis for the control of peripheral glands and tissues. Abbreviations: CRH = corticotropin-releasing hormone; GHRH = growth hormone-releasing hormone; GnRH = gonadotropin releasing hormone; ? = unknown; ACTH = adrenocorticotropic hormone; LH = luteinizing hormone; FSH = follicle-stimulating hormone; GH = growth hormone; TSH = thyroid stimulating hormone; AVT = arginine vasotocin; IT = isotocin; E_2 = estradiol; T = testosterone; P = progestogen; KT = 11-ketotestosterone; T_3 = triiodothyronine; T_4 = thyroxine (compiled from Oliveira and Goncalves, 2008; Zohar et al., 2010).

to ovaries is spread over a relatively longer period than in oviparous fishes (cf. Fig. 9). Besides, "many neuropeptides and neurotransmitters have been identified by immunohistochemistry in fibres penetrating the pars distalis *in vivo* and *in vitro*. This is true for GnRH, neuropeptide Y (NPY), dopamine (DA) and γ-aminobutyric acid" (GABA) (Kah and Dufour, 2011).

Vested with the main function of controlling synthesis and release of GTHs (gonodotropin) (Zohar et al., 2010), the GnRH (gonadotropin-releasing hormone) is the key player in the control of HPG-axis in fishes. The first to be purified among fishes is the salmon GnRH (sGnRH), a decapeptide that differs from mammalian GnRH (mGnRH) only by two amino acids in positions of seven and eight (Table 1). The sequence of GnRH known for six fish species are all highly conserved decapeptides with amino acid terminus (pGlu) at one end and carboxy-NH_2 terminus at the other. Barring the lamprey lGnRH 2, all forms have conserved positions at 1–4, 6 and 9–10 (Powell et al., 1994). It is well known that synthetic analogues of

Table 1. Primary structure of known forms of GnRH in selected craniates. The six variants found in teleosts are marked in bold.

Form of GnRH	1	2	3	4	5	6	7	8	9	10	Reference
Mammals (mGnRH)	pGlu	His	Trp	Ser	Tyr	Gly	Leu	Arg	Pro	Gly-NH_2	Burgus et al. (1972)
Chicken (cGnRH)	-	-	-	-	-	-	-	Gln	-	-	Miyamoto et al. (1982)
Frog (frGnRH)	-	-	-	-	-	-	-	Trp	-	-	Yoo et al. (2000)
Lamprey 1 (lGnRH1)	-	-	-	-	His	Asp	Trp	Lys	-	-	Sherwood et al. (1986)
Lamprey 2 (lGnRH2)	-	-	-	-	His	-	Trp	Phe	-	-	Kavanaugh et al. (2008)
Lamprey 3 (lGnRH3)	-	-	Tyr	-	Leu	Glu	Trp	Lys	-	-	Sower et al. (1993)
Dogfish (dfGnRH)	-	-	-	-	His	-	Trp	Leu	-	-	Lovejoy et al. (1992)
Salmon (**sGnRH**)	-	-	-	-	-	-	Trp	Leu	-	-	Sherwood et al. (1983)
Seabream (**sbGnRH**)	-	-	-	-	-	-	-	Ser	-	-	Powell et al. (1994)
Whitefish (**whGnRH**)	-	-	-	-	-	-	Met	Asn	-	-	Adams et al. (2002)
Medaka (**mdGnRH**)	-	-	-	-	Phe	-	-	Ser	-	-	Okubo et al. (2000)
Catfish (**cfGnRH**)	-	-	-	-	His	-	-	Asn	-	-	Ngamvongchon et al. (1992)
Herring (**hgGnRH**)	-	-	-	-	His	-	-	Ser	-	-	Carolsfeld et al. (2000)

GnRH are effective in inducing final maturation, ovulation and spermiation in many fishes (Peter et al., 1988). Incidentally, GnRH-like decapeptides are known to have existed some 600 Mya (million years ago) and appeared with their corresponding receptors in prochordates (Kah et al., 2007). In fact the possible presence of GnRH-like peptide with a potential role for reproduction in cnidarians suggests the ancient evolutionary origin of GnRH system much before the emergence of the pituitary gland and GTHs (Tsai, 2006; Twan et al., 2006). Three molecular variants of GnRH have been found in lampreys (GnRH 1, GnRH 2 and GnRH 3), whereas only two variants have been reported from primitive teleosts such as salmonids and cyprinids. A third form of GnRH 3 has been reported from more recently derived teleosts such as seabream. Frequent reports on the presence of three forms of *gnrh* from other teleosts are made, for example, *Fundulus heteroclitus* (Ohkubo et al., 2010). Briefly, cDNA sequencing or biochemical characterization studies have identified three forms of GnRH in most teleosts especially among perciformes (e.g., Senthilkumaran et al., 1999) and these three GnRH forms are generated by three different paralogous genes: *gnrh1, gnrh2* and *gnrh3* (Fernald and White, 1999). The *gnrh* genes of vertebrates share a common structure by the presence of four exons and three introns. Sequence analysis of the different *gnrh* genes has shown that coding regions are highly conserved (see Zohar et al., 2010). The presence of three GnRH forms raises many issues, especially the multiplicity of receptors. In fact five GnRH receptors have been characterized in *Takifugu rubripes* (Moncaut et al., 2005). Many fish species express two or three GnRH variants and GnRH exert pleiotropic actions via several classes of receptors (Kah et al., 2007). For more details on distribution and localization of the GnRH forms within the brain of teleosts, Kah and Dufour (2011) may be consulted. There are differences in localization and distribution of different GnRH forms within the brain resulting in multiple GnRH neuronal system (Oliveira and Goncalves, 2008). GnRH 2 cells are restricted to the dorsal synencephalon but that of GnRH 1 and GnRH 3 cell distribution overlaps in the olfactory bulbs ventral telencephalon and preoptic area (POA) (Gonzalez-Martinez et al., 2001, 2002). The neurohormones controlling the activity of the three different cell types are released directly by nerve endings located in close vicinity of their target organs (Kah and Dufour, 2011). GnRH 1 influences reproductive behavior through regulation of the pituitary gonadotropins and finally by sex steroids (Hofmann, 2006); GnRH 2 affects spawning behavior in the female (Volkoff and Peter, 1999) and GnRH 3 both sexual and aggressive behavior of the males (Ogawa et al., 2006).

Apart from the unique regulation of HPG-axis, there are a large number of neurotransmitters, neuropeptides (e.g., NPY), neurosteroids and other hormones like ghrelin, that participate in the neuroendrocrine regulation. Among the hormones, ghrelin is a well characterized, multi-functional hormone and is now known to regulate reproduction by affecting synthesis and release of reproductive factors from the brain and pituitary (Tene-Sempere, 2008). The presence of ghrelin and its receptor within the brain suggests the possible local and endocrine action of ghrelin in regulation of the modulators at the brain level. Ghrelin has direct effects on all three components of the H–P-G axis. Intra-cerebro-ventricular and intraperitoneal injection of synthetic octanoylated native ghrelin stimulates LH release from pituitary cells of goldfish both *in vitro* and *in vivo* (Unniappan and Peter, 2004). In the sexually mature female common carp too, the administration of human ghrelin stimulates LH release from dispersed pituitary cells (Sokolowska-Mikolajczyk et al., 2009). Incubation of rat ghrelin with the pituitary fragments of the grouper *Epinephelus coioides* has been found to stimulate the release of GH and also augment the GH mRNA transcription, suggesting that the ghrelin-like peptide and growth hormone secretagogue receptor (GHS-r) are involved in regulation of GH synthesis and release (Ran et al., 2004). Preproghrelin mRNA has been detected in the testis and ovary of zebrafish (Amole and Unniappan, 2009) and Atlantic halibut (Manning et al., 2008). The oscillations in the plasma ghrelin level in burbot *Lota lota* (Mustonen et al., 2002), tilapia (Parhar et al., 2003) and Arctic charr (Froiland et al., 2010), as functions of reproductive cycle and season, evidence the possible reproductive functions for endogenous ghrelin in fishes.

The kisspeptins are the products of the *kiss1* gene; the gene encodes for a 145 amino acid precursor protein that is cleaved to yield four biologically active peptides, each with 54, 14, 13 and 10 amino acid residues (see Kitahashi et al., 2009). Recently, *kiss1* has been identified from zebrafish (van Aerle et al., 2008) and medaka (Kanda et al., 2008). Kisspeptins (*kps*) and their receptors (*Gpr54*) are shown to be key players in regulation of reproductive functions. The 10 amino acid sequence in the *kp* is highly conserved among vertebrates (Zohar et al., 2010). The role played by kiss proteins, the peptide products of *kiss1* gene, and its receptor (*Gpr54*) as gate-keepers of GnRH as well as *gnrh1* and its receptor *degnrh2* on the onset of puberty has been briefed elsewhere (Pandian, 2011). Among the *kp* gene series, *kiss2* is more effective in stimulating GTH subunits in zebrafish (Biran et al., 2008). In medaka fish, ovariectomy and estradiol (e.g., E_2) replacement studies have shown that the ovarian estrogens are positively regulated by *kp* expression in a specific hypothalamic nucleus ventral tuberis (Kanda et al., 2008). Light regime has also been shown to affect the expression of *Gpr54* in Nile tilapia (Martinez Chavez et al., 2008), suggesting that the *kp* system may be one pathway for the photoperiod/melatonin (MEL) modulation of reproduction.

A good example for genome doubling is the presence of neural or brain aromatase gene *cyp19a1b* encoding p450 arom B highly expressed in the brain of males and females (Patil and Gunasekera, 2008) and ovarian aromatase gene *cyp19a1a* encoding p450 arom A in gonad (Chiang et al., 2001). Incidentally, eels possess a single *cyp19a1*, unlike other teleosts. Hence the eel is an important representative of phylogenetically ancient group of teleosts. The female Japanese eel *Anguilla japonica* requires 8.3 ± 1.6 yr to reach puberty. The delayed onset of puberty is associated with very low brain aromatase activity, which is in the range of K_m = 75 nM; Vmax = 1.14 fmol/min mg protein, against that (K_m = 5 nM; Vmax = 189.4 fmol/min mg protein) reported for the modern teleost like the goldfish (Zhao et al., 2001). Eels also require long treatment with repeated injections of gonadotropins to induce to maturation and spawning (e.g., *A. anguilla,* Lin et al., 1998; *A. japonica,* Jeng et al., 2002). The forebrain aromatase activity of the Japanese eel increases with gonado somatic index at all stages (Jeng et al., 2005). Treatments with steroids E_2 or T lead to elevated expression of *cyp19a1* in the brain of the female eel, perhaps due to E_2-specific induction. In contrast, pituitary hormogenate (CPH) results in the increased expression of *cyp19a1* and aromatase activity both in the brain and ovaries, indicating the steroid-independent control, probably from the effect of gonadotropins contained in the pituitary extract. In the CPH-treated female eels, the increased expression of receptor subtypes (estrogen receptor) *esr*-α and (androgen receptor) *ar*-α is in the brain, while *ar*-α and *ar*-β are in the ovaries. Steroid treatment induces an upregulation by E_2 and *esr*-α, but not *esr*-β expression in the brain, pituitary and ovary, while no autoregulation by T of its own receptors. These results reveal both steroid-dependent and -independent mechanisms in the regulation of *cyp19a1* and steroid receptor subtype expressions in the eels (Jeng et al., 2011).

In mammals, the p450arom B protein is detected in radial glial cells but not in neurons. These glial cells are non-neuronal cells, strongly involved in embryonic neurogenesis (Pinto and Gotz, 2007). In developing embryos, radial cells produce new neurons by asymmetrical divisions. Similar divisions are reported from adult zebrafish showing that radial cells are progenitors of the brain of adult fish. Hence, the radial cells persist in the adult brain of secondary gonochoric fishes like zebrafish, unlike in mammals (Pellegrini et al., 2007; cf. Pandian, 2012). The persisting glial cells in the brain of adult zebrafish seem to retain the brain non-sexualized and thereby facilitate the exceptional sexual plasticity, albeit the plasticity, as indicated by the progressively slower response to exogenous steroids, decay as ontogenetic process progresses in primary gonochoric fishes (Blazquez et al., 1998). Interestingly, Diotel et al. (2011) have provided for the first time detailed information on the sites of expression of the main steroidogenic enzymes in the adult brain of zebrafish. Employing high performance liquid chromatographic studies, they suggested a biochemical pathway (Fig. 3), through which the zebrafish brain converts [^{3}H]-pregenolone into a variety of radio-labeled steroids including testosterone and estradiol.

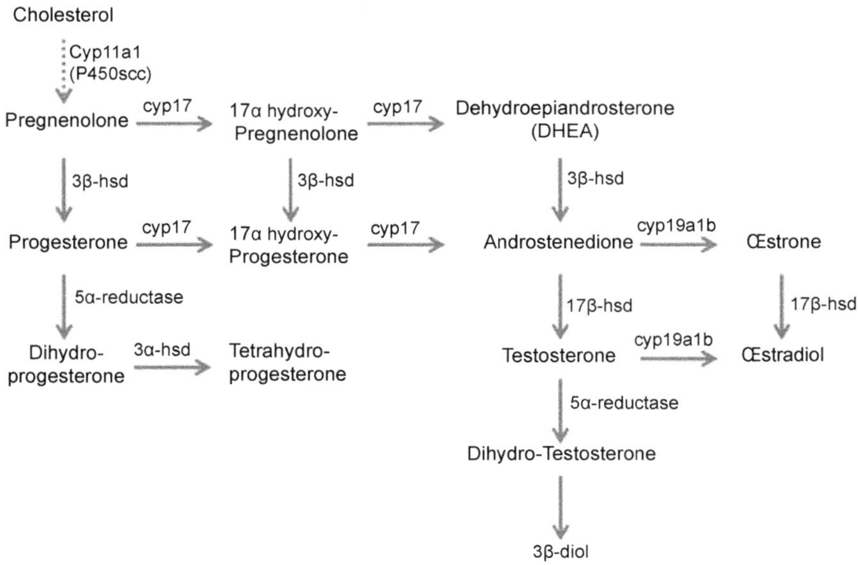

Fig. 3. Active biosynthetic pathways of neurosteroid formation in the brain of adult zebrafish. Full line arrows show active steroidogenic biosynthetic pathway in the brain of adult zebrafish. Dashed line arrow shows undetermined steroidogenic biosynthetic pathway (from Diotel et al., 2011).

Uniquely, the intense neurogenic activity linked to the persisting radial glial cells acting as progenitors and high aromatase activity supported by the strikingly high expression of *cyp19a1b* suggests that the steroids synthesized in the suggested pathway are able to modulate the activity of glial cells. By combining hybridization aromatase B immunohisto-chemistry with *in situ* hybridization, Diotel et al. (2011) further showed the presence of *cyp11a1*, *3β-hsd* and *cyp17* messengers in aromatase B-positive radial processes and thereby provided the first demonstration that the brain of the secondary gonochoric zebrafish can produce true neurosteroids.

Figure 2 shows a single ('food chain'-like) hierarchial pathway of each neuroendocrine system controlling the secretion and release of peripheral hormones; the pathway resembles a simple food chain. However, the *kiss2*, neuropeptides, neurotransmitters and others may also stimulate or inhibit the events at the GnRH and/or pituitary levels and thereby establish more than one pathway in the regulation of secretion and release of peripheral hormones; these multiple pathways seem to simulate a food web-like situation. Figure 4 shows some of the suggested pathways in the HPG-axis. For example, the secretion and release of LH can be stimulated by GnRH, *kiss2*, NPY through the pituitary and inhibited by DA and GnIH(?) at the hypothalmus level. These alternate pathways seem to play a role of check and balance.

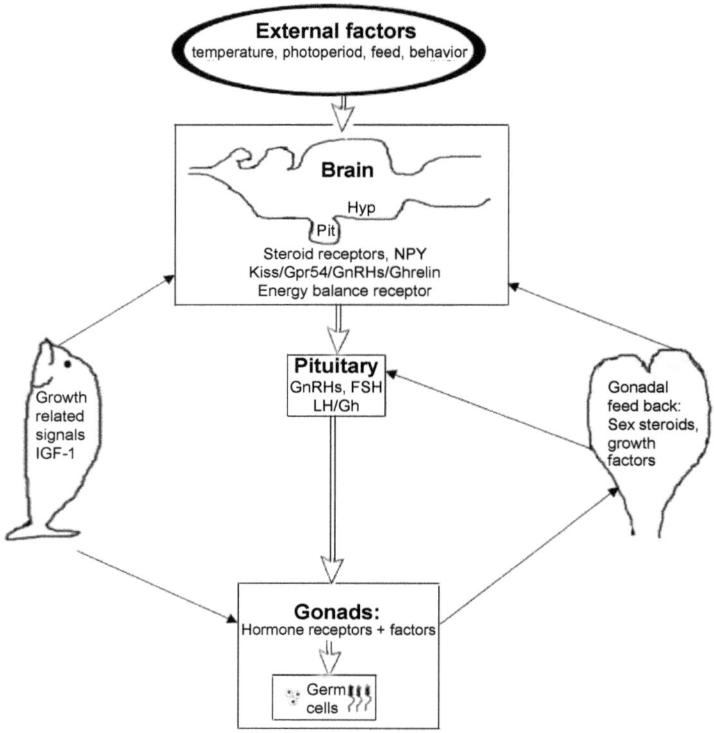

Fig. 4. Proposed 'food-web'-like neuroendocrine regulatory pathways in the brain-pitutary-gonad axis in teleost fish.

Peter et al. (1978) are the first to demonstrate the presence of a dopaminergic inhibitory neuroendocrine control of reproduction in *Carassius auratus*. Simple lesions on the brain of gravid female, maintained in environmental conditions that prevent ovulation, triggered large scale release of LH and ovulation. Dopamine (DA) may have multiple targets to regulate gonadotropic functions. The stimulatory action of GnRH is opposed by the potent inhibitory actions of DA. This dual neuroendocrine control of reproduction by GnRH and DA has been demonstrated in *C. auratus, Cyprinus carpio, Clarias gariepinus, Heteropneustes fossilis, Oncorhynchus mykiss, Oreochromis mossambicus, Mugil cephalus* and *Acipenser transmontanus* but its inhibitory role is absent in Atlantic croaker *Micropogonias undulatus* and *Sparus aurata*. "This dramatic discovery of the inhibitory role of DA on the neuroendocrine control of reproduction in *C. auratus*, besides the classical stimulatory role of GnRH, has opened new avenues for basic and applied research in teleosts. A special interest has been brought to fishes with the world wide development of aquaculture and the need to control gamete production" (Dufour et al., 2010). For instance, induction of artificial spawning of asp *Aspius aspius* was tested by administering analogues of LHRH (luteinizing hormone releasing hormone) and dopamine antagonists of two commercial products ovaprim and ovopel at the Czarci Jar Fish Farm, Poland by Targonska et al. (2010). As against no spawning in controls, the ovaprim group induced ≈ 80% ovulation and sustained 75% survival of embryos but the ovopel group induced 32% ovulation and 42% survival. The pharmological effects of agonists and antagonists of DA and its receptor DA D_2 as well as steroid levels on multiple endocrine control of reproduction are summarized in Table 2. Briefly, a new very successful method with a combination of GnRH agonist and DA D_2 receptor antagonist like domperidone is widely used in aquaculture as an alternative, more reliable and physiologically more relevant strategy to induce spawning, in comparison to treatment with pituitary extract, the so called hypophysation. Apart from this, DA is also known to reduce the production of GnRH and thereby leads to prepubertal blockade in the migratory European eel *A. anguilla* (see Dufour et al., 2010) and *M. cephalus* (Nocillado et al., 2007). But it plays no role on the onset of puberty in *Morone saxatalis* and *Pagrus major* (Kumakura et al., 2003). Hence the role played by DA in opposing GnRH role and prepubertal blockade appears to be species specific. DA inhibits the natural sex change in *Thalossoma duperrey*. Administration of DA or a combination of GnRH and DA D_2 receptor antagonist induces sex reversal in *Thalassoma bifaciatum*.

Dopaminergic activity is known to change with the developmental stage, sex and season. The activity is controlled by feedback mechanism from downstream peripheral components of the H-P-G axis (Montero and Dufour,

Table 2. Effects of agonists and antagonists of dopamine as well as steroid levels in multiple neuroendocrine control of reproduction in fishes.

Dopamine	
Apomorphine	An agonist, inhibits gonadotropin. Decreases the number of GnRH binding sites. Completely reverses the effect of domperidone
Quinpirole	An agonist, suppresses mRNA levels in GnRH and LH release
Dopamine D_2 receptor	
Pimozide	An antagonist, induces dramatic increases in synthesis and release of LH. Increases serum gonadotropin levels
Domperidone	An antagonist, increases GnRH binding capacity in a dose and time dependent manner
Steroids	
Estradiol	High levels of E_2 affects DA biosynthesis by regulating TH (Tyrosine hydroxylase) activity and degrades DA by modulating monoamine oxidase activity
Testosterone	Produce the same effect as E_2 and DA inhibitory tone

1996; Schmitz et al., 2005) and environmental cues, as well. E_2 seems to be the main regulator of DA activity; for example, E_2 may affect synthesis of DA in the Indian catfish *Heteropneustes fossilis* by TH (tyrosine hydroxylase) (Chaube and Joy, 2003) and by degrading DA through monoamine oxidase (Senthilkumaran and Joy, 1995). T (testosterone) is considered to exert most of its effect on local aromatizable E_2. Hence T produces the same effects as E_2 on DA inhibitory activity. However, T may also directly affect the DA activity. Weltzein et al. (2006) are the first to show the androgen-dependent modulation of *TH* expression in the central nervous system of a female vertebrate. They have treated the prepubertal female silver *A. anguilla* with E_2, aromatizable androgen T and 5-α dehydrotestosterone (DHT), a non-aromatizable androgen and evaluated the effects of these steroids on *TH* gene expression, a sensitive measure to evaluate DA activity (Goldstein et al., 1992), by qPCR and *in situ* hybridization; whereas E_2 shows no effect, T stimulates TH mRNA expression in the olfactory bulb and NPOav in the forebrain. Incidentally, T or DHT treatment produces identical stimulation in the forebrain, clearly indicating that both T and DHT are capable of directly affecting DA activity.

Through an impressive review, Miguad et al. (2010) comprehensively summarized the current status of our knowledge on the regulation of reproduction in temperate fishes by photoendocrines, especially serotonin. Reflecting the seasonal variation in photoperiod, the hormone melatonin is recognized as the biological time-keeper and is considered as a potent chronobiotic molecule associated with the temporal pattern of seasonal reproduction in vertebrates. It interacts with the reproductive cascade of fishes at a number of key steps through the dopaminergic system, clock-gene network and kisspeptin signaling that entrains circadian (a cycle of circa 24 hr) and circumannual (a cycle of circa one year) rhythms (Falcon et al., 2007, 2009). The seasonally changing photoperiod, a noise-free clear external clue, is used by most temperate fishes to initiate and co-ordinate gonadal development (Pankhurst and Porter, 2003). Only a few studies have addressed the question of melatonin-DA interaction in relation to reproductive functions. In *C. carpio,* melatonin inhibits DA release in the hypothalamus (Popek et al., 2005, 2006). Exogenous melatonin regulates the *TH* expression in the brain of the European eel by region specific localization of melatonin binding sites (Sebert et al., 2008). *In vitro* study of Maitra et al. (2005) showed that incubation of oocytes of *Catla catla* with melatoin for 4 hrs prior to the addition of maturation-inducing hormone (MIH) in the medium accelerates the oocyte maturation through the formation of cyclin B and cyclin dependent kinase Cdc2. Apparently, melatonin may be a potential physiological candidate involved in the maturation of oocytes in a sub-tropical annual spawner *C. catla*. It mediates the effects of environmental factors on the central nervous system by modulating the activity of DA neurons. Hence it can also have a positive or negative effect on reproduction through its interaction with DA system (Fig. 5).

Circadian axis: In a centralized model of circadian organization of mammals, the retina in the eyes captures photic signals and transfers them to the master clock in the brain via retino-hypothalmic tract. But in teleosts, the central master clock that connects the eye, brain and pineal complex is absent. Hence the circadian axis of fishes is not centralized. Fishes have two main identified photoreceptive organs: the eyes and pineal complex. Unlike mammals, but as in birds, the pineal organ of fishes is directly photosensitive (Migaud et al., 2006, 2007). The organ complex consists of the gland proper and a parapineal organ, and the latter is located in the

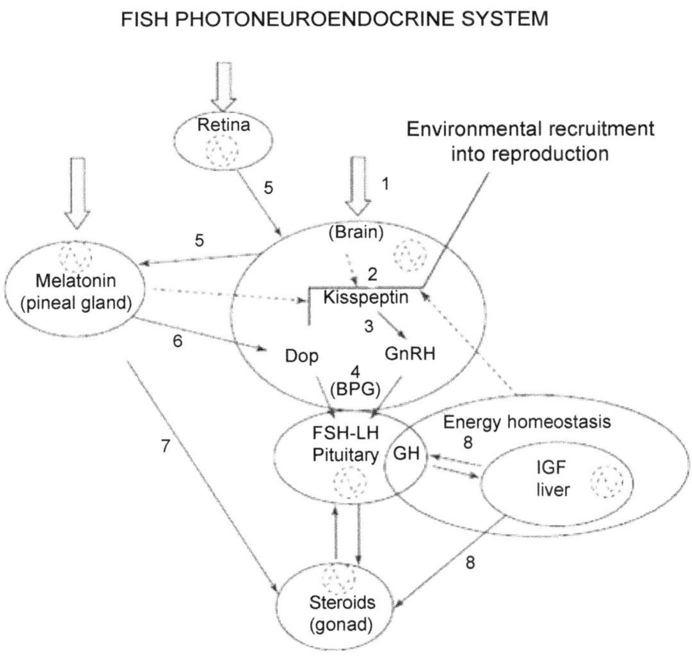

Fig. 5. Proposed photoperiodic control of reproduction in temperate fishes. Photic information perceived ⇒ via brain photoreceptors initiates the reproductive cascade through stimulation of the kisspeptin regulation on GnRH and BPG cascade. Light perceived through retina and pinial organ entrain pineal melatonin synthesis, which, in turn, interacts with reproductive cascade and number of key steps like the inhibition of FSH and LH singnaling via dopaminergic system and synchronization of the final oocyte maturation. Located throughout the system, the presence of endogenous clocks, and their role in reproductive cascade remains to be known (with kind permission of The Fisheries Society of the British Isles/Wiley Blackwell; from H. Migaud, A. Davie and J.F. Taylor. 2010. Current knowledge on the photoneuroendocrine regulation of reproduction in temperate fish species. J Fish Biol, 76: 27–68).

left side of the brain above the epithalmus (Eckstrom and Meissl, 2003). Pinealectomy studies have confirmed that the pineal organ is the main source of circulating melatonin present at night in the blood. Melatonin, an indoleamine derived from the essential amino acid tryptophan, is a highly lipophilic molecule that easily crosses the membrane of cells and cannot be stored within the pineal organ. The molecular machinery responsible for melatonin production and the mechanisms that regulate it differ between the retina and pineal photoreceptor cells (Falcon et al., 2007). Melatonin receptors or ^{125}I–melatonin–binding sites are present throughout the brain, especially the HPG-axis. Melatonin profiles in the blood accurately reflect the duration of the photoperiod and scotoperiod throughout the season with levels rising at night and falling during the day (Falcon et al., 2007). Photoreceptor cells draw tryptophan from the circulation, convert it into serotonin by means of the successive action of tryptophan hydroxylase and aromatic amino acid decarboxylase. Two other enzymes arylalkylamine N-acetyltransferase (AANAT) and hydroxyindole-O-methyltransferase contribute to the synthesis of melatonin from serotonin. Expression of AANAT, the rate limiting enzyme in melatonin synthesis, is downregulated by light, conferring melatonin a day and night production cycle (Dufour et al., 2010). Melatonin synthesis is mainly regulated by photoperiod, temperature, and other environmental factors such as salinity (Lopez-Olmeda et al., 2006) and feeding (Ceinos et al., 2008).

Tryptophan hydroxylase (Tph) is an initial and rate limiting homotetramic enzyme in the biosynthesis of serotonin (5-HT). 5-HT is well known to induce the release of GnRH and gonadotropin in teleosts. From a well designed study of Tph of the brain of Nile tilapia, Sudhakumari et al. (2010) for the first time implicated a role for Tph in sex determination and differentiation. The *Tph,* cloned from the tilapia brain, is 1,888 bp in length and encodes predicted protein of 462 amino acid residues. The tilapia *Tph* activity is confirmed from the conversion of L-tryptophan to 5-hydroxy tryptophan by recombinant enzyme after transient transfection of the cDNA clone in COS 7-cells. Identified for the first time in teleosts, the tissue distribution pattern has shown the presence/abundance of *Tph* in the brain, kidney, liver, gill, muscle and testis. A single transcript of around 2 kb is identified in the male brain but not in the female brain. RT-PCR analysis has revealed the exclusive expression of *Tph* in the male brain from 5 and 20 dph (day post-hatching), while it is from 25 dph in the female brain. Hence these results clearly show the sexually dimorphic expression of *Tph* in the brain until 20 dph. *In situ* hybridization has also revealed the exclusive expression of *Tph* mRNA specifically in the olfactory bulb, olfactory epithelium and telencephalon-preoptic area of male brain on the 11 dph. Immunocytochemical analysis has also confirmed the findings as *Tph-ir* is evident in nucleus pre-optics-periventriculus area of male. These results endorse that *Tph* expression is specific to the male brain during early development. Hence Sudhakumari and her team showed for the first time a definite role for *Tph* in sex differentiation during the critical labile period of (7–20 dph) gonadal sex determination (Nagahama, 2005) and first histological signs of sex differentiation (Ijiri et al., 2008).

2

Ovarian Differentiation

Introduction

In fishes sexuality ranges from gonochorism to unisexualism and self-fertilizing hermaphroditism to sequential and serial hermaphroditism (Pandian, 2010, 2011). Even among the gonochores, a few individuals of *Xiphophorus helleri* are known to switch from female to male (Lodi, 1979). Modes of reproduction also range from sexual to gynogenesis and to hybridogenesis and each of these modes may involve external fertilization and oviparity, or internal fertilization and viviparity, which itself ranges from zygoparity to true viviparity. In fact no other vertebrate group offers as much diversity as teleost fishes for the study of sex differentiation (Pandian, 2011). Hence this comprehensive account is restricted to oogenesis and related aspects alone; a descriptive account on the origin and differentiation of reproductive organs, system and related secondary sexual features may alone merit a title for a book.

2.1 Differentiation types

In fishes sex differentiation is a more labile and flexible process subjected to natural and artificial induction of sex change/reversal through genetic, endocrine and/or environmental manipulations. Based on the presence or absence of a non-functional intersexual phase during ontogenetic pathway of sex differentiation, gonochores are divided into primary and secondary gonochores. Primary or undifferentiated gonochorism occurs, when the differentiation process proceeds according to genetic sex determination. Hence the dimorphic sexual differentiation becomes recognizable by a number of markers; for example, the number of primordial germ cells (PGCs), appearance of ovarian cavity or p450 aromatase mRNA (Table 3). Whereas sexually dimorphic differentiation can be recognized by the *cyp19a1* as early as in 3–4 dpf (day post-fertilization) in Nile tilapia, the p450 aromatase mRNA can do it as late as when the fry has grown to 0.6 cm body length. The secondary gonochores commence with gonads that are already differentiated but remain non-functional as ovary (e.g., *Danio rerio*) or ovotestis (e.g., *Mycteroperca rosacea*) (see Pandian, 2012, *Sparodon durbanensis*, Buxton and Clarke, 1991). No information is yet available on the endocrine role on this initially non-functioning intersex gonad; however, steroidogenic cells have been immunohistochemically identified in the bipotential undifferentiated gonad of *Anguilla anguilla* (Grandi et al., 2010). In secondary gonochores, genotypic and phenotypic sexes differentiate in parallel lines for a short (e.g., *D. rerio*) or longer (e.g., *Dascyllus flavicaudus*) duration but only one sex, as genetically or environmentally determined, becomes functional (Fig. 6) and the other becomes latent with persistent useful (e.g., *Nemipterus bathybius*) or vestigial tissue, or may not persist. On the other hand, sex differentiation occurs

Table 3. Selected markers useful in recognition sexually dimorphic differentiation (for more information see Pandian, 2012).

Marker	Explanation and examples
PGCs	Dimorphic sexual differentiation becomes apparent on the 8th dph in *Oryzias latipes* and on the 15th dph in *Puntius conchonius* (see Fig. 1). However, in *Oreochromis niloticus* sexually dimorphic expression levels of *cyp19a1* becomes apparent between 3 and 4 dpf onwards, while that of PGCs only by the 15th dpf. Presumptive medaka male fry can be identified with 100 germ cells and no cells in the meiotic prophase. In the female fry, the number of germ cells are more and some of them have entered into the meiotic prophase (Satoh and Egami, 1972)
Vasa	Identification of sex chromosome and sex linked genetic marker has not been possible in *Danio rerio*. Based on gonad morphology and expression, Tong et al. (2010) could divide dimorphic sex differentiation in zebrafish as early as 10 dpf. On hybridizing germ cell marker with *vasa* gene as well as *zilli* and *zuwi*, one big and fatty and the other small slender juvenile gonads have been recognized from each other between 10 and 12 dpf
Ovarian cavity	In *Silurus meridionalis*, the proliferation of PGCs is commenced only in the 29th dph fry, clearly after the ovarian cavity is formed in the 12th dph fry
Stalked gonad	In poeciliids, stalk for stromal cells is absent in presumptive females but present in the males of poeciliids (Kramer and Kallman, 1985)
p450 aromatase mRNA	In *Paralichthys lethostigma* morphological and histological sexual differentiation becomes apparent in juveniles measuring 7.5–10 cm body length. But presumptive females of 0.6 cm body length show dramatically higher p450 aromatase mRNA levels up to 65, while presumptive males have a maximum level of 3 only (Luckenbach et al., 2005)

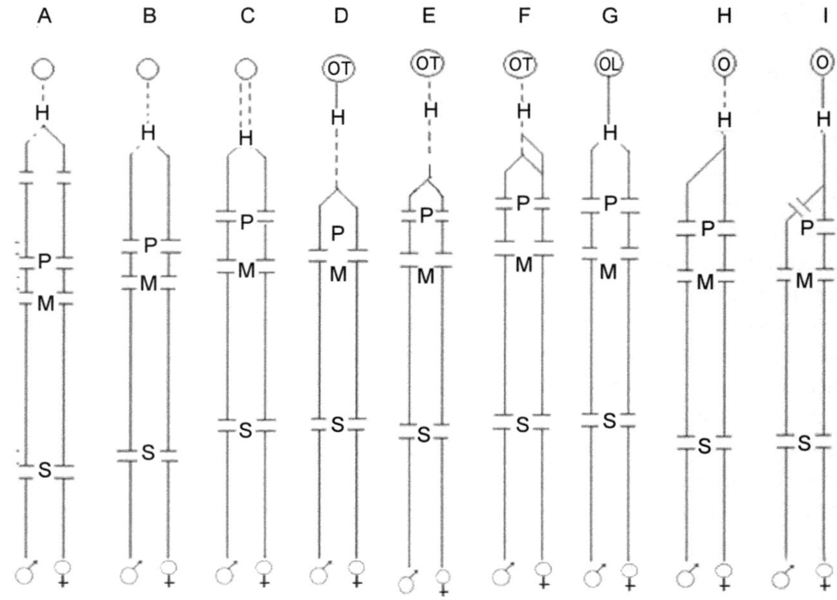

Fig. 6. Ontogenetic pathways of sex differentiation in selected gonochores. Primary gonochores: A = *Oryzias latipes*, B = *Oreochromis niloticus*, C = *Anguilla anguilla*, Secondary gonochores: D = *Pterogymnus laniarius*, E = *Sparodon durbanensis*, F = *Mycteroperca rosacea*, G = *Pseudorasbora parva*, H = *Danio rerio* and I = *Dascyllus flovicaudus*. OT = Ovotestis, OL = Ovarian lumen, H = Hatching, P = Puberty, M = Maturation, S = Senescence? (from Pandian, 2012).

through a contrasting juvenile hermaphroditic stage with the histological features of an early testes, but containing spermatogonia and oogenia (e.g., eels, Grandi and Colombo, 1997; however see also Grandi et al., 2010) or dominant testis (e.g., *S. durbanensis*). Final sex differentiation is induced by biotic (grouping and social interactions) or abiotic (temperature, food quality) environmental factors (Pandian, 2013). High temperature, quality feeding and groupings favor male differentiation (Holmgren, 1996). Despite these differences, endocrine mechanism of sex (final in the case of secondary gonochores) differentiation remains common to both primary and secondary gonochores.

2.2 Interrupted oogenesis

Using double colored *in situ* hybridization and real-time PCR, Chen and Ge (2011) investigated the ontogenetic expression patterns of *fshβ* and *lh*β subunits of the pituitary on ovarian differentiation of zebrafish. The expression of *fsh*β with its mRNA signal detectable (~ 2.3 cells/ pituitary) shortly after hatching (4 dpf) commences much earlier than that of *lhβ*. In contrast, *lhβ* expression becomes detectable at the time of sex differentiation (~ 25 dpf). In females, the first morphological sign of puberty commences with the first wave of follicle transition from primary growth to pre-vitellogenic stage at 45 dpf. Interestingly, the number of *lhβ* expressing cells is low (~ 5–6 cells/pituitary) prior to transition but dramatically increases during and after the transition. In contrast, the expression of *fshβ*, abundant prior to puberty, increases slightly. Hence *Lh* plays an important role in the onset of puberty of female zebrafish, as in mammals.

In teleosts, oogenesis is a highly complex process including five recognizable successive steps: I. Mitotic proliferation of oogonia, II. Transformation of oogonia into primary oocytes with the onset of meiosis (primary growth phase), III. Secondary growth phase characterized by significant oocyte enlargement due to the accumulation of yolk proteins (vitellogenesis) and lipids, IV. Maturation characterized by the resumption of meiosis, germinal vesicle break down, lipid and yolk coalescence and hydration, and V. Ovulation (Patino and Sullivan, 2002) (see Fig. 7).

Spermatogonia and oogonia display important differences in the commencement and completion of meiotic division. The entry of male germ cells into meiosis is an event associated with puberty and is a continuous process without any intermittent arrest. Besides, these germ cells complete the entire process of meiosis prior to spermiation. Conversely, meiosis takes place over a prolonged period of time in females to facilitate oocyte growth involving vitellogenesis prior to the entry into meiotic maturation; expectedly, the positive and negative interventions by the regulatory mechanisms are many, as it has been shown elsewhere (e.g., p 31). The oogonia enter meiosis much before puberty, but they are arrested at the diplotene stage of prophase I, which corresponds to G2 phase of cell cycle (Fig. 7). The prophase I-arrested oocyte is characterized by a nuclear structure known as germinal vesicle (GV); resumption of meiosis results in the GV break down (GVBD); it is followed by chromosomal condensation and formation of spindle that characterize the onset of meiosis I. Upon segregation of homologous chromosomes between the oocyte and expulsion of the first polar body, meiosis I is completed. It is followed by the transition to meiosis II, at which time, it remains arrested. Resumption of meiosis, often termed as final oocyte maturation, is stimulated by LH surge (see Senthilkumaran, 2011). Finally, the germ cell completes the entire process of meiosis, when the second polar body is released at the entry of a spermatozoa.

Briefly, the oogenetic cascade commences with the synthesis of GnRH in the hypothalamus and stimulates gonadotropic cells of the pituitary gland (Pankhurst, 1998). Activated gonadotropic cells, in turn, release GTH-I and GTH-II into circulation. GTH-I stimulates the ovary and testis.

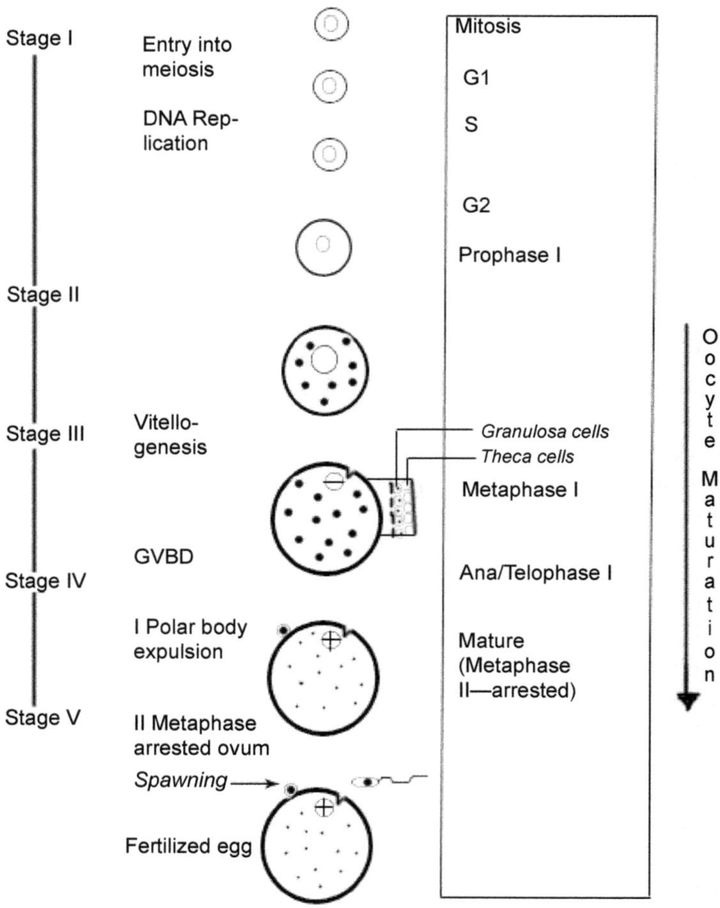

Fig. 7. Schematic illustration of oocyte development and differentiation in fishes. The sequence of oocyte stages is as follows: Stage I–primary growth; Stage II–cortical alveoli growth period; Stage III–early vitellogenic oocytes; Stage IV–late vitellogenic phase; and Stage V–mature/ ovulated oocyte, full of yolk. Note the constituent membranes of the oocyte and formation of micropyle through which sperm enters the egg.

Gonadal steroids are predominantly synthesized by the Leydig cells of testis and outer folliclar (thecal and granulosa) layers of ovary. The gonads synthesize the steroid precursor pregnenolone from cholesterol, followed by successive production of progesterone (P), 17-hydroxyprogesterone ($17P_4$), androstenedione ($\Delta 4$) and testosterone (T) (Devlin and Nagahama, 2002) (see Fig. 3). During vitellogenesis, aromatase converts a portion of the available T into estradiol-17β (E_2). But as ovarian maturation approaches, a switch to GTH-II synthesis in the pituitary leads to the decline of T and E_2 synthesis and simultaneous surge in the production of progestogens (see Frisch, 2004). However, these events are described more elaborately below.

The beginnings: Relevant information on different aspects of oogenesis in fishes has been compiled thanks to the excellent reviews by Lubzens et al. (2010); Senthilkumaran (2011); Urbatzka et al. (2011); Pandian (2011, 2012). Primordial Germ Cells (PGCs) are the progenitors of germ cell lineage and have the ability to differentiate into oogonia and spermatogonia. During early embryogenesis, they are differentiated from the totipotent blastomeres. While the other blastomeres proliferate and reach a 1,000–1,500 cells stage, the four celled PGCs at 32-cell stage suspend division. Subsequently, they undergo mitotic divisions; guided by chemokine-SDF1, they migrate by lophopodial movement and arrive in the gential ridges, where they are surrounded by germ cells supporting somatic cells (see Pandian, 2011). Recent findings suggest that the germ cells supporting somatic cells in the gonad autonomously activate male genotype but the germ cells activate the differentiation of female phenotype by sending signals to repress male pre-deposition and maintain sexual dimorphism. Thus it is the 'intimate cross-talk' and interplay between the PGCs and somatic supporting cells that seems to determine sex and subsequent differentiation (see Pandian, 2012). Incidentally, maternal genes are known to direct fertilization, egg activation, the first cell division(s) and initiation of zygotic transcription. The prospective antero-posterior axis of the embryo is already established during oogenesis and dorso-ventral axis, established during embryonic cleavage, also relies on maternal gene products. Thus, the maternal genes lay the foundation for embryonic development, on which the zygotic factors build upon (see Pandian, 2012).

Following some structural changes, the PGCs are transformed into oogonia and subsequently each oogonium proliferates by mitotic divisions and forms oogonial nests in association with granulosa-precursor cells. At this stage, the oogonium is surrounded by a monolayer of somatic granulosa cells, that secrete a basement lamina separating it from the ovarian stromal cells. Thus, the oogonium, surrounded by granulosa cells, basement lamina and theca somatic layer, constitutes the ovarian follicle and the oogonium is now called as primary oocyte (see Lubzens et al., 2010). The primary oocyte passes through a series of stages namely folliculogenesis and vitellogenesis, meiotic maturation and hydration, the latter is more conspicuous in marine pelagic fishes (than in freshwater fishes) before it is ovulated as a completely differentiated ovum. Hypophysectomy has been the earliest technique employed to know the hormonal mediation on oogonial divisions. It reduces the number and divisions of oogonia. However injection of pituitary extract ensures the normal oogonial divisions.

Prior to entry into the oogonial nests, the PGCs undergo proliferation through many rounds of mitotic divisions. Trends observed for the proliferation of PGCs provide the very first sign for sexually dimorphic differentiation (Fig. 1). The proliferation continues with intermittent divisions resulting in a larger number of PGCs in XX embryos. It is yet to be known whether the same holds true for female heterogametic fishes. However, due to the expression of *Dmrt1bY* in the male heterogametic *Oryzias latipes* or *Dmrt* in others, it is arrested on and off in XY embryos; consequently, the number of PGCs in XY embryos is less than those in XX embryos, except in *Silurus meridionalis* (see Pandian, 2012).

Folliculogenesis: The period from meiotic chromatin nucleolus stage to early cortical alveoli stage represents the primary growth phase of the oocyte. It is a GTHs-independent period, as hypophysectomy does not inhibit the folliculogenesis (Billard, 1992). It is characterized by synthesis of organelles and molecules that are to be used at later stages. For example, cortical alveoli, the membrane-vesicles of variable sizes, are synthesized and are initially localized near the Golgi complex. But they are eventually displaced to the oocyte periphery. Their contents are used for 'cortical reaction' at fertilization (Lubzens et al., 2010). GSDF (Gonadal Soma-Derived growth Factor), expressed in granullosa cells (Sawatari et al., 2007) may play a role in granulosa cell proliferation (Luckenbach et al., 2008).

In coho salmon *Oncorhynchus kisutch,* synthesis of *zp* (zona pellucida protein gene) transcripts and proteins involved in yolk incorporation and processing occur during primary growth phase, while increased expression of the cortical alveoli (CA) component and genes related to lipid incorporation occur concomitant with the appearance of CA. *3p*, vitellogenin receptor (*vldr*), cyclin E (*ccne*), a Dnaj transcript (*dnaja2*) and a ferritin subunit (*fth 3*) are significantly elevated at perinucleolar (P) stage, while a C-type lectin, retinol dehydrogenase (*rdh 1*) and co-atomer protein subunit (*cope*) are upregulated at the CA stage (Luckenbach et al., 2008).

Steriodogenesis is a complex process confined to specific steroidogenic cells facilitating the conversion of cholesterol to sex steroids (Payne and Hales, 2004). Cholesterol is the precursor for the synthesis of sex steroids and mineralocorticoids (e.g., aldosterone) and glucocorticoids (e.g., cortisol), as well (Hsu et al., 2006). The processes vitellogenesis and maturation are regulated by GTHs; GTH I (FSH) is involved in vitellogenesis, while GTH II (LH) triggers maturation and ovulation (Swanson et al., 1991; Nagahama, 2000). Steroidogenesis is regulated by GTHs that activate one or other signal pathway; the most important is the cyclic 3′ 5′ adenosine monophosphatase PKA pathway or protein kinase C (PKC) pathway (Stocco et al., 2005). The steroidogenic acute regulatory (StAR) protein transports cholesterol from the cytoplasm to the inner mitochondrial membrane (Stocco, 2001), where the first steroidogenic specific p450 side chain cleavage enzyme ($p450_{sec}$) converts cholesterol to pregnenolone (Arukwe, 2008) through one or another enzyme (Villeneuve et al., 2007). Synthesis of T occurs in theca cells; on its release to granulosa cells, T is converted into E_2 (Nagahama et al., 1994; see Urbatzka et al., 2011). For example, granulosa cells of *Fundulus heteroclitus* synthesize testosterone (Petrino et al., 1989). Hence synthesis of estradiol is a two-cell process. However, interstitial cells of the ovary may also produce androgen as in *Cyprinus carpio* (Epler et al., 1997). On diffusion into liver, E_2 binds and activates the estrogen receptor, which, in turn stimulates transcription of estrogenic genes including *Vtg*.

Vitellogenesis: The secondary growth phase coincides with vitellogenesis. During this phase, the oocyte incorporates vitellogenin (Vtg), protein, lipids, maternal mRNAs, vitamins and hormones. Consequently, oocytes are massively enlarged (Fig. 7). Vitamins have to be acquired through diet and the vitamin content of eggs is a measure of egg quality. Lipid soluble vitamins are transported by passive inclusion in the hepatically derived lipoprotein fractions in parallel to protein mediated transfer. Recruited from peripheral tissues and muscles, vitamin A and vitamin E are transported by vitellogenin (Vtg). For more details, Lubzens et al. (2010) may be consulted. Inclusion of hormones into the oocytes and their role on sex differentiation is discussed in an ensuing section.

During oogenesis, two vital processes, vitellogenesis and zonagenesis are involved in oocyte growth (Arukwe and Goksoyr, 2003). Zonagenesis is the process by which the zona radiata (vitelline envelope or chorion) forms around the developing oocyte as a thick, highly differentiated acellular zone. Two to four zona radiata proteins have been identified and characterized in many fishes, e.g., Indian fishes (Nath et al., 2007) and are incorporated into the egg shell, where they are crosslinked to form the zona radiata following fertilization and hardening (Oppen-Bernstein et al., 1990).

In response to GTH I-induced ovarian E_2 or other hormones (Babin et al., 2007), hepatocytes synthesize and release yolk precursor protein, vitellogenin (Vtg) into circulation, from which it is taken up for incorporation into the developing oocytes via specific *VtgR* receptor-mediated endocytosis (Sawaguchi et al., 2006). Vtgs are large blood-borne phosphoglycoproteins that serve as the major precursor proteins in oviparous female vertebrates. In fishes, Vtgs undergo proteolytic cleavage to form yolk proteins (YPs) such as lipovitellin (Lv), phosphovin (Pv) and β'component (β'c) prior to the deposition into the ooplasm (Hiramatsu et al., 2002). Lv is a large and lipid-richest protein, Pv is a much smaller unit with more than half of its amino acid residues in highly phosphorylated polyserine domains but β'c contains neither lipid nor phosphorus. For the first time in India, Nath and Sundararaj (1981) isolated a semi-purified Vtg (550 kDa) from E_2-treated male/female catfish *H. fossilis* and showed that alkali labile phosphorus (ALP) is a reliable index of Vtg. For more details on molecular size of Vtg of many Indian fishes and their information of YPs, Nath et al. (2007) may be consulted. Interestingly, the molecular weight of Vtg of *Salvelinus fontinalis* is also 562 kDa (Schafhauser-Smith and Benfey, 2002).

Very low density lipids are also drawn from plasma and vitellogenin and the process is facilitated by VLDL (very low density lipids) receptor genes *Vldro* and *Vldr* (Lubzens et al., 2010). Multiple forms of *Vtg 1a, Vtg 1b* and *Vtg 2* with different functions have been identified in *Clarias batrachus.* In all the fishes studied to date, E_2 is the most effective steroid for Vtg synthesis (Nath et al., 2007). On administration of E_2 during pre-spawning season, Vtg 1 induces vitellogenesis by influencing the release of GTH II and E_2, whereas Vtg 2 inhibits vitellogenesis through the induction of high levels of plasma E_2. Hence the circulating Vtg level regulates vitellogenesis by acting on the H-P-G-axis (Nath et al., 1997).

Vitellogenins (Vtgs) are phosphoglycoproteins present in the blood of all oviparous female vertebrates. Three different vtgs namely *vtgA, vtgB* and *vtgC* are incorporated into the oocytes of teleosts. Induced by E_2, they are synthesized in the liver (Tyler et al., 1997) but can also be induced by other hormones (Babin et al., 2007). *In vitro* experimental uptake of hepatic vtgs by rainbow trout ovarian follicles has been demonstrated (Tyler et al., 1991). The specific *vtgR* receptor of rainbow trout has been cloned and characterized. During vitellogenesis, its transcription is predominated.

Early vitellogenesis is characterized by increases in plasma FSH and E_2 and correlated with increased expression of ovarian FSH receptor (Swanson et al., 2003; Kwok et al., 2005; Kobayashi et al., 2008a). In vitellogenic salmonids, steroidogenic thecal cells supply androgen substrate to ovarian granulosa cells that express p450 aromatase and produce E_2 (e.g., Senthilkumaran et al., 2004). Both KT and T are also known to induce growth in pre-vitellogenic oocytes of *Gadus morhua,* which suggest a role for androgen in early follicular and oocyte growth. There is evidence of an interaction between the growth axis and reproductive axis. For instance, growth hormone (GH) potentiates the effects of E_2 in synthesis of vtg in many fishes (Lubzens et al., 2010). Numerous studies have reported that FSH, LH and partially purified heterologous GTHs stimulate E_2 production by vitellogenic ovarian follicles *in vitro* (e.g., Young et al., 2005). Several recent studies by Berishwilli et al. (2006); Nakamura et al. (2003) and Wuertz et al. (2007) indicate that the (growth factor) GF-II-signaling system regulates the steroidogenic activity of the ovarian follicles during vitellogenesis.

Experimental studies on unilateral ovariectomy (ULO) provide keys to understand whether the compensatory mechanisms initiate recruitment of additional oocytes for vitellogenesis and/or reduce follicular atresia. Following ULO and with availability of ingredients for egg production, the remaining ovary may opt for one or more of the following compensatory pathways: 1. Produce larger eggs from the existing pool of vitellogenic follicles; incidentally, it has implications to egg size, 2. Maximize the fecundity by reducing the number of atretic follicles and 3. Increase the number of dividing oogonia with concomitant production of stem cells. In iteroparous fishes, the compensatory ovarian hypertrophy is accompanied by options 2 and 3 but by options 1 and 2 in semelparous fishes (Pandian, 2012). The ULO studies by Garcia-Lopez et al. (2011a) have shown that within iteroparous fishes, the compensatory ovarian response may depend on the ovarian stage at the time of surgery. When subjected to ULO one month prior to the expected onset of vitellogenesis, i.e., early vitellogenic stage of sea bass *Dicentrarchus labrax,* all the perinucleolar and mid-late vitellogenic oocytes are fully compensated; however, the new early vitellogenic oocytes, that are originated in response to ULO, as in the Indian catfish *Heteropneustes fossilis* (Goswami and Sundaraj, 1968) and blue tilapia *Oreochromis aureus* (Dadzie and Hyder, 1976), do not pass through the primary and secondary growth phase. Hence the compensation is limited to option 2, in which maximization of fecundity is accomplished by reducing the number of atretic follicles. Hence the critical window for recruitment of follicles from the pool of early vitellogenic oocytes already existing in the ovary is opened until the early vitellogenic stage alone.

Incidentally, the dynamics of steroid synthesis and release by the ovarian tissue of ULO females differ from that of the intact control ovary. A two-fold increase in *StAR* transcripts, responsible for translocation of cholesterol into the mitochondria, a rate limiting step in steroidogenesis (Stocco et al., 2005), results in exceedingly large production of sex steroids by the remaining ovary. Targeted gene expression analyses across the brain-pituitary-ovary axis indicate that *Fsh/Fshr* and *GnRH 1* and *GnRH 2* signaling pathways act as the main axes orchestrating the ovarian compensatory growth resulting in unexcised ovarian lobe doubling its size equal to the intact paired ovaries. Steroid receptors, steroidogenesis-related proteins including StAR and IGF- related proteins, and members of the transforming growth factor β superfamily such as *amh* and *Bmp4* (bone morphogenetic protein 4) are all identified as potentially relevant players in the compensatory response process (Garcia-Lopez et al., 2011a). Incidentally, *Bmp15* and *Gdf9* (growth differentiation factor 9) are secreted by oocytes. *Bmp15* mRNA is detected pre-vitellogenic (i.e., perinucleolar) oocytes and its protein from lipid vesicle stage onwards; *gdf9* mRNA and Gdf9 protein are detected in primary nucleolar oocytes. Their functions are being elucidated (Garcia-Lopez et al., 2011b).

Stress response in fishes, like those in captivity in aquafarms, is mediated by the hypothalamus-pituitary-interrenal (HPI) stress axis and the autonomic sympathetic-chromaffin cell axis (Wendelaar Bonga, 1997). For most studies on the effect of HPI- axis, rainbow trout has served as the model to show that the activity of HPI affects HPG axis. Using *in vitro* and *in vivo* systems, several studies have demonstrated the stress-induced inhibitory effect of corticosteroids on biosynthesis of vtg; for instance, slow release of cortisol containing implant markedly reduces the vtg levels (Carragher et al., 1989). The administration of cortisol directly affects the liver and interferes with the estrogen-dependent induction of vtg production (Lethimonier et al., 2000, 2002) or indirectly by reducing the number of hepatic estrogen receptors (see Urbatzka et al., 2011), (however see also p 94).

Meiotic maturation: Based on 2.5% of genome is meiosis related genes in *Saccaromyces cerevisiae* (Goffeau et al., 1996), Schulz et al. (2010) suggested that there are about 900 potential meiosis-specific genes out of 37,460 genes in the zebrafish genome. Following growth including the vitellogenic inputs, the oocyte enters into meiotic process, which is morphologically indicated by germinal vesicle break down (GVBD). Electron microscopic studies (e.g., Le Menn et al., 2007) recognized that oocyte maturation is characterized by reduced endocytosis, GVBD, resumption of meiosis, formation of a monolayer of cortical alveoli under the oolemma, yolk platlet dissolution and hydration of pelagophil oocytes (Lubzens et al., 2010). Using *in vitro* and *in vivo* systems involving several fish species, extensive studies have shown that oocyte maturation is regulated by the following mediators: 1. LH, 2. Maturation-Inducing Hormone (MIH) and 3. Maturation Promoting Factor (MPF). Under the influence of GnRH, FSH regulates the oocytic growth through the mediation of E_2 and LH the final maturation partly by stimulating MIH, the 17α, 20β-dihydroxy-4-pregnen-3-one (17α, 20β-DP). Theca cells produce 17α hydroxy progesterone, which, in turn, is converted by granulosa cells to 17α,20β-DP in the presence of 20β-hydroxysteroid dehydrogenase (20β-HSD) (Nagahama, 1994; Sreenivasulu et al., 2012). Instead of 17α,20β-DP, 17α,20β-trihydroxy-4-pregnen-3-one (17α, 20β, 21P or 20β-S) is present in other perciforms like the Atlantic croaker *Micropogonias undulatus* (Trant et al., 1986) and *Dicentrarchus labrax* (Rocha and Reis-Henriques, 2000). Interestingly, both 17α,20β-DP and 17α,20β,21P are reported from protogynous hermaphrodites *Pagrus major* (Ohta et al., 2002) and *Halichoeres poecilopterus* (Matsuyama et al., 2002). C21 steroids like 17α,20β-DP, 11-deoxycorticosterone and diethylstilbesterol are effective inducers of GVBD.

A dramatic shift in the steroidogenic pathway from E_2 to 17α,20β-DP/17α,20β-21P occurs in the ovarian follicles prior to the commencement of meiosis. Hence the switch in the pathway is associated with decreased expression of p450 aromatase (Nakumara et al., 2005) and increased 20β, HSD activity in granulosa cells. From their observations consistent with 10 fish species, Patino et al. (2001) proposed the following two-stage model for oocyte maturation: 1. Ability of follicle cells to produce MIH and 2. Sensitivity or Oocyte Maturation Competence (OMC) to MIH through increases in MIH receptors on the oocyte membrane and communication among granulosa cells and between granulosa cells and oocytes. LH also induces the production of MIH by granulosa cells and sensitizes the oocytes to respond to MIH (Kagawa et al., 1998), by increasing the MIH receptor activity. Incidentally, adenylate cyclase activators and cAMP analogues also stimulate OMC but PKA (Protein Kinase A) inhibits it (Chang et al., 1999). Besides, IGF II (Insulin-like Growth Factor) (e.g., Bobe et al., 2004) GTHs and heterologous gab connections (e.g., Yamamoto and Yoshizaki, 2008) also induce the OMC. Many concomitant events are known to participate in the LH→MIH→MPF cascade, and endocrine and paracrine activities in preparation of ovulation of the differentiated oocytes. For example, simultaneous drops in expression levels of *cyp* and *shbgh*, following vitellogenesis in the the newly identified sex hormone binding globulin genes are reported (Bobe et al., 2006, 2008). The actions of GTHs may also be mediated or modulated by the local paracrine network consisting of activin, inhibin and follistatin system (see Urbatzka et al., 2011). Acting through the activin type I and type II receptors, activin A expedites maturation of post-vitellogenic follicles but its effects is blocked by follistatin in zebrafish (Pang and Ge, 1999). Activin B may play a role in ovulation. Both activins are strongly expressed in follicle cells but specific receptors are present only on oocytes. Hence the oocytes are the direct targets of activins. Expressed in follicle cells, the Epidermal Growth Factor (*Egf*) is another paracrine system that interacts with activin-follistatin by upregulation of activin A and B and downregulation of follistatin. It promotes oocyte maturation, as in goldfish (Pati et al., 1996) and suppresses atresia in pre-ovulatory follicles, as in rainbow trout (Janz and van der Kraak, 1997). Post-ovulatory follicles regress by apoptosis during ovarian recovery after spawning in repeat spawner *Sparus aurata* (Carnevali et al., 1999). The induction of OMC is important in fishes with asynchronous modes of spawning, for only selected clusters of oocytes are stimulated to mature at a given time (see Urbatzka et al., 2011).

Using specific anti-sera, radiometric assay and Western blot analyses, Sreenivasulu and Senthilkumaran (2009a,b) are perhaps the first to report the changes in enzyme activity and localization of enzymes aromatase and 20β-HSD in seasonally reproducing annual breeder *Clarias gariepinus*. Further they are also the first to clone a single form of *$p450_c17$* and relate the changes in the levels of its mRNA, protein and enzyme activity during preparatory and pre-spawning phases in the catfish. The key event of oocyte entry into meiosis is its induction by 17α, 20β-DP and the commencement of p450 17-11 expression marking the completion of vitellogenesis (see Zhou et al., 2007). The sequence of events, that follow the induction are: adenylate cyclase/phosphodiesterase activity → reduction in the meiosis arrester cAMP level → activation of MPF → entry of oocyte and progression of meiosis to MI and → meiosis II (Nagahama, 1997; Balamurugan and Haider, 1998; Haider, 2003; Basu et al., 2004; Nagahama and Yamashita, 2008). MPF is composed of two protein subunits namely a MPF kinase and a regulatory protein cyclin B. Receiving signal from MIH, cyclin B is synthesized from oocyte-stored mRNA to form the CAK, which subsequently activates the MPF. The MPF activation is the key to induce oocyte maturation (see Senthilkumaran, 2011).

Ovulation: The release of mature ovum, a metaphase II-arrested oocyte after the completion of the ovulatory process from the surrounding follicular cells is known as ovulation. Unlike in mammals, the negative feedback exerted by E_2 is a pre-requisite for piscine pre-ovulatory LH surge (Joy et al., 1998). Arachidonic acid and its metabolites including prostaglandins (PGs) are involved in induction of ovulation (Patino et al., 2003). Besides PGs, ovulation in fishes may also involve co-operation of other ovarian factors like proteases, progestins, eicosanoids, catecholamines and vasactive peptides (Goetz and Garczynyski, 1997). Among proteases, for example, serine proteases play an important role in the ovulatory process, e.g., rainbow trout (Bobe et al., 2006). During ovulation, the oocytes are expelled from the follicles through a localized rupture of follicular layers. Besides physical force due to increased size (see Lubzens et al., 2010), hydrolytic enzymes and metalloproteases are implicated in follicular rupture and ovulation (Ogiwara et al., 2005). Dopaminergic inhibition of ovulation/spawning of fishes is known. In captivity, *Mugil cephalus* successfully undergo vitellogenesis but do not spawn. Injection of the mullet with GnRH and a DA antagonist significantly increases spawning, more than those induced by GnRH alone (Aizen et al., 2005, see also Chapter 1.3).

Hydration: The process of oocyte hydration in pelagophil marine fishes, a unique phenomenon among vertebrates, facilitates acquisition of ≈ 75% of the final volume of the egg. The total water content of hydrated egg accounts for 90–95% of the egg weight, whereas benthophil species produce minimally hydrated demersal eggs that has no buoyancy, due to low water content (Pandian and Fluchter, 1968). Encountering the problem of space to accommodate the hydrated eggs within the body cavity, hydration of eggs is processed batch after batch, resulting in multiple/repeat spawning within a spawning season (see Fig. 9). The high water content in the hydrated pelagic eggs serve as a 'water reservoir' for the embryos to survive in the hyperosmotic sea water and as a 'buoy' to float and thereby disperse the eggs. The pronounced hydrolysis in the hydrated eggs has been considered as a source of free amino acids (FAA) that increase the osmotic pressure facilitating the influx of water into the eggs (Greeley et al., 1986). Yolk protein hydrolysis has been estimated to contribute 50% of the oocyte osmolality, while ions like K^+, Cl, Pi and NH_4^+ make up the balance (Finn et al., 2002). Besides generating FAAs, the cleaving yolk proteins also produce small peptides. During this process, the released yolk lobules coalesce with one another and eventually form a central mass of liquid yolk and the internal crystalline structure disassembles resulting in the characteristic transparency of mature ova (Cerda et al., 2007).

However, the discovery of the molecular water channels, the aquaporins (AQPs) found in all organisms (King et al., 2004) has prompted detailed studies into the molecular mechanisms involved in the hydration of the oocytes of marine pelagic fishes (Lubzens et al., 2010). These studies have shown that Aqp1b is encoded by a gene unique to teleosts, representing a neofunctionalized water channel adapted to oocytes of marine and catadromous fishes. For example, a novel water-selective *Aqp1b*, that is highly expressed in the ovary of gilthead sea bream *S. aurata* has been identified and characterized (Fabra et al., 2005). *Aqp1b* is synthesized at early vitellogenesis and transported to the vitelline membrane of the oocyte. The functional analysis has revealed that the sea bream *Aqp1b* C-terminus, unlike that in mammals, contains specific residues involved in phosphorylation-independent and -dependent mechanisms. Hydrolysis of yolk protein has been found to play a major role to create an osmotic driving force, while *Aqp1b* possibly facilitates the influx of water into the oocyte (Fabra et al., 2006). In contrast, *Aqp1b* has a completely different expression pattern in zebrafish and not found in the genome of medaka, that spawn strictly non-hydrated eggs (Tingaud-Sequeira et al., 2008).

Egg envelope: As mentioned elsewhere, the envelope of the oocyte is surrounded by an acellular vitelline membrane. Subsequently, three layers consisting of an inner granulosa layer, the middle basement membrane and an outer thecal layer. Usually a single spermatozoa enters through the micropyle. Following activation of the egg, the micropyle is closed to prevent polyspermy (see Pandian, 2012). Subsequent to fertilization, the egg envelope, facilitated by Z proteins is hardened to form the chorion. The vitelline membrane is known to show an antimicrobial function (Modig et al., 2007). Using chorion of *Paralichthys olivaceus* as specific substrate, hatching enzyme (PHE) has been isolated and characterized. The PHE has a molecular weight of 34.8 kDa and shows optimal choriolytic activity at the optimal pH of 7.0 and temperature of 35°C. It is extremely sensitive to Cu^{2+} and Ca^{2+} (Shi et al., 2006).

Reproductive cycles: "Whereas small fishes like the zebrafish continuously spawn throughout year, long living large fishes exhibit a definite reproductive cycle with clearly distinguishable preparatory, pre-spawning, spawning and post-spawning phases. Hence the sequence of events that trigger redifferentiation of ovarian components in cyclically reproducing fishes has attracted investigations. Seasonal reproductive cyclicity including regeneration of gonadal tissues is mediated by serum hormone levels, brain hormone content and brain enzyme activity, resulting in significant enrichment of neuroendocrine-related genes through the axis during spring-summer (monsoon in Asian countries)—winter. The storage of PGCs mostly singly or in small groups attached to the stroma of the ovary of polycyclic fishes is reported (e.g., viviparous clinid *Fuconimus mus*, Moser, 2007). Analyzing the multiple brain microarray data sets of female *Carassius auratus*, Zhang et al. (2009) have characterized global gene expression patterns as a function of the spawning cycle. A core set of 873 genes in hypothalamus are differentially expressed during the periods of pre-spawning (May), post-spawning (i.e., gonadal regression season) (October) and in the early gonadal redevelopment (December). Changes in expression pattern of these genes (including *isotocin, ependymin II, GABA* γ*2 receptor, calmpodulin* and *aromatase b*) are shared by telencepnalon. These genes are regulated by photoperiod. The hormones and their genes are functionally involved with G-protein coupled receptor, signaling the shift from gonadal regression to the pre-spawning stage" (see Pandian, 2012).

The team led by Dr B. Senthilkumaran has contributed much to our understanding of the cyclicity of reproduction in annual spawning African catfish *C. gariepinus*. In local ponds of South India, the fish exhibits distinct reproductive cycle with four distinguishable phases namely, preparatory, pre-spawning, spawning and post-spawning (Swapna et al., 2006). *Vasa* transcripts of this catfish are found in higher numbers in immature oocytes of stage I and II during ovarian recrudescence than in mature oocytes of stage III and IV. On *in vivo* treatment of oocytes II with human chorionic gonadotropin (hCG), *vasa* mRNA is up-regulated indicating its role in ovarian recrudescence (Raghuveer and Senthilkumaran, 2010). Further, 20β-hydroxysteroid dehydrogenase (20β-HSD) synthesizes 17β-dihydroxy-4-pregnen-3-one, a steroid required for resumption of prophase 1-arrested oocytes. Having cloned and characterized 20β-HSD from the catfish ovary, Sreenivasulu and Senthilkumaran (2009a) located the presence of 20β-HSD transcripts and protein in relatively higher concentration in mature ovary; further, immunocytochemical and immunoflourescence analyses with specific anti-body identified the presence of 20β-HSD in the follicular layer of the ovary. A rise in 20β-HSD enzyme activity has also been reported, indicating the role of 20β-HSD in the final oocyte maturation (FOM). Cytochrome p450 17α hydroxylase/c17–20 lyase (p450c17) is a key enzyme involved in the steroidogenic shift that occurs prior to oocyte maturation. *p450c17* is transiently expressed and converts progesterone to androstenedione through 17α-hydroxy progesterone. It is expressed at relatively higher levels in gonads. *p450c17* is expressed in the follicular layer of the ovarian follicle during preparatory and pre-spawning stages and thereby shows that the ovarian follicles resume capacity to produce maximum precursor steroid levels to induce the shift in steroidogenesis (Sreenivasulu and Senthilkumaran, 2009b). Rasheeda et al. (2010b) cloned both the ovarian *cyp19A1* (1941 bp) and brain *cyp19A2* (1786 bp), which are present in the ovary and brain, respectively. These isoforms efficiently catalyze the aromatization reaction by producing E_2 form testosterone and the *cyp19A2* expression is predominated in the brain and also in *cyp19A1* in the ovary. A good correlation of *cyp19A1* expression, corresponding enzyme activity and serum E_2 level suggests a crucial role of *cyp19A1* during the ovarian cycle. Thus the team of Sentilkumaran traced the sequence of ovarian recrudescence in the catfish with the commencement *vasa* expression → 20β-HSD synthesizing 17β-dihydroxy-4-pregnen-3-one, which resumes prophase 1 in the arrested oocytes → p450c17 expression finally elevating the serum E_2 level during the preparatory and pre-spawning phases of the reproductive cycle. In the channel catfish *Ictalurus punctatus* too, increasing abundance of the transcripts encoding 17α-hydroxylase/lyase ($p450_{c17}$), cholesterol side chain cleavage ($p450_{sec}$) and aromatase ($p450_{arom}$) at the onset of ovarian recrudescence and the precipitous decrease in all the three with the completion of vitellogenesis have been reported (Sampathkumar et al., 2000). Seasonal changes in the levels of expression of these genes are directly correlated with those in plasma levels of E_2 and T. A similar picture of seasonal changes in E_2 correlated with ovarian recrudencence has been reported in the seaperch *Psammoperca waigiensis* (Pham et al., 2010). A seasonal change in androgen receptor (*ar*) level has been reported with a five-fold peak during breeding season in *Carassius auratus* (Pasmanik and Callard, 1988). Larsson et al. (2002) also found a similar seasonal change in a nuclear *ar* during the breeding season in the brain of *M. undulatus;* they also confirmed that the brain *ar* expression is regulated by sex steroids.

Spawning periodicity: The described endocrine events within a reproductive cycle are to be tuned to different periods resulting in definite patterns of reproductive periodicity in fish. Expectedly, the large difference in reproductive investment between sexes has led to a more restricted period of sexual activity among annual single or multiple spawners (Nunez and Duponchelle, 2009). The restricted period among annual multiple spawners, however, ranges from 19 min (i.e., 3.2 eggs/hr) in Owens pupfish *Cyprinodon radiosus* to 8 d in zebrafish *D. rerio,* 15 d in fightingfish *Betta splendens* and to 19–60 d in *Tilapia zilli* (Fig. 8). Among the annual single spawners, single or multiple spawning occurs during the breeding season restricted to a short or longer duration. A number of factors like food availability, temperature, photoperiod and life history pattern, hydration level related pelagic/demersal eggs may regulate the reproductive periodicity. The pelagic eggs are relatively more hydrated than those of demersal eggs.

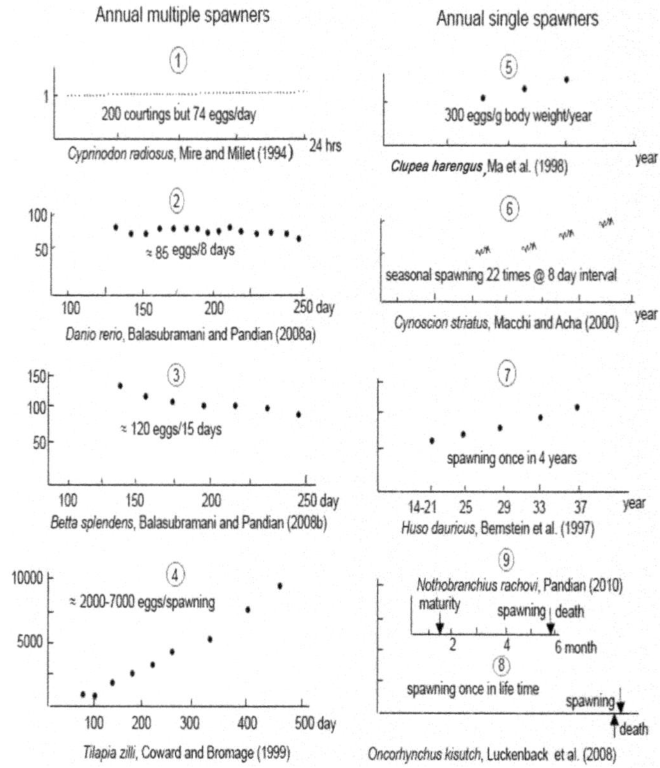

Fig. 8. Patterns of reproductive periodicity in selected teleost fishes.

Consequently, the pelagic spawner, though a seasonal breeder, may spawn eggs repeatedly within the season to accommodate the successive batch of hydrated eggs within the limited space available in the body cavity (see Fig. 9). Two different life history patterns of reproduction are known. In the semelparous pattern adults attaining sexual maturity die either while spawning or soon after. Species, in which adults in populations survive to breed again are iteroparous. There are also interesting examples to suggest that semelparity-iteroparity stands in a kind of a continuum in fishes. In populations of the American shad *Alosa sapidissima,* the repeat spawners progressively increase from almost 0% from the Florida coast (32°N) to 75% at the coast of New Brunswik (46°N). Hence, the shad are semelparous in Florida but is strongly iteroparous in New Brunswik (see Pandian, 2010).

Annual multiple spawners: A female *C. radiosus* is courted by a number of males as many as 200 times a day but ovulates a single egg/spawning. Only 38% courtings result in successful spawning of an egg. As the female has only a few maturing eggs at a time almost continuously for several months, the space problems within the body cavity, inherent in batch spawners, do not occur. Hence the pupfish produces 74 eggs a day, i.e., an egg for every 19 min or 3.2 eggs/hr. The laboratory reared zebrafish, an egg broadcaster with no parental care, matures at the age of 120 dph and spawns ≈ 85 eggs once every 7–8 d. Similarly, the fighting fish with paternal care of developing embryos in the bubble nests, releases ≈ 120 eggs once a fortnight. A 50 g weighing substrate spawner *Tilapia zilli* spawns about 2,100 eggs once every 19 d. With advancing age and increasing weight, the batch fecundity is increased to about 9100 eggs and interspawning period is prolonged to 60 d in a 450 g weighing fish.

Annual single spawners: Many long living fishes are seasonal spawners, i.e., spawn once a year or 4 y, or once in a life time. For example, *Clupea harengus* spawn once during the spawning season, as their demersal eggs undergo relatively limited hydration. Requiring a very high dose of hydration of eggs, the pelagic spawners do spawn seasonally, but during the season, they spawn repeatedly. The spawning frequency of the hydrated eggs within a breeding season is 15 times in *Brevoovtia aurea* and 22 times in *Cynoscion striatus* (Macchi and Acha, 2000) and 41 times/annum in *Serranus atricauda* (Garcia-Diaz et al., 2006). In sturgeons like *Huso dauricus*, the interspawning period is prolonged to 4 y. On attaining sexual maturity at the age of 14–21 years, it spawns once every 4 y. A climax of the restricted period of sexual activity, spawning occurs only once in a life time in long-living large *O. kisutch* and in *Nothobranchius rachovi*, a montane-pool inhabiting fish characterized by a short life span of 6 mon only (see Pandian, 2010).

2.3 Viviparity

Viviparous fishes have a much lower fecundity, by order of magnitude, than oviparous fishes (Grier et al., 2005). The mean fecundity is around 2,200 eggs for oviparous fishes but it is a tenth (230) for viviparous fishes (Goodwin et al., 2002). For example, the life time fecundity of a single female snook *Centropomus undecinalis* is 7–8 x 10^9 eggs; however, of these hardly 3,200 alone survive to reproductive adulthood. On the other hand, the mosquito fish *Gambusia affinis* is able to produce ≈ 800 neonates in her nine broods during its reproductive duration of its life (see Grier et al., 2005). Not surprisingly, more than 3% of all teleost species are viviparous (Wourms, 2005). As many as 577 species belonging to 135 genera and 14 families are viviparous fishes namely Latimeriidae (1 genus, 2 species), Hemiramphidae (4 genera, 10 species), Anablepidae (2 genera, 11 species), Goodeidae (17 genera, 39 species), Poeciliidae (28 genera, 208 species), Aphyonidae (6 genera, 8 species), Bythitidae (34 genera, 51 species), Parabroutulidae (2 genera, 2 species), Embiotocidae (13 genera, 24 species), Clinidae (21 genera, 80 species), Zoarcidae (1 genus, 3 species), Comephoridae (1 genus, 2 species) and Sebastidae (3 genera, 106 species).

Almost all the species belonging to the families Latimeriidae, Aphyonidae, Bythitidae, Parabroutulidae and Comephoridae are viviparous. Among teleosts, viviparity is estimated to have evolved from their respective oviparous clades as many as 12 times (Goodwin et al., 2002). Regarding maternal provisioning, there is a remarkable diversity and variations among viviparous fishes ranging from almost zero provisioning to species with intermediate degrees of provisioning and to species with an extreme of almost 100% maternal provisioning among closely related viviparous fishes (e.g., Poeciliidae, Pollux et al., 2009). In fact, with an array of different levels of maternal provisioning, the 577 viviparous species render the fishes as much a fascinating group as the 386 hermaphroditic fishes (Pandian, 2011). Hence, a little elaborate description on viviparity is provided in the context of yolk deposition and egg/foetus size of viviparous fishes.

Viviparity involves adaptations for internal fertilization, gestation and parturition. A fine distinction between internal fertilization, gametic association and gestation needs to be recognized. There are fishes like *Corynopoma riisei,* in which the sperm and eggs are found in ovarian lumen but the eggs are not fertilized until they are spawned and sperm are activated in water; this may at best be an "internal gametic association" (Burns and Weitzman, 2004). On the other hand, internal fertilization does occur in self-fertilizing hermaphrodite *Kryptolebias marmoratus* (see Pandian, 2011) and gonochores like *Horaichthys setnai* (Burns et al., 1995) but the fertilized eggs are not gestated. Hatching is not necessarily synchronized with parturition, for instance, hatching of larvae occurs several days prior to parturition in the rockfish *Sebastes flavidus* (MacFarlane and Bowers, 1995). Adaptations for internal fertilization and gestation have arisen from new functions of the cells that are common in all fishes and not at the participation of new cell types. Internal fertilization involves evolution of modified functions of Sertoli cells, which become columnar and secretory (Grier et al., 2005). The unspecialized follicular wall of lecithtrophic species remains very thin and plays no role in maternal provisioning. To suit intrafollicular (70 species of blennies, Clinidae, Fishelson and Gon, 2009 and poeciliids) and extrafollicular (Goodeidae) gestation, there are modifications in the female reproductive tract of viviparous fishes. For example, the germinal epithelial cells become secretory and vascularized, a new function associated with viviparity in the goodeid *Skiffia bilineata.* Intrafollicular gestation has eliminated the process of ovulation (e.g., *Heterandria formosa,* Uribe and Grier, 2011).

Evolution of viviparity is accompanied by several trophic adaptations to facilitate foetus development within the female's reproductive tract. Notably, some of these trophic adaptations have shifted the timing of the entire process of yolk deposition from pre-fertilization stage to post-fertilization period. Accordingly, lecithotrophs in oviparous and many viviparous fishes refers to the transfer of nutrient resources to the egg during oogenesis prior to fertilization. And matrotrophy in viviparous fish means reduction in pre-fertilization provisioning coinciding with an array of trophic specializations for the post-fertilization transfer of nutrients from the mother to the foetus by means of oophagy, histophagy or placentation (Trexler and DeAngelis, 2003). Placentotrophy is an important mechanism of matrotrophy (Pollux et al., 2009). A matrotrophic index (MI), defined as the dry weight of a neonate at birth divided by the dry weight of an egg at fertilization, has been developed to assess the levels of matrotrophy. Lecithotrophic species typically display a loss of 9–55% of dry weight on fetus metabolism (Trexler, 1997). Among the matrotrophic 279 poeciliid populations belonging to 132 species and 21 genera, more than 200 of them gain a little with MI of 0.5 to 0.75, while other matrotrophics like *Micropoecilia branneri* (Pires, 2007) gain a lot with MI of 63–99; MI of as high as 117 is reported for *Poeciliopsis latipinna* (Reznick et al., 2002). The smallest microlecithal egg of matrotrophic viviparous fish ranges in size from 350 µm in diameter, 0.014 mm^3 in volume and 2.25 µg in weight in the Australian clinid *Cristiceps aurantiacus* (Fishelson and Gon, 2009) to 400 µm in diameter in the poeciliid *H. formosa* (Uribe and Grier, 2011). The largest mesolecithal egg of lecithotrophic viviparous fish is perhaps that of *Poecilia monacha* with a diameter of 2.2 mm (Vrijenhoek and Schultz, 1974). This size range from 350 µm to 2.2 mm egg size in matrotrophic and lecithotrophic eggs of viviparous fishes may provide an idea on the quantum of maternal provision between fertilization and parturition. Incidentally, the process of oogenesis in *H. formosa* and *C. aurantiacus* may provide excellent opportunities to understand how vitellogenesis is almost eliminated during oogenesis in them and keys to understand the same in mammals.

According to the MI, the two closely related lineages of *Poecilia mexicana,* (i) one inhabiting surface waters (surface molly) and (ii) the second inhabiting sulfidic caves (cave molly) and a sister taxon *P. sulphurica* (sulfur molly) are all lecithotrophic. The transfer of radioactively labeled leucine from mother to foetus has shown low levels of post-fertilization transfer of maternal nutrients to the foetus (Riesch et al., 2010, 2011a,b). Fecundity of sulfidic cave mollies is 2 neonate/parturition, against 4 neonate/parturition in surface mollies; it remains unaffected by ambient light conditions, suggesting a heritable component to the reduction in fecundity (Riesch et al., 2009). However, the transfer of radio-actively labeled one or another nutrient has brought conclusive evidence for matrotrophy in many apparently lecithotrophic viviparous fishes like the rockfish *Sebastes.* In members of the rockfishes *Sebastes*, gestation occurs intralumenally, i.e., extrafollicularly for a period of about one mon. At 3 d, post-injection of [^{14}C]PC (phophosphatidyl choline), the presence of radiolabel is evident in embryos of *S. flavidus* and there is a progressive increase in it over the next 15 d of gestation period (MacFarlane and Bowers, 1995). Radiolabel is enriched 2 to 5.6-fold in embryos, as they develop from mid (embryonic shield stage) to late embryonic (pre-hatching) stage. After 18 days of gestation, the radiolabel level is 2396 disunits/g/min in comparison to 879 disunits/g/min in muscles. The incorporation of radiolabel into developing foetus from [^{14}C]PC in maternal serum *in vivo* has demonstrated the matrotrophic supply of phospholid in a viviparous fish. Further, the occurrence of the 70 kDa serum protein in all the embryogenic females of *S. flavidus* supports the potential for matrotrophy. Since vitellogenin contains 20% lipid, the vitellogenin may act as the carrier for the transportation of phospholipids as well as yolk protein.

The maternal placenta is a complex organ that mediates all physiological and endocrine interactions between the mother and foetus (Pollux et al., 2009). The mammalian placenta has been traced to an ancestor that lived 100 Mya (Pires et al., 2010). In viviparous fishes, the placenta is of recent origin and is estimated to have evolved some 64 Mya (Meredith et al., 2009). The placenta is found in Anablepidae, Clinidae, Embiotocidae, Goodeidae, Labrisomidae, Poeciliidae and Zenarchopteridae. Within Poeciliidae, for example, there have been several independent origins of placentation (Meredith et al., 2009, 2011). But different lineages have evolved that are structurally very different from one and another; for instance, the follicular placenta, through which a poeciliid mother transfers nutrients to her fetus, is formed by the close apposition of the embryonic absorptive tissues to the maternal follicular epithelium. The absorptive tissue of the fetus of *Heterandria formosa* consists of the highly vascularized surface of the greatly enlarged pericardial and reduced yolk sac, suggesting that the fetus surface area plays a role in acquisition of maternal nutrients. By contrast, the ventral portion of the maternal pericardial sac is greatly enlarged leading to the formation of large highly vascularized belly sac in *Poeciliopsis* spp. (see Pollux et al., 2009). However, the diverse poeciliidan placentas fulfill similar functions and are derived from the same pre-existing embryonic and maternal tissues (cf. Grier et al., 2005).

Superfetation refers to the presence of multiple broods at different developmental stages in a single female. It takes place in Poeciliidae, Clinidae and Zenarchopteridae (Pollux et al., 2009). Among poeciliids, placentotrophy and superfetation may have evolved independently. Extensive placentation has evolved at least four times within Poeciliidae (Pires et al., 2010) and placentotrophy has evolved in *Pamphorichthys hasemani* some 10 Mya but superfetation 9 Mya only. But they have originated together in branches leading to *P. hasemani* some 9 Mya, *Micropoecilia branneri* some 4 Mya and in *Poeciliopsis* later (Meredith et al., 2011). Poeciliids that are characterized by superfetation produce smaller broods and neonates, and parturiate more frequently than the lecithotrophs (Table 4).

Table 4. Reproductive potentials among selected poeciliids (compiled from Thibault and Schultz, 1978; Pires et al., 2010).

Species	Superfetation (level)	Age at maturity (day)	Reproductive life span (day)	Inter parturition interval (day)	Batch fecundity (no)	Egg weight (mg)	Mature embryo weight (mg)	Reproductive allocation (%)
Lecithotrophic poeciliids								
Poecilia reticulata	1	48	440	22	24	1.44	0.93	
Poecilia wingei	1	-	-	28	4	2.2	1.0	14.6
Poecilia picta	1	-	-	29	4	1.0	0.7	20.1
Poecilopsis monacha	2	82	380	12	12	2.08	1.26	
Matrotrophic superfetatious poeciliids								
Poecilopsis lucida	3	56	420	10	11	0.53	0.71	
Poecilopsis turneri	3	-	-	11	4	0.18	3.39	
Poecilopsis prolificia	5	65	190	7	4	0.16	0.56	
Poecilia branneri	5			5	4	0.01	0.7	5.9

Viviparity has been shown to impose severe locomotory constraint and cost to pregnant females (Plaut, 2002; Ghalambor et al., 2004). Because the same amount of muscular tissues has to accelerate and then maintain steady velocity of larger body mass (35% or more of total body weight, see Grier et al., 2005), pregnancy affects locomotory performance. In this connection, it is difficult to comprehend how the gravid semelparous Pacific salmon migrate against current. Nevertheless, a shift from matrotrophy to matrotrophic superfetation avoids the necessity for simultaneous transfer of reproductive allocation to all the fetus and successive broods provides scope for transferring the required nutrients to the growing fetus in one brood after the others (Fig. 9). Further, superfetation also enhances the

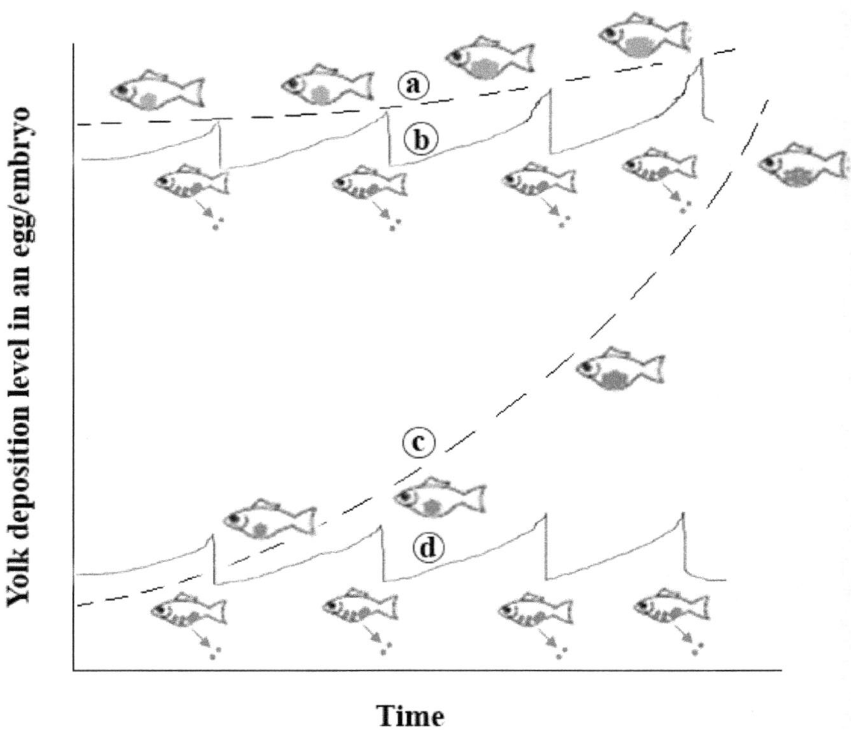

Fig. 9. Suggested effects of superfertation in oviparous/viviparous lecithotrophic and viviparous matrotrophic fishes. Note the relationship in the levels of reproductive allocation for yolk deposition. Also note water uptake (- - -) in non-superfetatious matrotrophics producing micro/mesolecithal eggs. Incidentally, the hydration, i.e., water uptake process necessitates in lecithotrophic oviparous pelagic spawners to repeatedly spawn during spawning season in an analogous similarity to matrotrophy with superfetation. (a) Lecithotrophy without superfetation, (b) Lecithotrophy with superfetation/hydration, (c) Matrotrophy without superfetation, (d) Matrotrophy with superfetation (compiled from Pollux et al., 2009 and others).

Color image of this figure appears in the color plate section at the end of the book.

swimming performance of pregnant females by reducing the reproductive burden without sacrificing the overall fecundity.

In viviparous fishes retention of fetus facilitates prolonged maternal nourishment, expanded options to spread out the reproductive investment and postpone the final commitment to neonate quality (Thibault and Schultz, 1978). Many internal and external factors seem to determine the neonate size and number. Schrader and Travis (2009, 2011) brought evidence for (i) asymmetric sibling competition and (ii) genomic conflict among fetus-mates, which may affect the neonate size but not egg size. Asymmetric sibling competition, a direct consequence of matrotrophic superfetation, arises when siblings with different competitive abilities share a limited resource, as in poeciliids like *H. formosa*. Consequently, the neonates from the first brood, in which the fetus predominantly compete with less developed siblings in the subsequent broods, are larger than those from subsequent broods (Schrader and Travis, 2011). Viviparous fishes can store sperm for periods from two mon (Takagi et al., 2008) to four mon (Love et al., 2002) and selectively fertilize eggs using the stored sperm from more than one male. Amazingly, 71–100% females in populations of marine viviparous fishes gestate fetus sired by more than one male (see Pandian, 2011). Consequent to the reduced relatedness among the fetus-mates, different fetus genotypes excert different levels of maternal investments and may ultimately affect maternal fecundity through higher frequencies of fetus abortion and lower number of fully grown neonates of *H. formosa* (Schrader and Travis, 2009). Neonate size strongly affects survival of offspring (Henrich, 1988).

Among external factors, actual and predictable nutrient resource availability is a key factor that ultimately determines the number and quality of neonates. Feeding the matrotrophic molly *Poecilia latipinna* on low (50% of high) and high rations, Trexler (1997) provided experimental evidence for the reduction in fecundity and possible resorption and abortion, which are reported in many poeciliids (e.g., Borowsky and Kallman, 1976) and clinids (e.g., Fishelson and Gon, 2009). Unfortunately, Trexler provided data on fecundity and neonate size in terms of relative brood size. Brood size of high and low ration-fed groups were 12% and 7.2% of body weight, respectively. This difference of 5.5% body weight is equivalent to 76% reduction in fecundity in the low ration-fed group. Aware of the reduced income of ingested energy, the low ration-fed molly increases the egg size; hence the 21% increase in blastula stage egg poses a question whether the decrease in fecundity is due to increased egg size. The embryonic metabolic cost is 8–9% and 33.7% in the high and low-ration groups. This clearly indicates a significant reduction in maternal provisioning in the low-ration group. Also the fat content in these group decreases from 24% to 14%. Presumably, the neonate quality is considerably reduced in the low-ration group, owing to resorption of nutrients. However, in the absence of exact number of neonates born from these broods, it is difficult to confirm whether there has been an abortion of some neonates.

Investigating how changes in low and high feeding regimes in lecithotrophic and matrotrophic poeciliids, Reznick and his group have made an excellent contribution. Experimenting on high and low feeding regimes in lecithotrophic *Poeciliopsis monacha* and matrotrophic *P. prolificia* for a period of 30 d or one gestation period, Banet and Reznick (2008) found no reduction in fecundity in the proximate brood that is ready to be parturiated. To achieve it, *P. monacha* sacrifices the total number of developing embryos, i.e., cumulative fecundity from 16 to 8, neonate mass from 224 mg to 179 mg and superfetation level from 2.7 to 1.8 (Table 5) but maintains the quality of neonate with ≈ 18% fat content. By contrast, *P. prolificia* sacrifices the quality of neonates with reduced lipid content from 12% to 8%, neonate mass from 1011 mg to 963 mg and cumulative fecundity from 19 to 13 but maintains the proximate brood number (= fecundity) around 12 and superfetation level at 3.9. Clearly, low ration regime in a matrotrophic poeciliid decreases the neonate quality reducing individual (weight) size, fat content and cumulative fecundity, perhaps by resorption, *albeit* with no abortion, as indicated by equal level of superfetation. But in a lecithotrophic poeciliid, it decreases only the neonate quality by reducing individual size; however the fact that superfetation is reduced from 2.7 to 1.8 and cumulative fecundity from 16 to 8 clearly indicates that there has been an abortion of at least one brood.

In an attempt to study the response of a superfetatious matrotrophic poeciliid *Phalloptychus januarius* (gestation period: 30 d), Pollux and Reznick (2011) designed the following four feeding regimes: High (15 d) + high (15 d), high + low, low + high and low + low. The fish groups fed on low

Table 5. Reductions in neonate number and quality of lecithotrophic *Poeciliopsis monacha* and matrotrophic *P. prolificia* following high and low feeding regimes for a period of one gestation period of 30 d (compiled from Banet and Reznick, 2008).

Dependent variable	*Poeciliopsis monacha*		*Poeciliopsis prolificia*	
	High	Low	High	Low
Neonate number before treatment (control) After treatment (experimental)	9.9 7.9	10.4 8.8	12 9	12 8
Neonate mass (mg) Control Experimental	185 224	162 179	676 1011	752 963
Neonate fat content (%) Control Experimental	18.2 11.8	17.5 11.8	9.4 12.3	10.9 8.0
Superfetation (level)	2.7	1.8	3.9	3.9
Cumulative fecundity (no)	16	8	19	13

+ high and low + low regimes reduced fecundity from 5 to 3 during the second fortnight, immediately reduced neonate size from about 1.5 g to 0.8 g and quality by decreasing lipid level to less than 25% of the control value in the low + low ration group (Fig. 10). In another interesting contribution, Reznick et al. (1996) also found immediate reduction in neonate size of both lecithotrophic *Poecilia reticulata* and *Priapichthys festae,* which allocate all resources before fertilization and matrotrophic *H. formosa,* in which all required provisions are made between fertilization and parturition. Apparently, resorption appears to be a common response of all poeciliids on low food availability, irrespective whether the poeciliid is a lecithotroph or matrotroph with or without superfetation. Since Pollux and Reznick (2011) have not provided data on the superfetation level in the poeciliid subjected to differently fluctuating feeding regimes, a generalization on abortion of a batch brood can not be made as yet.

Briefly, employing the capital breeding strategy, lecithotrophic oviparous and viviparous females meet the reproductive cost from their reserves, drawing down from their storage during oogenesis, while matrotrophic viviparous females, adopting an income breeding strategy, keep diverting a fraction of their ingested energy income to nourish their developing fetus. Consequently, yolk deposition level/egg size and fecundity are determined in the lecithotrophic species prior to fertilization based on prior food availability. But in matrotrophic species, the timing at which egg size and fecundity is determined, is decoupled with fecundity being determined after fertilization based on current resource availability rather than maternal reserves. Thus a change in food availability level is immediately reflected in yolk deposition level/egg size.

Incidentally, embryos of syngnathids are protected, osmoregulated and nourished in many species during the period of male pregnancy. The brood structures vary in complexity from (i) a simple unprotected ventral pouch area for gluing the eggs (ii) to individual membranous egg compartments, (iii) protection of eggs in a pouch, (iv) closed pouch and (v) to the most complex completely enclosed brooding pouch. Urophorids like *Hippocampus, Syngnathus* possess a tail pouch, whereas gastrophorids like *Nerophis, Dunckerocampus* have abdominal pouch (Wilson et al., 2003). Table 6 provides a comparative account on egg size and patrotropic levels in selected lecithotropic and patrotrophic syngnathids. Recalculations and estimations of the available data suggest that *H. kudo* is a lecithotroph; among patrotrophs, *Syngnathus typhle* is an obligate patrotroph, drawing more than 90% of its required nutrients from the paternal source. Ripley and Foran (2009) have demonstrated the embryonic uptake of paternally-derived nutrients by injecting L-lysine [^{15}N] and 16: O-palmitic acid l- [^{13}C] into the brooding males of *S. fuscus* and *S. floridae. N. ophidion* is a facultative patrotroph and draws less than 50% of its nutrient requirement from the

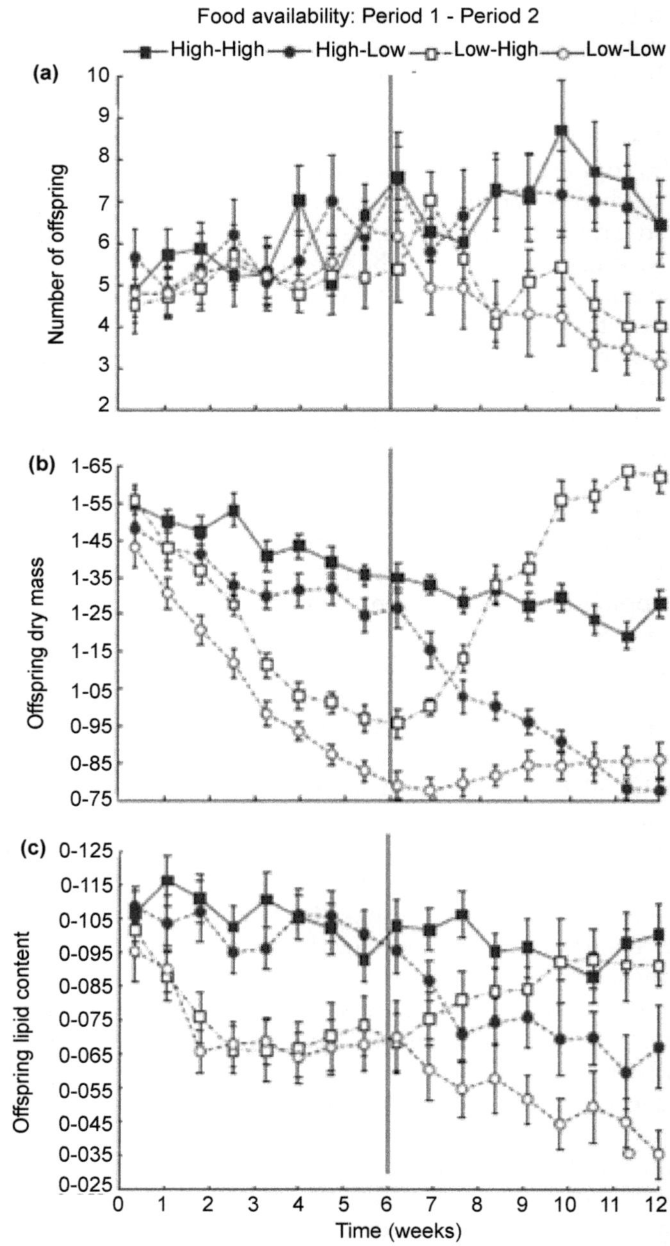

Fig. 10. Changes in neonate characteristics of *Phalloptychus januarius* during the course of experiment (a) Fecundity (no/5 day interval), (b) neonate dry weight (g) and (c) neonate lipid content (proportionate loss in dry somatic mass) (from Pollux and Reznick, 2011).

Table 6. Estimated egg size, fecundity and levels of patrotrophic contribution to the neonates of selected syngnathids.

Parameter	Lecithotrophic *Hippocampus kudo* (Thangaraj and Lipton, 2009)*	Facultative patrotrophic *Nerophis ophidion* (Berglund et al., 1986)	Patrotrophic *Syngnathus typhle* (Berglund et al., 1986)
Egg size	2.05 mg	4.65 J	23.74 J (1.1 mg)
Fecundity (no, mass [mg])	192/393	-	149/165**
Loss on embryonic metabolism (a) as per O_2 uptake (b) weight loss	 - 1.75 mg	 2.19 J 1.23 J	 21.2 J 1.21 J
Cumulative loss on metabolism	1.75 mg	3.42 J	22.41 J
Neonate size	0.3 mg	3.42 J	22.53 J
Neonate size + metabolism	-	6.84 J	44.94 J
Patrotrophic contribution (a) quantity (b) as% of egg size	 Nil Nil	 2.19 J 47	 21.2 J 93

*see Pandian (2010), ** from Berglund (1986)

paternal source; further *N. ophidion* can successfully be reared *in vitro* as much as the lecithotrophics like *H. erectus* (see Berglund et al., 1986). Surprisingly, the egg size of the facultatively patrotroph *N. ophidon* is smaller than that of the obligate patrotroph *S. typhle.*

More surprisingly, Sagebakken et al. (2009) demonstrated that the nutrients may also pass from "the missing or reduced embryos" to brooding males of *S. typhle.* It is known that there are reductions in the number of brooded embryos during the brooding period but the fate of the nutrients from these 'reduced embryos' has remained unknown. Using ^{14}C-labeled amino acids, Sagebakken et al. showed that the amino acids from these 'reduced embryos' are transferred to the brood pouch, liver and muscles of the brooding males but not to the half-siblings in the brood pouch. The possibility for a nutrients limited-male to use its embryos as a nutrient source opens up a scenario that strongly resembles filial cannibalism reported in other paternally caring fishes like the apongonids (see Pandian, 2010).

2.4 Quantum of yolk

Prior to opening an account on the quantum of yolk deposition in an oocyte of fishes, a distinction must be made between fishes with determinate and indeterminate fecundity. Among the former, the potential fecundity is fixed prior to the commencement of breeding season (*Clupea harengus, Gadus morhua,* Table 7) but it is not in the latter (e.g., *Merluccias merluccias*).

Table 7. Life history traits of marine fishes and their relevance to fecundity and spawning (compiled from Kjesbu et al., 2010).

Traits	Type I	Type II	Type III
Example	Hake *Merluccias merluccias*	Cod *Gadus morhua*	Herring *Clupea harengus*
Food	Piscivores	Carnivores	Planktivores
Reserves	Liver	Liver	Muscles, Viscera
Breeding type	Income breeder	Low capital breeder	High capital breeder
Breeder	Asynchronous	Synchronous	Synchronous
Fecundity	Indeterminate	Determinate	Determinate
Spawning area	Pelagic	Pelagic	Demersal
Spawning period	Months	Weeks	Days
Behaviour	Solitary	Solitary	Shoaling

Among the former, the stock of oocytes remains to be spawned during a breeding season declines, as successive spawnings occur (see Pandian, 2010). Hence fishes characterized by determinate and indeterminate fecundity shall have relevance to the quantum of yolk deposited in their oocytes. Considering the fishes characterized by these fecundities with their life history traits, Kjesbu et al. (2010) recognized three patterns for the European commercial fishes and the three patterns seem to hold for the other marine fishes too. A series of publications on 'Reproduction and Stock Evaluation for Recovery' (RASER) project brought out by Fisheries Research (2010) bridges the cellular events in oogenesis to the recruitment at population level and has more relevance to the fisheries reproductive biology. Enormous efforts are being made by the commercial fisheries scientists to identify the level of stock biomass that would permit sufficient recruitment to sustain the population of a stock in its constant level. In general, egg quality in terms of diameter and dry mass increases significantly with increasing female size (e.g., *Tilapia zilli*, Coward and Bromage, 1999; *M. merluccia*, Mehault et al., 2000). In a population, the larger and mid-aged females contribute a large fraction of eggs production (Korta et al., 2010). A 3-kg weighing carpenter sea bream *Argyrozona argyrozona* produces five-fold more eggs per kilogram body weight than a fish of 1 kg (Brouwer and Griffiths, 2005). The overexploitation of larger fishes has resulted in decreased spawning stock biomass and reduction in age at sexual maturity. For example, the reduced age at maturity of different stocks of the Atlantic cod *G. morhua* suggests the increased survival costs of reproduction associated with earlier maturity. The age-at-50% maturity has shifted from 10 yr in 1940 to 7 yr in 2000 in the Northeast Arctic cod; the same for body length has also decreased from 100 cm to 75 cm at the northern limit of the distribution in the NE Artic. Values for maturity-at-age and -length increase from 2 y and 40 cm in the southern warmer parts (George Bank and the Baltic) to

7 yr and 75 cm in the northern colder parts (Nash et al., 2010). The value for daily specific fecundity of the hate *M. merluccias* has been in the ranges of 8.1 and 14.1 eggs/g (ovary free) body weight during 1995–1998 and it has decreased to 4.0–5.0 eggs/g during 2001–2004 (Murua et al., 2010). Further the relatively younger but matured cods may also produce smaller eggs with decreasing survival and recruitment (Kjesbu et al., 2010). These findings have shown the negative effects of overfishing and the urgent need to implement conservation measures.

In fish eggs, the quantum of vitellogenin/yolk deposited in an egg is under the immediate control of endocrines GTH I (FSH) and E_2. However, not much is known on gene(s) responsible for the control of yolk deposition in the eggs, which, in turn, determines the egg size. The quantum of yolk deposition has to be considered with egg size and number (fecundity) together. In fishes, egg size is a highly variable trait and is controlled by a host of internal characteristics like the body size, pelagic or demersal spawning, mouth/pouch brooding or viviparous habit of the fish and external features like food availability and temperature. In general fishes produce smaller number of large demersal/brooded eggs or large number of smaller pelagic eggs. Within a species, body size is a determinant of egg size in many fishes, for example, *Oryzias latipes* (Fineman et al., 1975), *Tilapia zilli* (Coward and Bromage, 1999; see also Pandian, 2012); but in others like *Salvelinus leucomaensis* (Morita and Takashima, 1998), it may not be the determinant. Food availability is a major factor that determines the egg size. Duponchelle et al. (2000) reported a wide range of egg size in *Oreochromis niloticus* from nine reservoirs of Cote d'Ivoire. The size varies not only from one reservoir to another but also from year to year in a reservoir (Table 8). This wide variation in egg size is related to food availability.

Interestingly, Coburn (1986) summarized available information on mean egg diameter of mature oocytes of North American cyprinids belonging to 71 species from 12 genera; the oocyte size ranges from 0.64 mm in *Notropis buchanani* to 2.0 mm in *Campostoma anomalum.* Mean size is strongly conserved in several taxa; however, with increasing mean size, there are wider variations in the egg size among species. Cutting across species of different phylogenetic lines, habitats and niches, Coburn (1986) found a strong correlation between egg size and vertebral count (Fig. 11); the correlation suggests that for every 0.17 mm increase in egg size, one additional vertebra is formed. The origin of these eastern North American cyprinids is dated to the Pleistocene. The relative constancy of egg size and vertebral count in these cyprinids over such an extended period involving climatic shifts and others suggests the probable conservation of this relationship. This preliminary finding of such a high correlation by Coburn is certainly interesting. Two hypotheses are suggested: 1. Vertebral number is not casually linked to egg size. But, both egg size and vertebral

Table 8. Regulation of yolk deposition and hence egg size by environmental and internal factors.

Factor	Example/explanation
Pelagic spawners	Prolific producers of small (0.54–0.64 mm, see Pandian, 2010) eggs
Demersal spawners	Smaller number of large (e.g., 1.46 mm *Tilapia zilli*, Coward and Bromage, 1999) eggs
Smaller fishes	Produce large but smaller number of eggs, e.g., For a 3-fold increase in body size, 20% decrease in batch fecundity of Hawaiian apogonidian cardinal fishes (see Pandian, 2010)
Montane spawners	Lecithotrophic Cuban poeciliids inhabiting mountain streams produce large eggs and smaller broods but those in low land wetlands small eggs and large broods (Ponce de Leon et al., 2011)
Mouth brooders	Produce relatively smaller number of large eggs
Pouch brooders/ Viviparous fishes	Produce smaller number of eggs (e.g., 17 pouch-brooded egg/g body weight, 8 neonates/g body weight)
Body size	A major determinant of egg size in many fishes (e.g., *Tilapia zilli*) but not in others (e.g., *Salvelinus leucomaensis*)
Food availability	In *Oreochromis niloticus* (Duponchlle et al., 2000), egg size is 4.5 mg in 1995 but as much as 7.9 mg in 1996 at the Kossou reservoir; it is 7.5 mg in 1995 at the Aryame reservior
Temperate fishes	Finnish population of *Barbatula barbatulus* produce 2.5 times larger eggs than that in southern England (see Pandian, 2010)
Sexually dimorphic eggs?	Within a single brood, larger males and smaller females are produced by the Trinidadian guppy *Poecilia reticulata* in populations with an excess of females (Barbosa and Magurran, 2010)

count may vary in response to food availability and environmental factors like temperature, respectively. 2. The vertebral count is dependent on egg diameter. The correlation coefficient of 0.66 found in the study supports the second hypothesis. However, it is not known whether the difference in egg size within a species or population can induce changes in the number of vertebra. The first batch egg in *Danio rerio* measures 1.09 mm and the size increases to 1.21 mm at the mid reproductive age of about 100 d (for cod see Mehault et al., 2010). The estrone-treated XX, XY and YY females produce egg measuring 1.24, 1.21 and 1.71 mm, respectively (Fineman et al., 1975). The body size of MT-treated *Betta splendens* is reported to be half the length of the untreated male (Kirankumar and Pandian 2002). But it is not known whether the decrease in body size is due to the decrease in the number or size of vertebra.

The formation of vertebra is a process dependent on space for increasing cell number required for the formation of additional vertebra and rate of development. Space or cell number can be related to egg size but the rate and duration of somite formation are dependent upon an interaction between genetic and environmental parameters. It is then important to know whether the morphogenetic control of vertebral count reside in the egg size.

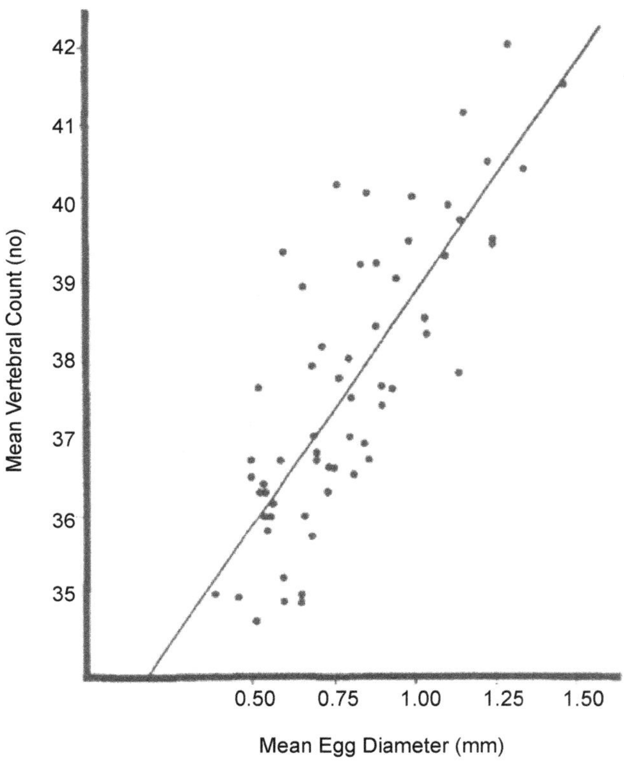

Fig. 11. Relationship of vertebral count to mean egg diameter of 71 cyprinid species of North America (from Coburn, 1986).

A number of fishes with different ploids simultaneously produce eggs of different sizes. For instance, the unusual F_3 hybrids (*Carassius auratus gibelio* ♀ x *Cyprinus carpio* ♂) simultaneously produce haploid (13 mm), diploid (17 mm) and tetraploid (20 mm) eggs (Liu SJ et al., 2007). Estimates of vertebral counts in progenies arising from these haploid and diploid eggs produced by the same tetraploid parents may provide an excellent opportunity to know whether the vertebral count is dependent on egg size.

In teleosts, the size of eggs produced by an individual female is not necessarily a species-specific trait; it may vary geographically, seasonally and even ontogenetically; egg size is variable even within a female and among females of a species. To this day, egg size has been suggested to be genetically fixed in *Cyprinodon pucosensis* only (Garrett, 1982). The theory predicts that mothers should produce offspring of a size that maximizes maternal fitness, as they have a finite quantity of resources that has to be allocated for reproduction, besides sustaining their own body growth

(Bownds et al., 2010). Hence the primary choice for a female is between reproductive effort and growth on one hand, and egg size on the other, which is under independent selection. For over 60 yr fishery biologists have sought to understand the selective process that drives the variation. Two predictions have been central to the study of how reproductive effort is reflected in egg size: 1. Egg size is inversely related to egg number and 2. Given optimal size of egg, egg number (batch fecundity) increases with female body size, as the available space within the body cavity to accommodate the ripe ovary increases (see Pandian, 2010). Incidentally, the space limitation within the body cavity inherent in batch spawner does not arise in fishes like the Owens pupfish *Cyprinodon radiosus*, which yolks only a few oocytes at a time continuously over a period of time (Mire and Millett, 1994).

Patterns of egg size and fecundity have been extensively studied in many fishes, especially in salmonids. Salmonids produce a small number of large eggs compared to other teleosts. Rapid maternal growth is associated with small but numerous eggs, whereas slower maternal growth is associated with fewer large eggs (e.g., Tamate and Maekawa, 2000). Anadromous salmonids present particularly good opportunities to test whether the slow growing female should produce fewer but large eggs; their earlier growth rate may be slower in limited freshwater than that after their migration to the sea, where opportunities for growth are better. Analyzing the temporal variations in reproductive effort of the semelparous *Oncorhynchus kisutch* and *O. tshawystcha* at the University of Washington Hatchery over three to five decades, Quinn et al. (2004) concluded that the egg size of these salmonids does not vary with growth rate either during freshwater or marine phases and is probably determined much later in life, corresponding to the number of remaining oocytes and the energetic constraints at maturity (see also p 29). This may be a contrast to the iteroparous *Simochromis pleuropilus*, in which resource availability at the juvenile stage seems to determine the egg size (Taborsky et al., 2007). Nevertheless, finding no evidence for the older/larger females to produce larger eggs with intrinsic higher fitness, Marshall et al. (2010) cautioned that larger eggs and hence larger offspring may fare better than smaller ones, when reared in the *same environment* but a *different environment* may elicit selection for different optimal sizes of eggs and hence offspring.

Further offspring from smaller eggs may grow faster and catch up with the size of those arising from larger eggs. As the somatotropic axis is a key regulator of early growth in vertebrates, Segers et al. (2011) have brought the very first evidence for potential link between maternal gene like (growth hormone receptor) *GHR* in smaller eggs and its elevated expression levels during the stages from alevin (8 dpf) to juvenile (38 dpf) facilitates faster growth of offspring arising from smaller eggs of *S. pleuropilus*. The expression level of growth hormone receptor *GHR* is higher in large eggs (19 mg each) than in smaller eggs (17 mg each) until 2 dpf. However subsequent to hatching on 5 dpf, the pattern is reversed from 8th to 38th dpf, i.e., the expression level of *GHR* in alevins and juveniles arising from smaller eggs is significantly higher than that in alevins hatched out from large eggs. There is also no significant difference in the expression levels of *IGF-1* and *IGF-2* among the offspring of larger and small egg classes and there are positive correlations between *GHR* and *IGFs*. This enables the young ones from small eggs to catch up fully in size within eight ws with conspecifics from larger eggs.

There is growing evidence that maternal stress imposes deleterious effects on her developing eggs/embryos. It may be adaptive for mothers to hormonally program (Weaver et al., 2004) their offspring to be suited for a particular environment. Exposed to the threat of predation during oogenesis, the stickleback *Gasterosteus aculeatus* produce larger eggs (2.8 mg) with higher cortisol (2.2 ng/ml) content than the control mother (egg size : 2.4 mg; cortisol : 1.5 ng/l, Giesing et al., 2011). Increasing cortisol level in eggs seems to be a common feature of mothers suffering stress. Increased daily chasing during oogenesis is known to cause elevated cortisol levels in both maternal plasma and eggs of coho salmon (Stratholt et al., 1997). Briefly, the increased egg and fry size provide the fry an advantage in a high predation environment (Taborsky et al., 2007). While the predatory risk during oogenesis affects egg size in some fishes like *G. aculeatus,* Taborsky (2006) has advanced it to juvenile mothers receiving limited or unlimited food supply. The juvenile mothers of the cichlid *Simochromis pleuropilus* raised with a little food source produce larger eggs (9 mg dry weight each) than those provided with unlimited food supply (8 mg dry weight/egg). Evidently, food resource availability to juvenile mothers seems to be a more critical determinant than that to maturing mothers. Incidentally, there are others like the salmonids and coral reef fishes, in which maternal stress causes the mothers to produce small low-quality eggs (Campbell et al., 1994; Contreras-Sanchez et al., 1998; McCormick, 1998, 1999). Apparently, the quantum of yolk deposition and hence the egg size is likely to be controlled by a number of genes.

Fragmentary information available on xenogenesis provides an insight into the genetic determination of egg size. Following the transplantation of PGCs of rainbow trout into the masu fry, the recipient adult masu differentiates a 'hybrid gonad' generating eggs derived from both the endogenous and transplanted PGCs. Consequently, the donor-derived xenogenic rainbow trout fry are hatched on the 34th dpf (day post-fertilization) but those of masou between 36th and 41st dpf (Takeuchi et al., 2004). This observation implies that the hybrid gonad simultaneously produces eggs of different sizes requiring different incubation periods.

Similarly, Yamaha et al. (2001) transplanted a 'sandwich' from the lower blastoderm presumably containing PGCs of goldfish into the blastoderm of sterile triploid crucian carp. Not only was fertility recovered (see also Yamaha et al., 2003) but also the xenogenic chimera, on attaining sexual maturity, simultaneously produced haploid goldfish eggs (1.52 mm each) and triploid gynogenic crucian carp eggs (1.67 mm each). Having a hybrid gonad with both PGCs derived from the endogenous triploid crucian carp and exogenous diploid goldfish, the chimeric xenogenic triploid crucian simultaneously produced smaller haploid goldfish eggs and larger triploid crucian carp eggs. Conceivably, the respective PGCs seem to possess gene(s) responsible to fix the approximate egg size in these chimeric fishes.

Incidentally, Saito et al. (2008) generated xenogenic pearl danio transplanting its single PGC into blastoderm of danio and produced all male chimeras. Subsequently, some of these xenogenic chimeras were feminized by estrogen administration. Expectedly, these female chimeras produce the donor-derived pearl danio's eggs of the size of 700 μm, typical of that (706 μm) of danio but not like that of pearl danio (816 μm). If PGCs are the determinants of the egg size, then it is not clear why the pearl danio PGC-derived danio, on feminization, produces eggs, whose size is characteristic of danio. Information available on this aspect is fragmentary to make any generalization. While awaiting more information from the ongoing researches of Dr T Saito, it is suggested that the scientists working on this frontier area may also provide the desired information to know whether PGCs determine the egg size.

Available bits and pieces of information seem to provide circumstantial evidence for the genomic control on the quantum of yolk deposited and egg size in fises. Table 9 lists the recorded multiploid eggs produced by gonochoric and hybrid fishes. Among fishes characterized by different ploidy levels, there are a few rare cyprinids and a cobitid that simultaneously produce eggs of different sizes (Table 10) and thereby provide an excellent opportunity to show that ploidy level in the eggs seems to control the quantum of yolk deposited in eggs of different sizes. Among them, the cobitid loach *Misgurnus anguillicaudatus* simultaneously produce haploid and diploid eggs as well as haploid and triploid eggs; triploids produce diploid and triploid eggs, and tetraploids produce diploid and tetraploid eggs. An interploid hybrid between 2n ♀ and 4n ♂ loach simultaneously produces 49% smaller haploid eggs, each measuring 1.06 mm and 51% of larger triploid eggs of 1.41 mm size of each. The pentaploid loach also simultaneously produces haploid smaller eggs and triploid larger eggs (Matsubara et al., 1995). Hence there is an increase in egg size with increasing ploidy level, albeit the increase is not proportionate. In the triploid instead of the expected size of the egg of about 1.59 mm, the observed increase is 11% only. The increase in egg size of the hybrid triploid between *Carassius*

Table 9. Recorded multiploid eggs produced by gonochorics and their hybrid fishes. PE indicates unreduced eggs produced through premeiotic endomitosis; POAg indicates the unreduced eggs through post-ovulatory ageing (from Pandian, 2011).

Classification of ploidy in eggs
I 2n undreduced eggs *1a Bisexuals*
2n *Misgurnus anguillicaudatus* PE
3n *M. anguillicaudatus* PE
4n *Cobitis biwae* PE
2n *Carassius gibelio* PE
1b Hybrids
3n (*Oryzias luzonensis* ♀ x 2n *O. latipes* ♂) PE
3n (*O. latipes* ♀ x 2n *O. curvinotus* ♂) PE
2n (*Fundus heteroclitus* ♀ x 2n *F. diaphanus* ♂) PE
3n (*Salmo salar* ♀ x 2n *S. trutta* ♂) PE
3n (*S. trutta* ♀ x 2n *S. salar* ♂) PE
3n (*C. auratus gibelio* ♀ x *Cyprinus carpio* ♂) PE
3n (*C. carpio* ♀ x 2n *C. auratus gibelio*) PE
3n (*Lepomis gibbosus* ♀ x *L. cyanellus* ♂) PE
3n (*L. gibbosus* ♀ x *L. macrochirus* ♂) PE
II 3n unreduced eggs *2 Bisexuals*
3n *M. anguillicaudatus* PE
Carassius lansgdorfi PE
Tinca tinca POAg
III 4n Unreduced eggs *3 bisexuals*
4n *M. anguillicaudatus* PE
4n *C. biwae* PE
4n *C. auratus lansgdorfi* PE

auratus cuvieri ♀ and *Cyprinus carpio* ♂ is also 14% only (Zhang et al., 1992a). In tetraploid *Megalobrama ambycephala*, the increase egg size is about 25% of the expected size (Liu et al., 2007). The unusual F_3 hybrid (*C. auratus gibelio* ♀ and *C. carpio* ♂) simultenously produces haploid (13 mm) and diploid (17 mm) euploid eggs and 4n (20 mm) aneuploid eggs, i.e., the increase in size of (aneuploid) diploid egg of this hybrid line is also about 14% of the expected increase. Clearly, increase in genome size from diploid to triploid cyprinids and cobitids increases yolk deposition and thereby increases the egg size from 11 to 14% of the respective expected size. However, the increases from haploid to diploid (euploid) egg in *M. ambycephala* and diploid (aneuploid)

Table 10. Recorded multiploid eggs simultaneously produced by gonochores and their intraspecific and interploid hybrids (from Pandian, 2011).

1a Haploid-diploid eggs
2n *M. anguillicaudatus*
3n (*C. auratus cuvieri* ♀ x *C. carpio* ♂) ♀ F_1 x 2n *C. auratus cuvieri* ♂
1b Haploid-triploids eggs
3n ♀ *M. anguillicaudatus* ♂
2n ♀ x 4n ♂ *M. anguillicaudatus*
3n ♀ *Squalias alburnoides*
3n xenogenic crucian carp
2a Diploid-triploid eggs
2n ♀ x 4n ♂ *M. anguillicaudatus*
2n ♀ x 3n ♂ *M. anguillicaudatus*
2b Diploid-tetraploid eggs
4n *Megalobrama ambycephala*
3a Multiple ploid eggs
4n *M. anguillicaudatus* n, 2n, 3n, 4n eggs
Multiple 3n carp 1.5n, 2n, 3n eggs

egg size of the hybrid (*C. auratus gibelio* ♀ and *C. carpio* ♂) are 12–14% of the respected expected size. Hence increase in ploidy increases the egg size but not proportionately (see also p 98).

2.5 Maternal hormones

Unfertilized eggs contain a wide complement of maternally contributed endogenous hormones. Generally, developing embryos display declining levels of these hormones during the yolk sac absorption stage and increasing levels during hatching (Hines et al., 1999). Yamamoto (1969) proposed that the endogenous sex steroids act as natural inducers of gonadal differentiation in fishes. Cloning and sequencing DNA encoding steroidogenic enzymes of tilapia, Nakamura and Nagahama (1985) synthesized corresponding peptides and used them to raise polyclonal antibodies against four steroidogenic enzymes essential for the biosynthesis of all major steroid hormones including androgens and estrogens (Kobayashi et al., 1996; Chang X-T et al., 1997). Of them, the cytochrome p450 aromatase antibodies stain theca cells and granulosa cells in the ovary but not the Leydig cells in the testis. Thus steroid producing cells (SPCs) in the presumptive ovaries at the undifferentiated and differentiating stages express the steroidogenic enzymes for 17β-estrodiol (E_2) biosynthesis from cholesterol. These observations have been considered as evidence in support of Yamamoto's proposal (see also Pandian, 2010). However, the SPCs first appear after the completion of gonadal differentiation in *Oryzias latipes* (Satoh, 1974; Kanamori et al., 1985),

Oncorhynchus mykiss (van den Hurk et al., 1982), and *Anguilla japonica* (Chiba et al., 1997). Besides, 3β-hydroxysteroid dehydrogenase, a key enzyme for steroidogenesis, is detected after differentiation in medaka (Hishida, 1969). Besides, Petkam et al. (2002) elucidated the metabolic pathways used by *O. mykiss* from hatching to yolk absorption stage. Based on conversion of tritium labeled substrate, the quantum progesterone (P_4) metabolism is more than androgens and estrogens. Except for the limited two-directional conversion of the androgens T, A_4 (androstenedione) and the estrogens E_1 and E_2, no other steroid of biological potency is synthesized.

However the onset of steroidogenesis occurs prior to the initiation of gonadal differentiation in *O. rhodurus* (Nakamura and Nagahama, 1993). In *O. kisutch*, the gonad remains undifferentiated between hatching and 59–70 dpf but by 77 dpf the ovarian differentiation becomes evident (Feist et al., 1990). In coho salmon too, definite increase in T and KT levels following the initial decline until the 30th dpf is reported. For example, the non-eye and eyed embryos, in which the gonads remain still undifferentiated, the increases are from 413 to 611 ng/embryo for T, 106 to 223 ng/embryo for KT, while the levels of progesterone (P_4) and 17α-hydroxy-20β-dihydroprogesterone (DHP) remain around 320 and 450 ng/embryo, respectively. Only E_2 precipitously decreases from 500 to 18 ng/embryo. Hence coho salmon is capable of synthesizing T and KT either before or concomitantly with the onset of gonadal differentiation and its patterns of steroid level have "distinguished themselves dimorphically before gonadal differentiation". A similar sexual dimorphism of T during early development has also been documented in tilapia (Rothbard et al., 1987). The findings of Govoroun et al. (2001) have some implications on the role played by the endogenous steroids in gonadal differentiation of *O. mykiss*. These authors have examined the expression of key steroid enzyme genes during natural differentiation and androgen-induced differentiation in male and female gonads at 55 dpf, i.e., two weeks before histological differentiation of gonads, developed from (XX) eggs that have been fertilized by sperm of sex reversed neomale (XX) or supermale (YY). The following inferences have been made: 1. The steroid enzymes required for a complete steroidogenesis are available at the early stage in both sexes of rainbow trout and thus estrogens in females and 11-oxygenated androgens in males can be effectively synthesized. 2. Steroid biosynthetic capacity preceeds histological gonadal differentiation (Hines et al., 1999). Immune-histochemical analysis using $p450_{sec}$, 3βHSD and $p450_c17$ specific antibodies shows that the enzymes are present prior to the ovarian differentiation but not in the testis before histological differentiation. Thus gonadal differentiation is potentially active before histological differentiation. 3. Their position in this cascade of events is before the stage, in which the fate of the gonadal sex is irreversibly determined, making them key players in the control of gonadal phenothypic differentiation. 4. The exogenous (11β-hydroxyandrostenedione) 11βOHΔ4 acts as a substitute for the endogenous 11-oxygenated androgens. The inhibition of *p450arom* in females treated with androgens mimics is observed in the male differentiating testis, indicating that the role of 11-oxygenated androgen in testis differentiation may be to inhibit *p450arom* gene expression and hence estrogen synthesis (Govoroun et al., 2001). In the light of such contrary information, the function of endogenous sex steroids in early sexual differentiation remains unclear.

It is in this context, the contributions by Iwamatsu et al. (2005, 2006a, b) are interesting, especially their experimental designs: (i) For experiment *O. latipes* is selected; unlike in salmonids and cichlids, sex is determined in *O. latipes* by a single gene *Dmy/Dmrt1bY* (Matsuda et al., 2002; Nanda et al., 2002) at the developmental stage 36 (= 7 dpf), i.e., prior to hatching (on the 10th dpf); incidentally, Yamamoto (1969) too postulated his concept on steroids as natural inducers of sex differentiation from his studies on *O. latipes*, (ii) In the selected S-rR medaka strain, male and female embryos can be recognized by orange and white color, respectively and (iii) Measurements of 17-β estradiol (E_2) and testosterone (T) in oocytes, eggs, embryos as well as fluids of the ovarian cavity may help to know whether the steroids are synthesized by the embryos or simply absorbed from the steroidogenic granulosa cells in the follicle.

From the results reported by Iwamatsu et al. (2005), the following is summarized: (1) The E_2 content of fully grown follicular oocyte is 23 pg/oocyte (equivalent to 25.6 ng/ml) in comparison to 23.3 ng/ml in the ovarian fluid. Steroid profile of egg essentially parallels to that of ovarian fluid of *O. kisutch* (Feist et al., 1990). Apparently, the endogenous E_2 in mature oocyte of medaka results from diffusion of the steroid E_2 from the steroidogenic follicular cells to the ooplasm and yolk of the pre-ovulatory oocytes during oogenesis. Steroids are lipophilic and easily cross cell membranes. (2) On ovulation, the E_2 content precipitously decreases to 6.0 pg/egg (6.7 ng/ml), further down to 4.1 pg/egg after 6–12 hr post-fertilization (hpf) and to 3.2 pg/gastrula and finally to 0.9 pg/embryo (16 somite stage) and remains at the same level at hatching. In *O. kisutch* and *O. rhodurus* also, the E_2 decreases to less than 48 ng/embryo (Feist et al., 1990). However it is likely that E_2 is not synthesized in medaka embryos but is rather utilized on embryonic metabolism. (3) More importantly, there is no difference in E_2 content (0.5 pg/embryo) between orange colored male and white colored female of embryos at the age of 5–10 hpf. This suggests that steroids are not sex determining factors but they are rather the results of sex differentiation (see Hines et al., 1999). From their studies on tilapias, Rothbard et al. (1987) also concluded that "the steroids stored within the eggs do not allow differentiation into either sex during initial period". Clearly, the endogenous E_2 seems not a natural inducer of sex differentiation in *O. latipes*. Hence, it is likely that *O. latipes* in which sex is decisively determined by a single gene and sexual dimorphism becomes evident prior to hatching, adopts a pattern of sex differentiation, while others like *O. kisutch, O. rhodurus,* in which sex differentiation is apparent only after hatching follow a different mechanism of sex differentiation. (4) Experiments using radiolabels by Piferrer and Donaldson (1994) indicated that the yolk can retain the incorporated exogenous E_2. The medaka eggs incubated for 24 hr after fertilization in salines containing 10 ng E_2/ml induces functional sex reversal from male genotype to female genotype in 78% of the treated embryos. This suggests the overriding role of exogenous E_2 on the genetic sex determination, when late gastrula (1dpf) embryos are treated at high exogenous E_2 conconcentration and also confirms the findings of Yamamoto.

The role of androgen on sex differentiation is a more complicated, as aromatizable androgens like T, 17α-methyltestosterone (MT) can be converted into estrogen. Hence, Iwamatsu et al. (2006a) measured the T contents of eggs and embryos and examined sex differentiation of embryos exposed to selected levels of exogenous T and MT as well as non-aromatizable 17α-methyldihydrotestosterone (MDHT). From 4.7 pg/oocyte (7.8 ng/ml), T precipitously decreases to 2.7 ng T/egg (3.6 ng/ml) and subsequently to 1 pg/fertilized eggs and 0.33 pg/2-d old embryo and remain at that level until hatching. On incubation of 2-dpf eggs in media containing 100 ng T/ml and 500 ng MT/ml, E_2 increases from the control level of 0.5 pg/embryo to ≈ 0.9 pg/embryo and ≈ 1.1 pg/embryo, respectively on the 8th dpf. Expectedly, there is no increase in E_2 level on exposure to 100 ng MDHT/ml during the period between 2 and 10 d-post-fertilization. Embryos incubated at 10–1000 ng T/ml undergo sex reversal including paradoxical reversal from male genotype to normal fertile female phenotype. Treated at 0.1 ng MT/ml, the embryos also undergo sex reversal. The paradoxical effect of androgens on sex differentiation results from the conversion of androgens to estrogen (Piferrer and Donaldson, 1991; Piferrer et al., 1994a). Surprisingly, the exposure to 0.1–1.0 ng MDHT/ml also induces sex reversal from female to male phenotype. From the results reported by Iwamatsu et al. (2006a), the following may be inferred: (1) The levels of E_2 and T in the egg and embryo of medaka are not critically adequate to induce natural sex reversal, (2) The embryo has the machinery to convert exogenous aromatizable androgens to estrogen, (3) The developmental stage 36 (= 7 dpf), when *Dmy/Dmrt1bY* is expressed in XY embryo (Matsuda et al., 2003), is insensitive to exogenous sex steroids (Scholz et al., 2003). Irrespective of the expression of Y chromosomal *Dmy,* the steroids alter the cascade of expression of downstream acting gonad-specific genes such as *Dmrt1.* Thus sex steroids seem to exert a priming effect on the sex determining pathway at an early stage of development long before the initiation of gonadal differentiation. Incidentally, the exposure of fertilized eggs of medaka to sub-lethal concentrations of aromatase inhibitor (AI) Fadrozole and/or methylandrostenediol (MA) has shown the estrogen-independent ovarian differentiation and a key role played by estrogen receptor activation in ovarian differentiation of medaka (Kawahara and Yamashita, 2000). Furthering their experiments, Iwamatsu et al. (2006b) showed that *in vitro* treatment of immature oocyte with non-aromatizable androgen and MDHT during the period immediately before the pre-ovulatory maturation is adequate to induce complete sex reversal of genetic females to functional males. A series of these experimental observations suggests the existence of a sex determinant in the oocyte sensitive or labile to androgen and manipulability of sex prior to fertilization.

To decisively understand the role of endogenous sex steroids, Rougeot et al. (2007) estimated the levels of T, E_2 and KT in unfertilized eggs, embryos and lavae/juveniles of the European perch *Perca fluviatalis*. The perch eggs have been fertilized by XY male in the 'sexual' group and XX sex reversed neomale in the 'all female' group. This design of experiment is to have a better understanding of the role played by the endogenous sex steroids in the presence and absence of Y chromosome. For a broader and better understanding, the mean values reported by Rougeot et al. (2007) are compiled in Table 11. According to the authors, the differences reported for T, E_2 and KT do not significantly differ in these groups during the period from unfertilized eggs to larval stage, except for E_2 and KT during the larval stage as well as KT in the egg stage. At the completion of the experiment, the sex ratio remains 0.5 ♀ : 0.5 ♂ in the 'sexual' egg group and 1♀ : 0 ♂ in

Table 11. Mean levels (pg/g) of sex steroids testosterone (T), 11-ketotestosterone (11-KT) and estrogen (E_2) present in unfertilized 'sexual' and 'all female' eggs, embryos (fertilization to hatching) and larval/juvenile stages of the European perch *Perca fluviatalis* (compiled from description of Rougeot et al., 2007).

Developmental stage	'Sexual' (XX♀/XY♂)	All-female (XX♀/XX♂)
	Testosterone	
Egg	1634	1513
Embryo	1331	1281
Larva	542	727
	17β-Estradiol	
Egg	497	550
Embryo	494	254
Larva	726	156
	11-ketotestosterone	
Egg	1513	629
Embryo	760	718
Larva	432	0

the 'all female' egg groups. Besides, the following may be inferred: 1. The perch eggs are endowed with maternally-derived T, E_2 and KT, and these 'maternal' endogenous sex steroids can be used as precursors to induce the initial sex differentiation of the brain and/or gonads. 2. The presence of E_2 and KT in embryos and larval stages before the histological differentiation of the gonads, suggests extra-gonadal synthesis of these sex steroids. 3. The high levels of KT in 'sexual' eggs suggest that this steroid is particularly important for male sex differentiation in the perch. 4. In the perch, gonadal differentiation is perhaps controlled by KT to E_2 ratio: an excess KT, as in 'sexual' egg group prevailing during embryo-larval/juvenile stages, may induce the development of testes and excess E_2 present in the larval/juvenile stage may induce the development of ovary (Rougeot et al., 2007). However, it is still not clear whether the maternally-derived KT induces the testicular development in the presumptive males and/or the maternally derived high level KT serves in the egg as precursor that ultimately leads to testicular development.

2.6 Sexual receptivity

Investigations on gonadectomy and hormonal manipulations have revealed the important role played by steroidal hormones on reproductive behavior. Incidentally, gonadectomy and hormone therapy experiments have now become a standard procedure to test the effect of gonadal hormones (Goncalves and Oliveira, 2011). Ovariectomy results in the loss of sexual receptivity in female *Poecilia reticulata* (Table 12); the receptivity is not restored on receiving progesterone, cortisol + corticosteroid or MT. But it is restored in an ovariectomized and hypophysectomized female, when administered with E_2 (Liley, 1972). This clearly indicates that the ovarian hormone E_2 is essential to render the female receptive to mating. Ovariectomy also results in decreased courtship display in cichlids *Hemichromis bimaculatus* and *Sarotherodon melanotheron* (see Oliveira and Goncalves, 2008).

In externally fertilizing species, prostaglandins (PGs) produced by mature oocytes signal a ready to spawn state and thus promote sexual receptivity and spawning (see p. 80-81). On the administration of PGF2α, the females of *Barbionymus gonionotus, Carassius auratus* (Cyprinidae), *Macropodus opercularis* (Anabantidae) and *Cichlasoma bimaculatum* (Cichlidae) commence spawning activity (Table 12). In fact the male goldfish injected with PGF2α displays female-like sexual behavior not different from PGF2α-injected female (Stacy, 1976). Progestogen (PGF2α) is suggested to enter the circulation and act on the brain circuit to promote spawning behavior (Stacey and Peter, 1979). Response to PGF2α administration is similar in ovariectomized and intact females, suggesting that spawning behavior may be induced by PGF2α produced in the oviduct during ovulation (Kobayashi and Stacey, 1993).

Table 12. Effects of ovariectomy and hormonal manipulation on female's reproductive behavior in fishes. + = increase, – = decrease, 0 = no effect, ND = not described, * = ovariectomized (compiled from Oliveira and Goncalves, 2008, with kind permission of Science Publishers and Prof RF Oliveira).

Species and sex	Hormone	Sexual receptivity	Courtship display	Spawning/ oviposition
Hemichromis bimaculatus ♀*		ND	–	ND
Sarotherodon melanotheron ♀*		ND	–	ND
Carassius auratus ♀*		ND	0	0
C. auratus ♀	PGF2α	+	+	+
C. auratus ♂	PGF2α	+	ND	ND
Macropodus opercularis ♀	PGF2α	ND	ND	+
Cichalosoma bimaculatum ♀	PGF2α	ND	ND	+
Barbionymus gonionotus ♀	PGF2α	+	+	+
Salaria pavo ♀	AVT	ND	+	ND
Poecilia reticulata ♀*		–	ND	ND
P. reticulata ♀*	E_2	–	ND	ND
P. reticulata ♀*	MT	0	ND	ND
P. reticulata ♀*	Progesterone	0	ND	ND
P. reticulata ♀*	Cortisol + Corticosterone	0	ND	ND
*Thalassoma bifasciatum**	KT	–	+	0

GnRH is also implicated in the female's receptivity. In *Astatotilapia burtoni,* the size of POA-GnRH neurons is smaller in brood-attendant females, in comparison to females that are immature and post-reproductive state (see Oliveira and Goncalves, 2008).

2.7 Stress and spawning

Most food- and ornamental-fishes reared in captivity suffer from one or other reproductive dysfunction: failure to undergo final oocyte maturation (FOM), ovulation and/or spawning. In a commendable contribution, Zohar and Mylonas (2001) summarized the problems, traced the causes and suggested solutions. The dysfunctions may result from combination of captive-induced stress and the absence of appropriate environmental triggers. The stress may lie on hypothalamus-pituitary-interrenal [HPI] stress axis or on hypothalamus-pituitary-gonadal [HPG] reproductive axis (Meghan et al., 2011). Reproductive problems are more serious in female brooders (Zohar and Mylonas, 2001). On the failure of spawning, ovulated eggs are retained in the abdominal cavity and are subsequently resorbed in salmonids. In others, like catfishes, the ovulated eggs may undergo post-ovulatory ageing and on fertilization, hatch out as triploid progeny, due to duplication of the haploid eggs or by facilitating the entry of two sperm (see Pandian, 2011). To induce ovulation and spawning the earliest method used is to inject freshly ground pituitary containing LH. Subsequently, purified GnRH, which induces the fish's own gonadotropin from the pituitary, has been employed. Highly useful synthetic agonist of GnRHa and its incorporation into polymeric sustained release delivery system, which releases the hormone over a period of days and weeks, have been developed (Table 13). This has ensured multiple spawning in fishes characterized by asynchronous or repeat spawners.

Hypophysation, which is still practiced in some developing countries, involves a more or less standardized procedure with a small priming dose (10–20% of total) followed by a large resolving dose of 2–10 mg/kg body weight administered 12–24 hrs apart. The major drawbacks in hypophysation are: (i) variability in pituitary LH content, (ii) side effects that may be caused by other hormones present in the pituitary and (iii) potential transmission of diseases from donor to recipient. On the other hand, the use of GnRH pepitides to induce spawning has important advantages: 1. GnRH and its agonists are small decapeptides that do not trigger an immune response and can be re-used with no reduction in their efficacy. 2. By acting at a higher level of H-P-G axis and inducing the release of endogenous LH, the GnRH 'repairs' the entire sequence of endocrine dysfunction. 3. It can be synthesized in its pure form, and hence avoids the scope for transmission of diseases and 4. Most importantly, because of the structural similarities, the use of GnRH is generic and the same GnRHa has been successfully

Table 13. Production, administration and advantages of GnRHa delivery systems (compiled from Zohar and Mylonas, 2001).

Components and production	Administration-advantages
Cholesterol-cellulose (CC). Varying the C-C ingredients in matrix 'fast' and 'slow' implants that retain elevated plasma GtH level for 8 d and 8 wk, respectively	Solid, cylindrical pellets of 3 mm diameter that can be implanted intramuscularly with a scalpel. Easy and inexpensive to fabricate but the disadvantage is that the releasing duration is extremely variable
Lactic acid-glycolic acid (LGA). Biodegradable microspheres (5–200 μm) produced by double emulsion solvent evaporation method, in which GnRHa is dissolved in microscopic water droplets, followed by entrapping in a matrix of LGA	Microspheres are suspended in viscous vehicle and injected into muscle with an 18-gauge needle on a volume to weight basis. Hence the method is suitable for small and large fishes. It produces immediate release and lasts for a few months. No harmful residuals are left in 'retired' broodstock
Ethylene-vinyl acetate (EVAc). Solid, monolithic non-degradable implant. GnRHa is mixed with an inert bulky agent and the mixture is entrapped in EVAc matrix	Fabricated discs of 2 mm and administered intramuscularly using an implanter. Unlike the others, it has a long shelf-life up to 3 yr, when stored in desiccator at –20°C

applied to a wide range of fishes. The effective doses are 1–15 mg GnRH/kg or 1–100 μg GnRHa/kg. Three types of GnRHa delivery systems have been developed (Table 13). These systems involve changes in chemicals used in the matrix and have their own inherent advantages and limitations (Zohar and Mylonas, 2001).

Table 14 lists the advantages of the delivery systems in the context of dopamine antagonists (see Table 2), reproductive season and responsiveness of oocytes of brooder fishes, that have been kept in captivity for long time. The following may be noted: 1. Luteinizing hormone releasing hormone analogue (LHRHa), as in *Clarias gariepinus, Centropristis striata,* or GnRHa, as in others, serves as equally good inducers of spawning, though a

Table 14. Effects of LHRH/GnRH and dopamine antagonists on oocyte maturation and spawning in fishes.

Species & Reference	Reported observations
Clarias gariepinus Richter et al. (1987)	Pimozide/LHRHa induces pre-ovulatory surge, oocyte maturation and ovulation
Centropristis striata Balinsky et al. (2005)	Implantation of 50 μg LHRHa in vitellogenic females (oocyte size > 550 μm) induces spawning within 36 hr
Epinephelus marginatus Marino et al. (2003)	GnRHa administration via the CC controlled delivery system is an effective method to ensure quality eggs
Bidyanus bidyanus Levavi-Sivan et al. (2004)	30 μg sGnRHa/kg fish induces spawning more efficiently than hCG or 20 μg mGnRHa + domperidone/kg fish
Sphoeroides annulatus Duncan et al. (2003)	Implantation of slow-release ethylene vinyl acetate (EVAc) co-polymer containing the identical LHRHa is as good as multiple injections of LHRHa in inducing spawning
Solea senegalensis Agulleiro et al. (2006)	A single sustained release from a 50 μg GnRHa loaded/kg implant is as efficient as three weekly injections of 5 μg/kg in inducing multiple spawnings
S. senegalensis Guzman et al. (2009)	A single sustained release from a GnRHa microsphere (LGA) implant induces more frequent spawnings, i.e., 18–22 times than GnRH-injected females from farm
Gadus morhua Garber et al. (2009)	A single sustained release from an implant loaded with LHRH ethylamide (150 μg LHRHa/kg + sGnRHa 150 μg ovaplant/kg) induces ovulation in females and spermiation in males within 20 d of implantation
Labeo rohita Dasgupta et al. (2009)	A single injection of 10 μg sGnRHa + 5 mg domperidone/kg fish during monsoon season induces 2.5 times more spawned good quality eggs than that induced during pre-monsoon season
Hippoglossus hippoglossus Vermeirssen et al. (2000)	GnRHa-implanted males spermiate four weeks prior to the control
Clarias batarachus Manickam and Joy (1989)	Injection of 50 ng LHRHa + 0.5 μg pimozide/g induces spawning of quality eggs in 86% females
Synbranchus marmoratus Ravaglia et al. (1997)	Treatment with sGnRHa + domperidane induces sex reversal but not ovulation

combination of both of them is required for *Gadus morhua*. 2. In many species, the need for dopamine antagonists is not obligatory (e.g., *Solea senegalensis*, *G. morhua*). 3. The use of delivery systems like the cholesterol-cellulose (CC) (e.g., *E. marginatus*), lactic acid-glycolic acid microsphere (LGA) (e.g., *S. senegalensis*) and Ethylene-vinyl acetate (EVAc) (e.g., *Sphoeroides annulatus; G. morhua*) eliminates the need for multiple injections and are equally efficient in inducing single (e.g., *S. annulatus*) and multiple (e.g., *S. senegalensis*) spawning(s) as well as spermiation (e.g., *G. morhua*) within a short period of treatment. 4. To respond positively to the inducers, the oocyte must have matured to a minimum size during the pre-ovulatory season, say, early spring in *S. senegalensis* and pre-and-monsoon seasons (*Labeo rohita*). For example, an oocyte size of 500 µm is the minimum required for positive response of *S. senegalensis*. On injection of 10 µg sGnRHa + domperidone, 100% mature rohu females that contain a minimum of 55% oocytes at stage II (migrating germinal vesicle) respond positively (Dasgupta et al., 2000). 5. Reproductive season is synchronized with seasonal changes in photoperiod, temperature, water and food availability. The synchronization is tuned with internal processes, that regulate the spawning season. While there are many examples from temperate fishes, the publication by Dasgupta et al. (2009) stands alone to show the effect of monsoon, a vitally important event in the seasonality of Asian countries on induce spawning. The responsiveness of pre-monsoon female rohu *Labeo rohita* (PMR) in terms of latency period, spawning frequency, production of quality eggs and hatchling size is significantly lower, when compared to the response parameters in monsoon rohu (MR).

All the benefits accruing from hormonal and/or photoperiodic induction of ovulation may also induce specific defects in egg quality. Since 2004 the effect of hormonal induction on egg quality has begun to receive attention (e.g., Arabaci et al., 2004). Following hormonal induction of ovulation using intraperitoneal GnRH analog injection or a specific photoperiod regime designed to advance spawning, Bonnet et al. (2007) undertook a detailed microarray-based analysis of egg quality of *O. mykiss*. Survival of normal alevin at yolk sac absorption stage is 84% for the eggs ovulated naturally; in contrast, it is significantly lower 65 and 37% for the eggs ovulated following hormonal and photoperiodic inductions, respectively. The microarray analysis of egg transcriptome after natural and manipulated ovulations has led to the identification of 26 genes. Of them, a dramatic increase of Apolipoprotein CI (APOCI) and tyrosine protein kinase (HCK) is observed in eggs ovulated following hormonal induction (Fig. 12). Besides, microarray and RT pCR analyses show that the abundance of prohibition 2 (PHB2) egg mRNA reduces the developmental success. In view of its commercial importance, more publications are required to solve the problem of increased defects in eggs ovulated following hormonal induction.

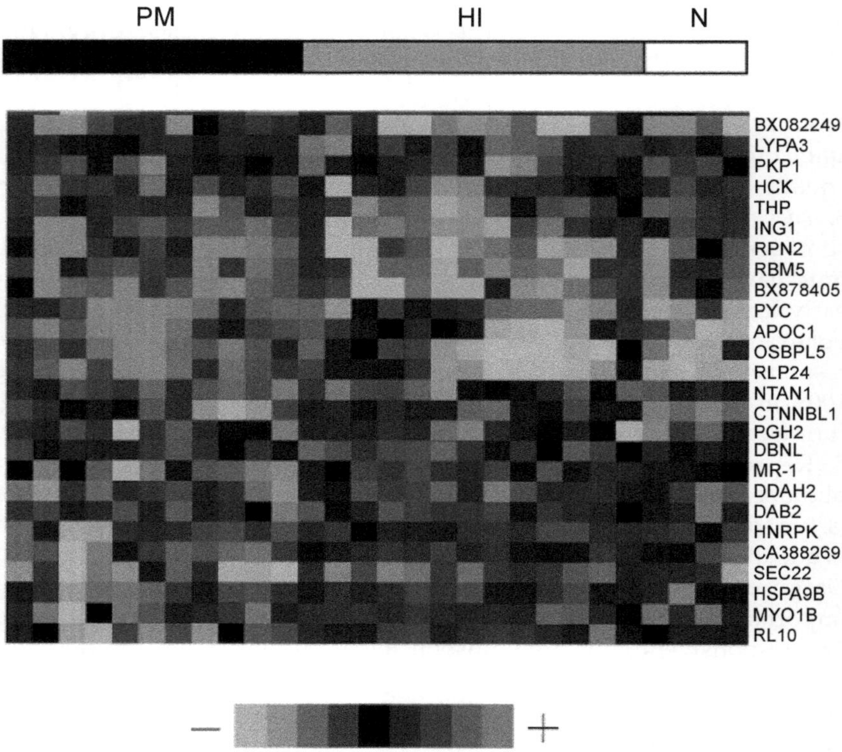

Fig. 12. Mean linkage clustering analysis of the 26 differentially abundant genes in eggs collected after photoperiod-manipulated ovulation (PM), hormonally-induced ovulation (HI) and natural ovulation (N). Each row represents a gene and each column represents an egg RNA sample. For each gene, the expression level within the sample set is indicated using a color intensity scale. Red and green are used for over and under abundance respectively, while black is used for median abundance (from Bonnet et al., 2007).

Color image of this figure appears in the color plate section at the end of the book.

2.8 Departures from norm

In fishes 130 F_1 fertile natural hybrids and 150 artificially produced F_1 hybrids are reported (Argue and Dunham, 1999), but natural triploidization is limited to seven species only (Pandian, 2011). Our knowledge on the effect of triploidization is based mostly from fishes artificially induced triploidization. Parental females of the reciprocal cross between *Poecilia sphenops* and *P. velifera* maintain the same interparturition period and fecundity, and remain unaffected by fertilization with a heterospecific male (George and Pandian, 1997), implying that the egg number of the female remains a maternal trait. Information on egg size of other fertile hybrid fishes is not immediately available. However, there are indications that the disparity in egg sizes of chosen species for crossing can be an important factor resulting in success of hybridization. For instance, the egg size of *Poeciliopsis viriosa* and *P. monacha* is 1.4 mm and 2.2 mm, respectively and yet successfully hybridize; however, that of *P. viriosa* is so small (<1.4 mm) and does not readily hybridize, as it contains inadequate yolk to produce a viable hybrid, when crossed with *P. monacha,* whose embryo is nourished by large ova of 2.2 mm (Vrijenhoek and Schultz, 1974).

In general hybridization leads to male sterility and triploidization in female sterility (Pandian, 2011). However, fertile hybrid females produce a fewer eggs; for example, the relative fecundity of the hybrid female between *Clarias gariepinus* and *Heterobranchus longifilis* is just 12 eggs/g body weight, as against 70 and 106 eggs/g body weight of the parental species (Legendre et al., 1992). Hybrid female fertility is recovered by switching to gynogenesis, as in fertile hybrid female between *Oryzias latipes* (2n = 48) and *O. curvinotus* (2n = 48, Sakaizumi et al., 1992). In fact *Oncorhynchus mykiss* (2n = 60) is amenable to diploid and triploid hybridizations with *Salmo salar* (2n = 60), *S. trutta* (2n = 40), *Salvelinus fontinalis* (2n = 42), *O. kisutch* (2n = 60), *O. masou* (2n = 60) and *O. keta* (2n = 74). However, its compatibility with the genome-based chromosome number, structure or DNA content of any one of the hybridizing species is not easily reconcilable (see Pandian and Koteeswaran, 1998). A second route to neutralize chromosomal incompatibility is triploidization. The cross between *Clarias macrocephala* ♀ (2n = 54) and *Pangasius sutchi* ♂ (2n = 60) results in the production of gynogenics (2n = 57) and triploids (3n = 54+30 = 84, NaNakorn et al., 1993a,b). In all such events, departure from normal oogenesis occurs. Premeiotic endomitosis or elimination of meiosis I is the most common route of departure (Fig. 13). However, genomic addition of a single set of chromosomes due to post-ovulatory ageing is the other departure pathway but not so frequently used route. Most unfortunately, none of the authors, who have successfully induced such gynogenics and/or triploids, have reported the egg size of such depauperates (e.g., Argue and Dunham, 1999; Scribner et al., 2001).

Fortunately, Piferrer et al. (1994b) estimated a 25% decrease in fecundity due to a 30% increase in gynogenic, i.e., diploid oogonial diameter, implying that a diploid egg may have 25–30% increase in its size. Interestingly, Arai et al. (1999) have successfully made interploid crossings in *Misgurnus anguillicaudatus* between 2n ♀ x 4n ♂, 4n ♀ x 2n ♂ producing triploids, 4n ♀ x 4n ♂, 6n ♀ x 2n ♂ and 2n ♀ x 6n ♂ producing tetraploids as well as between 6n ♀ x 6n ♂ producing hexaploids (Table 15). Fortunately they have also provided desired information on fecundity and egg size of these interploid hybrids. In the light of findings by Piferrer et al. (1994b), higher percentage reductions in fecundity of tetraploid and hexaploid are expected. Indeed, the reductions in fecundity are 53% and 62% in tetraploids and hexaploids,

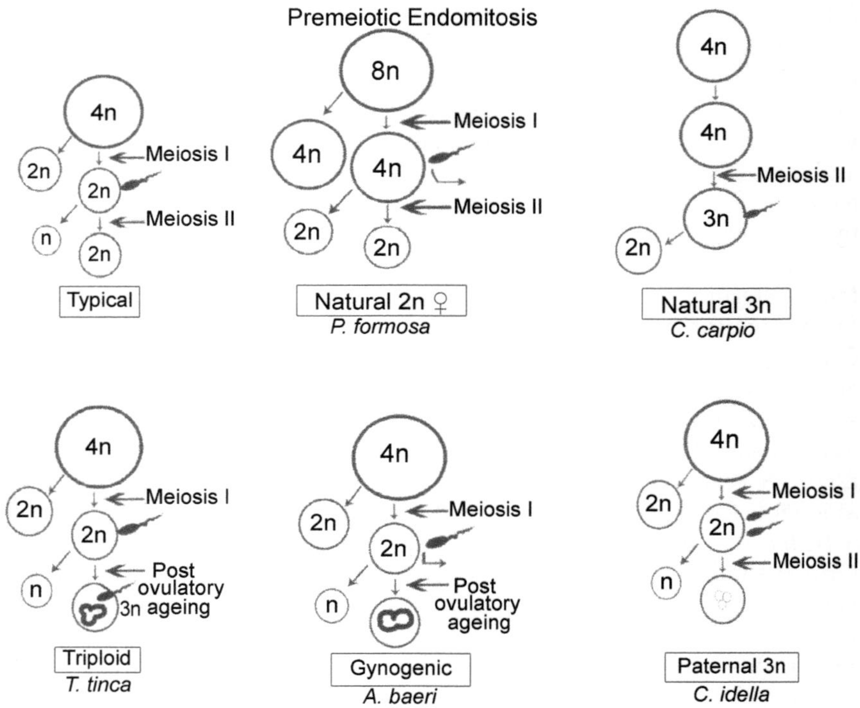

Fig. 13. Oogenesis and its modifications in natural and induced gynogenic diploids-triploid fishes. Note the premeiotic endomeiosis in 2n♀ *Poecilia formosa*, the elimination of Meiosis I in *Cyprimus carpio*, the addition of n chromosomes due to post-ovulatory ageing in *Tinca tinca*, gynogenic *Acipenser baeri* and the entry of 2 sperm in paternal triploid of *Ctenopharyngodon idella* (from Pandian, 2011).

Table 15. Ploidy, body weight, fecundity and survival of *Misgurnus anguillicaudatus* (compiled from Arai et al., 1999).

Parameter	Diploid	Tetraploid	Hexaploid
Body weight (g)	17.6	14.02	12.01
Reduction in weight (%)	0.0	20.00	31.02
Fecundity (no)	4,106.0	1,834.00	1,544.00
Reduction in fecundity (%)	0.0	53.03	62.04
Relative fecundity (no/g)	238.0	140.00	127.00
Egg size (mm)	1.0	1.25	1.44
Hatching (%)	25.4	30.08	54.08
Survival as% of hatching	1.0	20.04	45.02
Net progeny production (no/♀)	669.0	374.00	698.00

respectively. But the reductions in body weight are just 20% and 31% in tetraploids and hexaploids. Due to doubling and tripling of genome size, somatic cells seem to suffer less than those of gonadal cells. Further, the egg size increases from 1 mm in diploid to 1.25 mm (12.5%) in tetraploid and 1.44 mm (14%) in hexaploid, instead of the expected proportionate increases to 2.0 mm and 3 mm respectively. These findings of Piferrer et al. (1994b) and Arai et al. (1999) suggest that the increase in ploidy/genome size of the fish does increase the yolk deposition resulting in increased egg size but the increase is, however, not proportionate.

3

Testicular Differentiation

Introduction

Sexual reproduction involves the maintenance of males within a species and is so costly that a dozen unisexual fish species have all females only. Yet the overwhelming evolutionary success of sexual reproduction among microbes, plants and animals including fishes witnesses the importance of males in ensuring recombination to generate genetic diversity (see Pandian, 2012). In teleosts, mode of reproduction ranges from sexual to gynogenesis and to hybridogenesis and each of these modes may involve external fertilization and oviparity or internal fertilization and viviparity. Internal fertilization is facilitated by a specific sperm storage organ, the seminal vesicles and intromittent organ like the gonopodium to transfer sperm into the oviduct. Among the oviparous fishes too, the sperm delivery ranges from simple broadcasting in the vicinity of female's spawning site to delivering them into the respiratory chamber of a host of animal like the mussels, ascidians, sponges and crabs, in which the stalked eggs, on fertilization, enjoy protection and ventilation at the cost of the host (see Pandian, 2010). A climax among the oviparous fishes seems to be the females performing 'sperm drinking', while spawning, and 'the ingested sperm' requires adaptive protection, while passing through the gut and being delivered safely inside the modified pelvic-pouch, where the freshly laid eggs await the sperm (Kohda et al., 1995; see also Pandian, 2010). Teleosts are uniquely characterized by acrosomeless spermatozoa and micropylar eggs. Hence hybridization among closely related species is rather ubiquitous. Eggs of many teleosts are known to tolerate the entry of heterospecific sperm and in a few cases the egg genome accommodates the genome of heterospecific sperm in diploid status, as in *Oryzias latipes* x *O. curvinotus* and in triploid status, as in *Oncorhynchus mykiss* (2n = 60) x *Salmo trutta* (2n = 40).

3.1 Spermatogenesis

It is a highly organized and coordinated developmental process, during which a small number of diploid Spermatogonial Stem Cells (SSCs) proliferate and produce a large number of differentiated 'spermatid' carrying a haploid recombined genome (Schulz et al., 2010). Despite the extraordinary diversity and plasticity in sex determination and differentiation processes in fishes, the major structural organization and endocrine system of spermatogenesis remain highly conserved. As in other vertebrates, the structural organization consists of 1. Tubular and 2. Intertubular compartments. The latter contains the steroidogenic Leydig cells, blood/ lymphatic vessels, macrophages, mast cells, neural and connective tissue cells (Koulish et al., 2002; Nobrega and Quagio-Grassiotta, 2007). Regarding the tubular compartment, there are two basic types (anastomosing and lobular) of testicular architecture in teleosts (Grier and Taylor, 1998). "In the anastomosing tubular type, SSCs are distributed throughout the testis, spermatocytes do not move during maturation process, and anastomoses are observed between the spermatogenic tubules" (Schulz and Miura, 2002). This type of testis is common from the basal lineages such as Coelacanthiformes, Acipenseriformes and Cypriniformes as well as basal Euteleosts such as Salmoniformes (Parenti and Grier, 2004). The lobular testes contain some germinal compartments that terminate blindly at the periphery of the testis. They are classified as either *unrestricted lobular,* in which cysts containing spermatogonia are located along the length of the lobules (e.g., *Mugil cephalus*) or *restricted lobular,* in which spermatogonial cysts are located at the periphery of the lobule (e.g., *Fundulus grandis*) (see Knapp and Carlisle, 2011). However, intermediate forms are also reported from Perciformes (e.g., *Oreochromis niloticus,* Vilela et al., 2003), Pleuronectiformes (e.g., *Solea senegalensis*, Garcia-Lopez et al., 2005) and Gadiformes (e.g., *Gadus morhua,* Almeida et al., 2008).

Functionally type A spermatogonia are divided into (i) undifferentiated (A single A_s), that are the stem cells being continuously present and (ii) differentiated spermatogonia (A_1–A_4; A_{diff}) that are present at specific stages of the epithelial cycle and are committed to form mature sperm. The number of spermatogonial generation is genetically determined. The undifferentiated type A spermatogonia (A_{und}) give rise to differentiated spermatogonia (A_{diff}) (Fig. 14) which still share some morphological characteristics of A_{und} but with a greatly reduced potential for self-renewal. Further maturation associated with changes in morphological aspects gives rise to more rapidly dividing type B spermatogonia (Schulz et al., 2010). On entry into the maturation

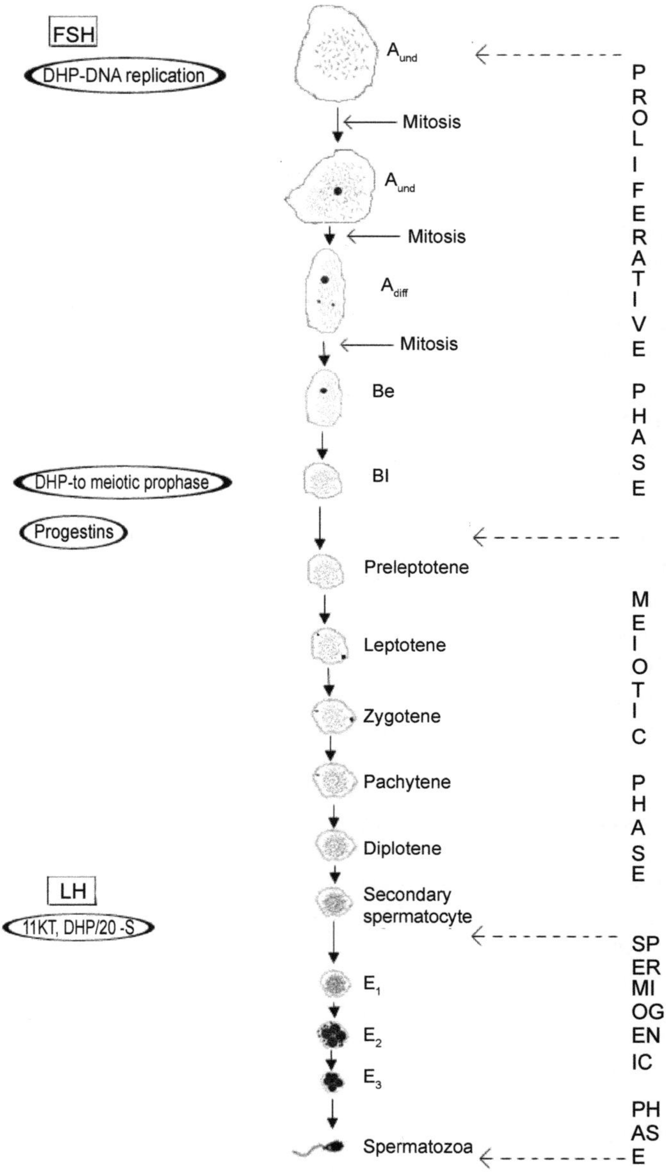

Fig. 14. Schematic representation of spermatogenesis in fishes. A_{und} = undifferentiated spermatogonium, A_{diff} = differentiated spermatogonium, B_e = early sperpatogonium type B, B_l = late sperpatogonium type B, E_1 = early spermatid, E_2 = intermediate spermatid and E_3 = final spermatid. Note the progressive reductions in cytoplasmic and nuclear volumes. On left, the hormones responsible for the differentiation are listed (source: Schulz et al., 2010 and others).

pathway, the spermatogonia type B typically undergo a species-specific number of mitotic divisions to produce the primary spermatocytes. Following the final mitotic division, the preleptotene spermatocytes enter the lengthy prophase of the first meiotic division and proceed through the leptotene, zygotene, pachytene and diplotene stages (Fig. 14). In this first meiotic division, the maternal and paternal chromosomes pass through segregation and recombination during the ensuing meta-, ana- and telophase, resulting in the production of secondary spermatocytes. "These short-lived cells quickly commence the second meiotic division without a DNA synthesis phase, giving rise to genetically unique haploid spermatids" (Schulz and Miura, 2002). Briefly, meiosis I-reductional, where the homologous chromosomes are segregated into secondary spermatocytes and meiosis II-equatorial, where the sister chromatids segregate resulting in four haploid cells harboring one copy of each chromosome (see Fig. 14). An aim of meiosis is to generate genetic diversity through the two events: the first recombination and the second segregation of homologous chromosomes. With respect to nuclear condensation, early spermatid E_1, intermediate spermatid E_2 and final spermatid E_3 can be characterized (Fig. 14). Secondly, a key event of spermatogenesis is the 80–90% reduction in cellular volume (Sprando et al., 1988) and nuclear contents, in a striking contrast to that in oogenesis. Morphometric studies of Schulz et al. (2005) have shown a progressive 93% reduction (from 2300 to 160 μm^3) in cellular volume of the germ cell including 85% reduction (from 550 to 80 μm^3) in its nuclear volume of tilapia. During the corresponding period, the number of germ cells (from 1 to $\approx$ 120 cells) as well as Sertoli cells (from 1.4 to 3.6 cells) per spermatogonial cyst increases. The haploid spermatids mature as spermatozoa through the process known as spermiogenesis (Knapp and Carlisle, 2011).

Following entry of the SSCs into spermatogonia, the spermatogenic cysts/spermatocyts (Schulz and Miura, 2002), which are completely enveloped by Sertoli cells (Le Gac and Loir, 1999), are formed (Nobrega et al., 2009). The cells resulting from the mitotic divisions of a single primary spermatocyst remain interconnected by cytoplasmic bridges due to incomplete cytokinesis. Desmosomes present between outer layer of germ cells and surrounding Sertoli cells together with the interacellular bridges serve as medium for intracellular signal propogation among the clone members of germ cell (see also Sprando and Russel, 1987). Hence all the germ cells within a cyst have originated from a single SSC and develop synchronously, producing a cyst that contains 'clones of germ cells' in the same stage of spermatogenesis (Le Gac and Loir, 1999). The different stages of spermatogenesis can be distinguished by cell size, nuclear histology and number of cells present in a cyst (Schulz et al., 2005, 2010). Using BrdU technique, the percentage of the BrdU-incorporated cells containing nascent DNA during S phase of the cell cycle as well as the duration required for a cell cycle can also be estimated (Schulz et al., 2003). The cystic spermatogenesis, resulting in formation of 'clones of germ cells', is typical of aquatic fishes and amphibians (anamniotes) and is associated with external fertilization. Schulz et al. (2010) postulated that the cystic spermatogenesis is more efficient than the non-cystic spermatogenesis that occurs in the terrestrial amniotes, characterized by internal fertilization. Viviparity involving internal fertilization is recorded in 577 fish species belonging to 135 genera and 14 families (see p 37). Many publications describe the spermatogenic and aspermatogenic regions of the testicular tubes and formation of spermatozeugma (e.g., Burns et al., 1995, 1998; Pecio et al., 2001; Grier et al., 2005) but it is desirable to have an idea on whether or not the viviparity and associated internal fertilization have switched from the cystic to acystic spermatogenesis in some viviparous fishes.

Sertoli cell number determines the spermatogenic capacity. In amniotes, they proliferate until puberty. Hence the adult testis of amniotes contains a fixed number of Sertoli cells that support waves of spermatogenesis. But the cyst-forming Sertoli cells of fishes and amphibians retain their potency to proliferate also in adults (Schulz et al., 2010). The main functions of Sertoli cells are to support survival and development of germ cells, secrete fluid that generates the tubular lumen and to rapidly phagocytize by apoptosis the residual bodies, discarded by spermatids and residual sperm. That the cystic spermatogenesis is more efficient than the non-cystic one, is indicated by 60%–80% residual loss of germ cell in the rat (Hess and Franca, 2007), in comparision to ~ 30% in fish (Matta et al., 2002). Notably, the asynchronous terminal differentiation of germ cells in fishes reveals that each spermatogenic wave occurs independent of others.

"The complex process of spermatogenesis in fishes can be divided into three major phases (Nobrega et al., 2009): 1. Proliferative mitotic phase. The number of mitotic divisions that a SSC undergoes prior to forming spermatocyte is: 10 in Anguilliformes, 8 in Characiformes, 8 or 9 in Cypriniformes, 6 or 8 in Salmoniformes, 11 in Gadiformes, 7–10 in Perciformes, 10 in Beloniformes and 14 in Cyprinidontiformes (Miura, 1999). Spermatogonial cells of the allogenic rainbow trout *Oncorhynchus mykiss* is estimated to undergo mitosis 27 times instead of 7–8 times (Okutsu et al., 2006); the allo-androgenics of rosy barb *Puntius conchonius* (Kirankumar and Pandian, 2004) and Buenos Aires tetra *Hemigrammus caudovitatus* (David and Pandian, 2006a) also undergo increased number of mitotic divisions, as evidenced by an order of greater number of spermatozoa". Incidentally, Schulz and Miura (2002) ruled out the endocrine role on the number of germ cell divisions, as this is apparently determined genetically. It is not then clear how allogenesis and interspecific androgenesis induce this radical change in the number of mitotic divisions. 2. Meiotic phase in which the genetic material is duplicated, recombined and segregated and 3. Spermiogenetic phase, during which spermatids are transformed into spermatozoa (Nobrega et al., 2009). "The duration of spermatogenesis is species specific and lasts from 12 d in medaka *O. latipes* to 14 d *Oreochromis niloticus,* 14.5 d in guppy *Poecilia reticulata* and to 21 d in *P. shenops* (Vilela et al., 2003). These durations are relatively shorter than those reported for mammals. In fishes, a single Sertoli cell is able to support around 100 spermatids, which are about 10 times higher than those of mammals. Moreover, as indicated earlier the number of Sertoli cell is also more than doubled during spermatogenesis. Briefly, these characteristics of the piscine spermatogonial system provide a larger number of sperm within shorter duration to sustain external fertilization in fishes" (Pandian, 2010).

The credit for having projected an apparently unnoticed but a seminal publication by Shibata and Hamaguchi (1988) goes to Schulz and Miura (2002). Justifiably, Shibata and Hamaguchi have clearly distinguished the early slowly dividing spermatogonial germ cells as undifferentiated generations of cells (see Fig. 1) from the subsequently rapidly dividing cells producing differentiated generations of cells. Using the BrdU technique, the duration of a cell cycle in early spermatogenesis is five times longer than in late spermatogonia of the African catfish *Clarias gariepinus* (Schulz et al., 2003). " In medaka, for example, the early spermatogonial generations up to the 5th of 8 in total mitotic divisions can develop into oocyte in response to estrogens, but the male developmental path is irreversibly fixed after the 5th division in XY embryo" (Schulz and Miura, 2002) (Shibata and Hamaguchi, 1988). Hence the irrevocable dimorphic sexual differentiation is indeed commenced from the 6th mitotic division that occurs at the 36th developmental stage in medaka (see Fig. 1). In the Japanese eel too, the hCG treatment of immature males stimulates proliferation of early spermatogenesis in all the treated individuals; however, it ceases after the 4th or 5th mitosis in the so called 'non-responders' (Miura et al., 1997). This clearly suggests that a regulatory check point separates the early undifferentiated from subsequent differentiating spermatogonia (Schulz and Miura, 2002).

Gonadotropins: Our understanding of endocrine aspects of testicular differentiation is based on publications limited to about 23 species belonging to 13 families among 10,598 species belonging to 162 families considered by Parenti and Grier (2004) and as shown in Fig. 3.3 of Knapp and Carlisle (2011). Not surprisingly, our understanding of the endocrine testicular differentiation in fish, in comparison to that of ovarian differentiation, is in its infancy. Nevertheless the vast numerous publications in this area have been reviewed from time to time by Schulz and Miura (2002); Miura and Miura (2003); Lahnsteiner and Patzer (2009b); Schulz et al. (2010);

Scott et al. (2010) and Knapp and Carlisle (2011). Thankfully, the following comprehensive account on endocrine testicular differentiation is based on these reviews and other original contributions. Gonadotropins (FSH, LH) are the primary pituitary hormones involved in regulation of spermatogenesis. They act by stimulating growth factor release or the gonadal biosynthesis of sex steroids.

The BPG axis orchestrates the activities of the reproductive system by adjusting its level according to the environmental conditions on one hand, and to the development and physiological state of the individual on the other hand. Accordingly, the hypothalamus regulates, on receiving environmental cues, the pituitary secretions. The pituitary hormones FSH and LH are the key players in regulation of testicular functions. FSH regulates the Sertoli cell activities like structural, nutritional and sustenance of germ cells (Schulz et al., 2010). Hypophysectomy involving the ablation of the pituitary is known to disintegrate spermatocysts in immature eel (Khan et al., 1986) and to block spermatogonial differentiation at the 7th (of 14) mitotic division and reduce the number of early spermatogonia in the guppy (Billard, 1969). LH regulates the production of sex steroids, meiotic maturation of germ cells and spermiogenesis. The FSHr of fishes show preference for FSH but is also activated by LH; however, the cross-activation of FSHr requires high LH levels (e.g., channel catfish, Sampathkumar et al., 2000; zebrafish: So et al. 2005; African catfish: Garcia-Lopez, 2008; Japanese eel: Kazeto et al., 2008). Hence the available data (e.g., Ohta et al., 2007; Garcia-Lopez et al., 2008) on receptor pharmacology and localization suggest that steroidogenesis in Leydig cells is directly regulated by LH and FSH, while the functions of Sertoli cells are mainly regulated by FSH. More importantly, both receptor specificity and site expression indicate that the physiological activities of piscine gonadotropins are not clearly separated; that calls for more detailed studies on gonadotropins of fishes. Publications on the dynamics of plasma level of FSH is limited to only males of salmonids. In pubertal chinook salmon FSH level is transiently elevated coinciding with spermatogonial proliferation. But the plasma LH level increases in association with spermatogenic maturation, further increases at the time of spermiation but declines prior to the commencement of the spawning season (Campbell et al., 2003). Other signaling molecules, like activin Bβ is expressed in Sertoli cells of *Anguilla japonica* and stimulates FSHβ but suppresses LHβ in the pituitary of goldfish (see Ge, 2005). Sertoli cells are also known to produce activin (Miura and Muira, 2001), gonadal soma derived factor (GSDF) (Sawatari et al., 2007) and AMH (Miura et al., 2002), which regulate germ cell proliferation.

Insulin-like growth factor 1 (IGF-1) is a key somatotropic hormone that may convey growth status to the reproductive endocrine system. In his review, Reinecke (2010) brought clear evidence to show that *IGF-1* serves as a local regulator of spermatogenesis in fish and suggests an entire Gh-IGF-1-gonad axis involved in this process. For example, *IGF-1* regulates mainly growth, differentiation and reproduction by selectively promoting mitogenesis and differentiation (Duan, 1998; Reinecke et al., 2005). Its mRNA and/or peptide occur in spermatogonia, spermatocytes, Sertoli and Leydig cells of rainbow trout (Le Gac et al., 1996), tilapia (Reinecke et al., 1997) and sea bass (Perrot et al., 2000), suggesting autocrine and paracrine functions. In agreement with this suggestion, the *IGF-1* type *igf1r* receptor has been identified in spermatogonia and primary oocytes of rainbow trout and sea bream. In testes of Japanese eel and Nile tilapia, *IGF-1* stimulates spermatogenesis induced by KT (Reinecke, 2010). Lynn et al. (2009) also found significantly positive correlations between liver *1GF-1b* and *esr2a* mRNA levels as well as liver *1GF-1b* and *cyp19a1a* levels in seasonal spawning yellow perch *Perca fluviatalis*. *IGF-1* increases steady-state mRNA levels of *gnrh1* and four gonadotropin subunits and may act alone or with GnRH to increase pituitary FSH in *O. kisutch* male over one year before puberty. This finding of Luckenbach et al. (2010) suggests that *IGF-1* may prime pituitary gonadotropic cells of prepubertal coho salmon to respond to GnRH by stimulating synthesis of GnRHR and FSH during the onset of puberty.

Thyroid hormones: In teleosts, thyroid is not a discrete compact organ but the thyroid follicles are diffusively dispersed. Cyr et al. (1988) showed that the seasonal reproductive cycle is associated with well defined chronological sequence of changes in thyroid hormone levels. The following findings have catalyzed much of the desired research in this area: 1. Thiurea as an effective thyroid depleting drug (Davidson et al., 1979). 2. The cloning of the thyroid stimulating hormone (TSH) receptors from the gonad of *Morone saxatalis* (Kumar et al., 2000) and the testis of *C. gariepinus* (Vischer and Bogerd, 2003) and 3. Release of GnRH and its stimulation increase the number of Sertoli and germ cells per cyst in *O. niloticus* on administration of goitrogen-6-*n*-propyl-2-thiouracil during the testis development (Matta et al., 2002). Evidence accumulated on the association of GnRH and catecholaminergic system in fishes (e.g., Parhar et al., 2000; Chaube and Joy, 2003) has increased the possibility for thyroid hormone action on gonadal function through hypothalamic-hypophyseal axis. Using the strategy of thiurea-induced thyroid hormone depletion in *C. gariepinus*, Swapna et al. (2006) have shown the shrinking seminiferous tubules with fewer spermatozoa. These changes are accompanied by reduction in serum and tissue level of T and KT, and qualitative changes in *cg*GnRH and LH, as well. Together, these results show a significant role played by thyroid hormone in testicular function in the catfish and modulation of sex steroids either directly or through the hypothalamic (GnRH)-hypophyseal (LH) axis.

Steroid hormones: In teleost males, as in other vertebrates, the Leydig cells are the major sites, from which steroid synthesis proceeds by converting cholesterol through a series of enzymatic steps to pregnenolone and subsequently through different sub-pathways to a number of androgens, estrogens, progestogens and corticosteroids (Norris, 2007). For more details on pathways synthesizing steroid hormones, for example, the progestogens 17, 20α DHP, 17, 20β DHP, Kime (1993); Scott et al. (2010); Knapp and Carlisle (2011) may be consulted. In fishes, the dominant testicular steroids are the 11-oxygenated androgens like KT (Borg, 1994). T is also synthesized but its derivative dihydrotestosterone is rarely found. KT-treated juvenile African catfish precociously commences spermatogonial proliferation followed by meiosis, as well; the fish also develops secondary sexual charactaristics (Cavaco et al., 1998). This observation irrevocably confirms the role of KT in masculinization. Using long term testis tissue culture system, Miura et al. (1991b) demonstrated that KT promotes successful completion of the entire sequence of spermatogenesis in the Japanese eel (see also Miura et al., 1996). This is an important prelude for establishing similar long term testis tissue culture system for successful rearing of spermatozoa (see Hong et al., 2004) of culturable protogynic fishes like the groupers, in which males are scarce. In contrast, T inhibits the natural onset of spermatogenesis in the catfish, as it blocks the KT-induced spermatogenesis and development of seminal vesciles (Cavaco et al., 2001). Hence T may play a feedback role in inhibiting FSH release and steroidogenesis (Schulz and Miura, 2002).

Spermatogenesis is initiated, when spermatogonial mitosis switches from the slow self-renewal pathway to rapid proliferation. The main function of FSH is to stimulate spermatogenesis-inducing steroids like KT (Miura et al., 1991a). Activin, a dimeric growth factor composed of two activin β subunits, induces proliferation in spermatogonia (Miura et al., 1995) (Fig. 16). DHP or 20β-S is a major progestin, an indispensable hormone to initiate spermatogonia to enter the meiotic prophase (Miura et al., 2006). On the other hand, trilostane, an inhibitor of 3β-dihydroxysteroid dehydrogenase inhibits the *Fsh*-induced initiation of spermatogenesis. AMH is also an inhibitor of spermatogenesis (Schulz et al., 2010) (Fig. 15).

In the context of the important role played by KT in induction of spermatogonial proliferation, it is interesting to note the role of *p450 (11β)*. The key steroidogenic enzyme 11β-hydroxylase, encoded by *p450 (11β)*, is involved in the production of 11β-hydroxytestosterone in the process of KT biosynthesis in the testes (Kusakabe et al., 2006). During spermatogenesis, the expression level of *p450 (11β)* is low at the early spermatogenesis but subsequently increases sharply to the peak (Maugars and Schmitz, 2008). Incidentally, its expression can be influenced by chemicals, which can induce sex reversal. By *in situ* hybridization, the *p450 (11β)* expression is detected in steroid producing cells such as Leydig cells in the testis (Kusakabe et al.,

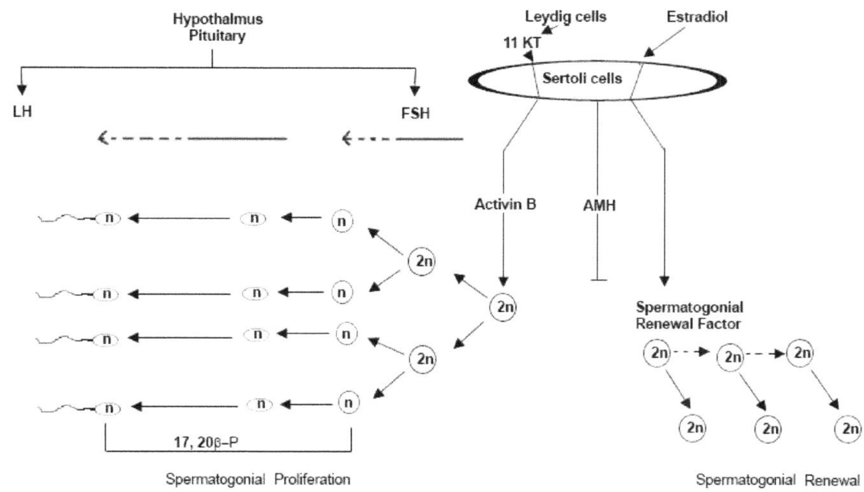

Fig. 15. Schematic representation of hormonal influence on spermatogonial proliferation (on the left) and spermatogonial renewal (on the right) in fishes (source: Schulz et al., 2010; Knapp and Carlisle, 2011).

2006) and spermatogonia during early stages of spermatogenesis (Vinas and Pifferer, 2008). Interestingly, two alternatively splicing isoforms: (i) *p450 (11β)*-1 and (ii) *p450 (11β)*-2 have been cloned from *O. niloticus* by Zhang et al. (2010). They are 1,614 and 1,227 bp in length with open reading frames encoding proteins of 577 and 408 amino acids, respectively. They are expressed in the testis from 50 dph (day post-hatching) onwards but not in the ovary. They are strongly expressed in sex reversed XX testis, following fadrozole and tamoxifen treatment but are completely inhibited in 17β-estradiol-induced XY ovary. These data suggest a role for the *p450 (11β)* in spermatogenesis.

The enzyme 11β-hydroxysteroid dehydrogenase (11β-HSD2) is required for the conversion of 11β-hydroxytestrosterone into KT. Rasheeda et al. (2010b) cloned the full length cDNA of *11β-HSD2* from the testis of *C. gariepinus* and showed its high sequence homology to that of the zebrafish and eel. It is prominently expressed in the testis and is accompanied by serum level KT elevation. Administration of hCG during pre-spawning and resting phases results in peak elevation in *11β-HSD2* transcripts within 24 hr of induction only in the testis of the pre-spawning phase. Sreenivasulu et al. (2012) found that the expression of *20β-HSD,* detectable in male *C. gariepinus* fry at 50 and 100 dph, peaks in the fry at 150 dph. During the testicular cycle, mRNA and protein of 20*β*-HSD reach the highest levels during the spawning phase. The 20*β*-HSD protein is prominently present in interstitial cells and spermatogonia/spermatocysts but is totally undetectable in haploid spermatid and sperm. Together with Rasheeda et al. (2010b), the ontogenetic study of Sreenivasulu et al. (2012) suggests a pivotal role for *11β-HSD2/20β-HSD* as marker for determination, differentiation and development of testis.

Progestin: The major progestin in teleosts is not progesterone but 17, 20β-dihydroxy-pregnen-4-en-3-one or 17, 20β-trihydroxy-pregnen-4-en-3-one. Two years following the discovery 17, 20β-P in blood plasma of female *Oncorhynchus nerka* (Idler et al., 1960), its presence in males of *O. nerka* and *Salmo salar* (Schmidt and Idler, 1962) were also discovered. Its role in stimulation of spermiation has catalyzed the appearance of over hundred publications during the last 30 ys (Scott et al., 2010). Spermiation occurs, when close contact between developing germ cells and Sertoli cells is broken by opening the cyst and facilitating the continuation of the cystic lumen with that of spermatogenic tubule (O'Donnel et al., 2000). Besides, 17, 20β-P also has a role in initiation of meiosis and maturation of gametes; injection of 17, 20β-P induces spermiation a few months ahead of the normal spermiation period (Miura et al., 2007b). The testicular origin of 17, 20β-P has been established in many fishes, especially salmonids (see Scott et al., 2010). Its absence in some fishes like *A. japonica* has been traced to its precursor 17-P; on addition of 17-P, the synthesis of 17, 20β-P is enhanced, suggesting the presence of 20β-hydroxy steroid dehydrogenase (20β-HSD), the enzyme required for conversion of 17-P to 17, 20β-P (Miura et al., 1991b). 17, 20β-P plays a major role as an hormonal pheromone.

A more important role of 17, 20-β-P can be explained by taking a short side-step to have a look at hormonal pheromones. Living in aquatic medium, where the olfactory organ is readily exposed to soluble conspecific odorants, fishes have evolved pheromonal functions for released sex hormones. Intriguingly, derived from a homologous and conserved endocrine system of vertebrates, the hormonal pheromones may be incapable of conveying the species-specific and reproductive event-specific information. Closely related species have similar pheromone system; for example, *Cyprinus carpio* and *Carassius auratus* share common steroidal pheromone systems; both species are sensitive to 17α-20β-dihydroxy-4-pregnen-3-one, 17α, 20β-P-sulphate and androstendione (AST). Similarly, sympatric species *S. salar* and *S. trutta* display responses to ovarian fluid and urine of conspecific and heterospecific females. Nevertheless, fishes can also discriminate between specific mixtures of compounds and avoid hybridization (see Burnard et al., 2008). The hormonal steroids including their precursors and metabolites and prostaglandins, collectively called hormonal pheromones (Stacey, 2011), that are released into water are known to promote the desired reproductive synchrony among conspecifics. The hormonal pheromones eliciting diverse physiological (primer) and behavioral (releaser) effects on conspecifics are known atleast in a dozen fish species belonging to Cyprinidae, Salmonidae, Cichlidae and others. But, the olfactory extraordinary responsiveness to a host of pheromonal hormones at pico molar threshold has been demonstrated in more than hundred species. Many fish species are capable of discriminating a variety of hormonal pheromones through multiple, highly specific olfactory receptor mechanism (Stacey, 2011).

The communication system between the signaler and receiver mediated by hormonal pheromone is better understood in goldfish than in other fishes (Kobayashi et al., 2002; Stacey and Sorensen, 2006, 2009). The pre-ovulatory female goldfish releases three steroid odorants, namely 17, 20β-P, its 20β sulfated metabolite 17, 20β-P-s and AST (Sorensen et al., 2005). Other pheromones like prostaglandin $F_{2\alpha}$ ($PGF_{2\alpha}$), synthesized by eggs in the oviduct of female are also released (Stacey and Peter, 1979). Males detect AST and 17, 20β-P only in the immediate vicinity of the female but 17, 20β-P-s is detectable at greater distances (Sorensen et al., 2000, Scott and Ellis, 2007). During each spawning pulse, a female urination creates 40 odor patches each in an active volume 10 l of water (Sorensen et al., 2000; Appelt and Sorensen, 2007). $PGF_{2\alpha}$ is released both across the gills and in the urine but 15K-$PGF_{2\alpha}$ is released almost exclusively in urine pulses. They are detected by separate olfactory mechanisms at nM and pM thresholds, respectively (Sorensen et al., 1988), although they elicit more or less the same response from the male. The objective of preovulatory pheromonal communication seems to attract a more suitable male with better sperm quality. The pheromonal 17, 20β-P-induced increase in LH is known to enhance spermiation, hydration and sperm motility (Pankhurst, 2008). Within 30 min of priming, i.e., exposure to 17, 20β-P, the males remain aroused for 10 hr, and the aroused male produces more spermatozoa with greater mobility (Defraipont and Sorensen, 1993). Studies of group and isolated male goldfish indicate that males inhibit milt production from other males, when exposed to AST. Males can also discriminate AST of other males from that of females (Fraser and Stacey, 2002). For more details on other cyprinids, salmonids and cichlids, Stacey (2011) may be consulted.

Spermatogonial renewal: "Gonadotropins (FSH and LH) are the primary pituitary hormones involved in regulation of spermatogenesis. They act by stimulating growth factor release or the gonadal biosynthesis of sex steroids. These hormones, in turn, regulate renewal of Spermatogonial Stem Cell (SSC) and sperm production from spermatogonial cells (Schulz and Miura, 2002). In the absence of gonadotropins, Sertoli cells produce a growth factor, an orthologue of anti-Mullerian hormone, which directs stem cell divisions to the self renewal pathway. Alternatively, gonadotropin or KT stimulation induces spermatogonial proliferation" (Fig. 15).

"It is now established that E_2 regulates the renewal of SSCs (Miura et al., 1997, 1999) (Fig. 15). 1. E_2 receptor expression is prominent in Sertoli cells. 2. In *Anguilla japonica,* mitosis of SSCs is promoted by the implantation of E_2 but is suppressed by tamoxifen, an E_2 antagonist (Miura and Miura, 2003). 3. In the Japanese huch *Hucho perryi, in vitro* treatment of testicular tissues with E_2 promotes renewal of SSCs (Amer et al., 2001)". 4. *In vitro* 10 pg/ml of E_2 is adequate to induce spermatogonial renewal divisions in cultured testicular tissues. Incidentally, E_2 receptors are expressed throughout the BPG axis and in the male reproductive organs (O'Dennel et al., 2001). 5. In immature testis of Japanese eel, a 'spermatogenesis-preventing gene' has been identified. The gene has been cloned (*eSRS21*) and it shares about 40% amino acid sequence similarity with mammalian *AMH/MIS.* Suppression of its expression by hCG or KT stimulation results in initiation of spermatogenesis. Electroporatic transfer of some of these genes of *A. japonica* has evidence to confirm that *eSSRs34* gene induces ssc renewal and *eSSR21* induces spermatogenesis inhibiting substances (Miura et al., 2007a). Hence, the credit of establishing the presence of estrogen, a female hormone in males and its implication to regeneration of spermatogonial cells goes to Dr T Maiura and his colleagues.

3.2 Xenogenesis and differentiation

Fishes produced by transplantation of PGCs and their derivatives SSCs and oogonial stem cells (OSCs) to homospecific recipients are called allogenics and those by the transplantation to heterospecific recipients are xenogenics. With relatively better availability of SSCs in terms of number and duration (see Pandian, 2011), the SSCs transplantation may provide important keys to many issues on genetic and endocrine control of gonadal differentiation. Hence a re-examination of xenogenics has become necessary. A comparison of relevant results reported on xenogenics by different authors led to the following inferences:

1. Transplantation of PGCs or their derivatives SSCs or OSCs into early developmental stage of recipient generate F_0 chimeras characterized by morphological (exception: body color in rainbow trout, Okutsu et al., 2006, and medaka, Shimada and Takeda, 2008) and anatomical features of the recipient but with donor-derived complete or partial gonad.
2. For the transplantation, the authors used grafts ranging from a single PGC (e.g., Saito et al., 2008) to 5–30 PGCs (e.g., Takeuchi et al., 2004), a layer (Shimada and Takeda, 2008) or sandwich inserted with a major fraction of the donor (Yamaha et al., 2001) or sandwich inserted with the entire basket of PGCs from the donor (Yamaha et al., 2003), SSCs (e.g.,

Okutsu et al., 2006) and OSCs (Wong et al., 2010); the developmental stage of the recipients also ranges from blastula of hybrid carp (Yamaha et al., 2001, 2003), fertile alevin (Takeuchi et al., 2004) or sterile alevin (Okutsu et al., 2006) and fry (Wong et al., 2010)—all of them possessing undifferentiated gonad to mature artificially sterilized testis (Lacerda et al., 2010) or alevin, whose gonad is committed to become the testis (Wong et al., 2010).

3. Yamaha's transplantation technique involves simple insertion of donor's sandwich of PGCs into the recipients's blastula, after removing the recipients's sandwich containing some PGCs (Yamaha et al., 2001) or the donor's sandwich containing none of the PGCs from the recipients (Yamaha et al., 2003) (Fig. 16). Briefly, the technique may result in partial 'sterilization' as in 2n goldfish (Yamaha et al., 2001) and complete 'sterilization' of the recipient's blastula of hybrid carp (Yamaha et al., 2003). In the absence of the recipients's PGCs, the entire gonad of the chimera is completely colonized by the donor's PGCs, as in 2n sterile (carp x goldfish) hybrid (Yamaha et al., 2003). Consequently, the 2n female and male chimeras produce n egg and n sperm, as in goldfish (Fig. 16B), but in the presence of recipient's PGCs, n eggs of the recipient goldfish and 3n unreduced clonal eggs of the crucian carp (Fig. 16A).
4. Transplantations of a single PGC of pearl danio, goldfish and loach into the early blastulae of zebrafish generated only fertile male chimeras, which are amenable to endocrine feminization (Fig. 17A, B, C). Perhaps a single PGC is not adequate to neutralize the germ cells supporting somatic cells, which are predisposed toward male development. This suggestion is confirmed by transplantation of a graft containing a

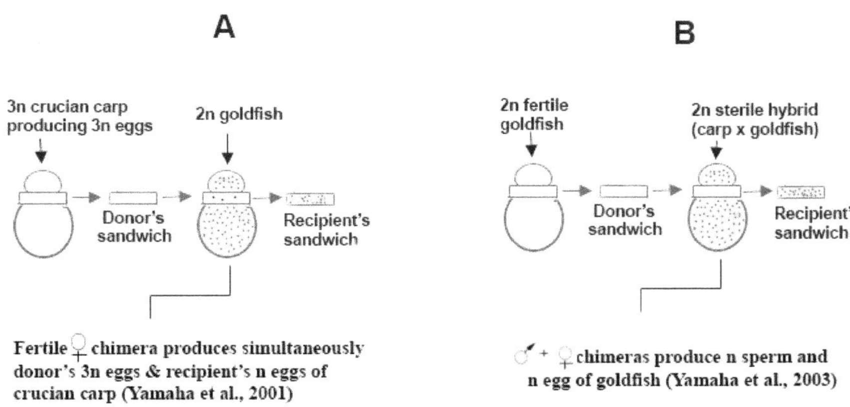

Fig. 16. Schematic representation of Yamaha's PGCs transplantation technique by removing a sandwich from the donor and inserting it in the recipient involving the removal of recipient's sandwich. Essentially the technique involves sterilization of the recipient's blastula.

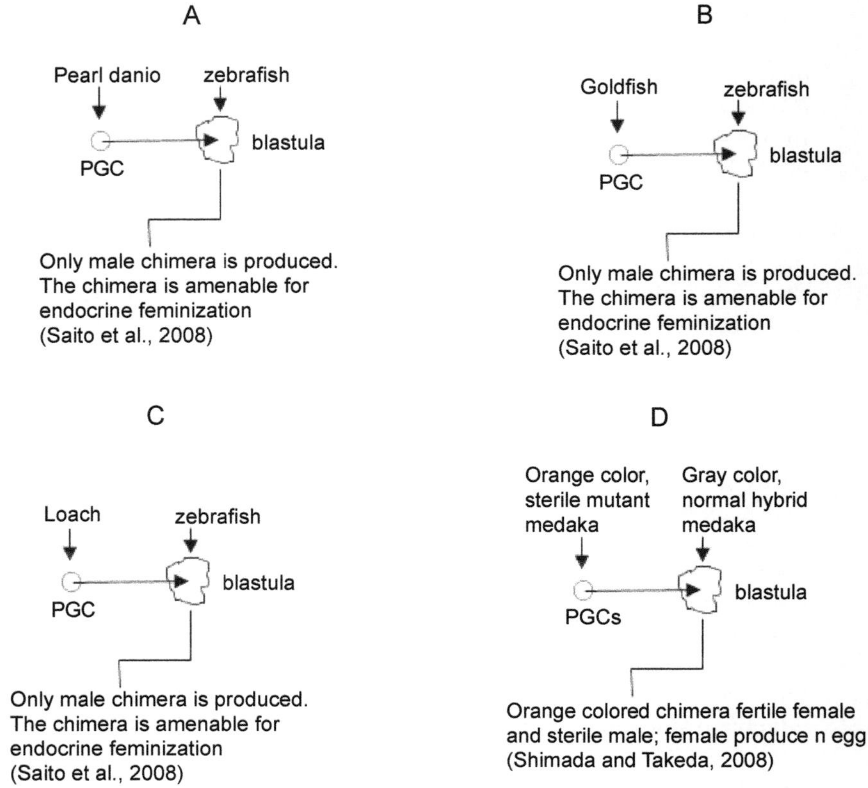

Fig. 17. Schematic representation of Saito's method of transplantation of a single primordial germ cells into the recipient blastulae resulting in all male chimeric progenies. In contrast the transplantation of a PGC layer into the recipient blastula produces female and male chimeras.

layer of PGCs into sterile fry of zebrafish by Wong et al. (2010) (Fig. 17) and adequate number of PGCs into medaka blastula by Shimada and Takeda (2008) (Fig. 17D) as well as into the alevin of masu by Takeuchi et al. (2004) (Fig. 17). Hence to generate an almost equal number of male and female chimeras, transplantation of more than five PGCs are required.

5. On transplantation of PGCs of rainbow trout into alevin masu salmon, both the recipients and transplanted PGCs colonize the gonads of the male and female chimeras (Takeuchi et al., 2004). The 'hybrid ovary' simultaneously produces fertilizable eggs both masu and trout (Fig. 18A). It remains to be known whether the male chimera with the presumable 'hybrid testis' too can simultaneously produce fertilizable sperm of both salmon and trout. In the absence of the recipients's

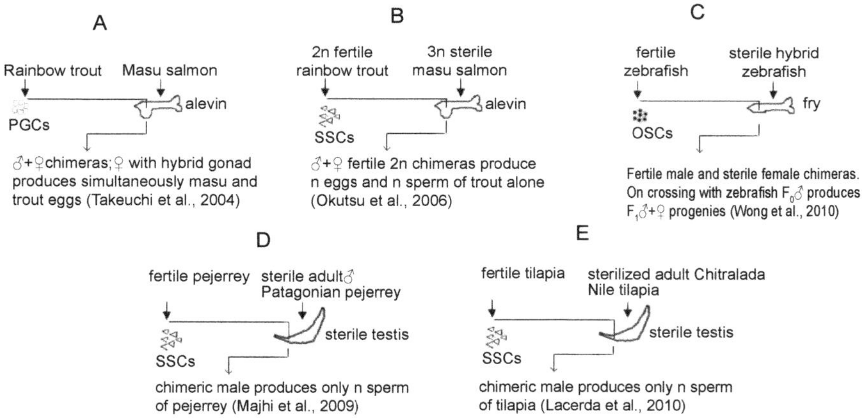

Fig. 18. Schematic representation of PGCs (A), SSCs (B) and OSCs (C) transplantations from donor to recipients with undifferentiated gonad. Contrastingly, SSCs are transplanted into sterilized testis of mature males (D and E) resulting in production of male chimeras alone.

PGCs, the transplanted SSCs of rainbow trout into alevin of 3n sterile masu, the 3n fertile chimeric masu salmon produces either n sperm or n egg of the trout alone (Fig. 18B). Only the donor-derived gametes are produced by the chimeras (Okutsu et al., 2006). In zebrafish, the transplantation of OSCs into the hybrid (of pearl danio x danio) fry also produces both fertile male and female chimeras, although the female chimera is sterile, for yet to be known reasons of the failure of endocrine differentiation (Fig 18C). These experiments of Takeuchi et al. (2004) (Okutsu et al., 2006) and Wong et al. (2010) demonstrate the retention of bisexual potency by the PGCs and their derivatives, SSCs and OSCs, despite the SSCs and OSCs are drawn from mature males and females that have earlier spermiated/spawned once or more.

6. The transplantations of SSCs of pejerrey and tilapia into artificially sterilized mature testis of hybrid pejerrey or chidralatha tilapia, respectively generate only male chimeras, producing the donor-derived sperm alone (Fig. 18D, E). Sex of a fish is initially determined by the intimate 'cross talk' (Tanaka et al., 2008) and interplay between germ cells and germ cells supporting somatic cells (Pandian, 2011). The experimental knockdown of *Cxcr4*, the chemo-PGCs-attractant receptor gene has shown that the supporting somatic cells are predisposed toward male development and the presence of PGCs is essential for sustenance of sexual dimorphism in medaka (Kurokawa et al., 2007). PGCs and their derivatives OSCs and SSCs retain their bisexual potency *per se* and the supporting cells are also considered to retain bisexual potency (Nakamura et al., 2003). However, studies on transplantation of germ cells, endocrine sex reversal and gonadectomy clearly indicate

that these somatic cells lose bisexual potency, when the fish attains sexual maturity (see Pandian, 2012). Incidentally, Ogawa et al. (2008), who have hormonally induced 'post-sex differentiational' sex reversal in *Cyprinus carpio*, a primary gonochore (Devlin and Nagahama, 2002), have also clearly stated that "the bisexual potency of the ovary is maintained in the common carp at least before puberty". In primary gonochores gonadal differentiation implies the loss of bisexual potency. However, the potency is retained by the somatic cells of hermaphroditic fishes (Pandian, 2012). Evidently, the bisexual potency has been lost by the somatic cells of hybrid pejerrey and chidralatha tilapia as well as the masu alevin committed to become male. Hence the gonadal differentiation in gonochores implies the loss of bisexual potency by the germ cells supporting somatic cells.

3.3 Spermiogenesis

It is a developmental process through which spermatids are differentiated into spermatozoa. It includes a series of morphological changes, namely nuclear condensation, formation of flagellum/flagella, elimination of cytoplasm including some of its organelles and rearrangements of other cellular organelles along spermatozoan cytoplasm (Jamieson, 1991). Based on the orientation of flagellum to nucleus and nuclear rotation, three types of spermatozoa are recognized (Schulz et al., 2010): Type 1 is characterized by a perpendicular flagellum in relation to nucleus with its rotation, in Type 2, the flagellum is laid in parallel to the nucleus but without its rotation and in Type 3 the flagellum is central without nuclear rotation (Quagio-Grassiotto and Oliveira, 2008). These flagellar types are useful in establishing phylogenetic relationships among species (Jamieson, 1991).

"In coupling with diverse modes of fertilization in fishes, the spermatozoa display various morphological adaptations in size (spherical to elliptical), shape (15 µm in *Catla catla* to 24 µm in *Labeo rohita* within Indian major carps alone, Verma et al., 2009) and ultra-structure. The spermatozoan structure of more than 300 fish species spread over 100 families has been described (Jamieson, 1991). For example, the sperm cell of *Hemigrammus caudovittatus* consists of a small spherical head, a short collar and a single flagellum of 18 µm length (David and Pandian, 2006b). Typically, the head lacks an acrosome and accommodates electron dense nucleus; the collar with or without cytoplasmic channel (e.g., Pomocentrids, Lahnsteiner and Patzner, 2009b, *Clarias gariepinus*, Mansour et al., 2002) contains a few mitochondria. With this basic morphology, the 'aquasperm' is adapted for external fertilization (e.g., *Oncorhynchus tshawytscha*, Fig. 19a left panel). Its spermatogenetic process is characterized by a short duration of the spermiogenetic cycle and reduced energy expenditure. Consequently, these

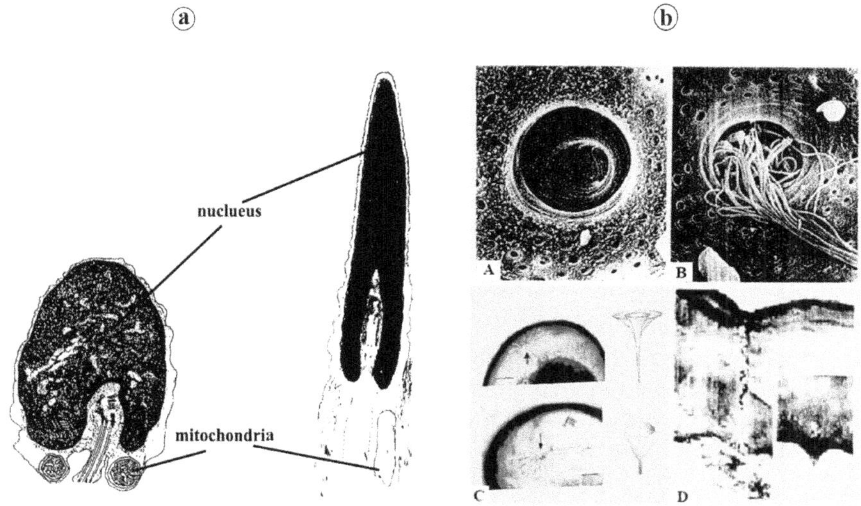

Fig. 19. (a) (Left panel) Longitudinal sections of aquasperm of oiviparous *Oncorhynchus tshawytscha* (left) and interosperm of viviparous *Poecilia latipinna*. Note the mitochondria in *O. tshawytscha* sperm and the large modified ones in *P. latipinna* (from Jamieson, BGM. 1991. Fish Evolution and Systematic Evidence from Spermatozoa, Cambridge University Press). (b) (Right panel) Electron microscopic views of (A) egg showing micropyle and (B, D) entry of sperm through the micropylar canal in an egg of *Alcichthys alcicornis* (from Munehara et al., 1989). Apical view of unfertilized eggs of (C) *Gymnocorymbus ternetzi* (lower) and *Hemigrammus caudovittatus* (upper) and schematic views of their respective micropylar canals in eggs (from David, 2004).

enable the externally fertilizing fishes to produce a large number of sperm cells (see Kweon et al., 1998)".

"On the other hand, the teleosts that employ internal mode of fertilization have specialized spermatozoa with a relatively larger and elongated head containing nucleus and collar containing extensively modified mitochondria (Fig. 19a left panel). This kind of 'interosperm' with its elongated head ensures the passage of the sperm cell through more viscous fluids of the female reproductive tract (Gardiner, 1978) and may also facilitate side-by-side alignment and thus the formation of sperm bundles (Burns et al., 1995). An enlarged collar containing numerous mitochondria prolongs the life of the sperm cells within the female reproductive tract" (Pandian, 2010). Interestingly, the absence of acrosome in piscine spermatozoa (Fig. 19b: A-D right panel) coincides with the presence of micropyle in the eggs. Since entry of sperm is made through the micropyle during fertilization, fertilization in fishes may not necessarily be a species specific event. For more details on the consequent hybridization and polyploidization, Pandian (2011) may be consulted; similarly, for more details on the number and architecture of micropyle and acrosome, Pandian (2012) may be consulted.

In semicystic spermatogenesis (Mattei, 1993) that occurs in Blenniidae (Lahnsteiner and Patzner, 1990), Opheliidae (Mattei, 1993), Scorpaenidae (Munoz et al., 2002), Soleidae (Garcia-Lopez et al., 2005) and Corydordinae (Spadella et al., 2007), the release of germ cells occurs in the spermatid stage. Individualization of spermatids from clones of different germ cells proceeds through spermiogenesis in the tubular lumen (Schulz et al., 2010). Even after completion of the spermiogenesis, sperm of many species (e.g., Salmonids) are incapable of fertilizing eggs. The developmental process involving physiological (but not morphological) changes from non-motile spermatozoa to that of being capable of vigorous motility and fertilization is referred to as 'capacitation'. This capacitation is accompanied by a fluid producing efferent duct, causing sperm hydration. The seminal fluid containing hydrated sperm suspended in the efferent duct is known as milt (Schulz and Miura, 2002).

Naturally mature or artificially matured males that have been treated with ovaprim are amenable for stripping the semen. The semen yield, for example among the Indian major carps, ranges from 6.6 ml/kg body weight in *Labeo calbasu* to 8.9 ml/kg *Cirrhinus mrigala* and the sperm count from 2.6×10^{10}/ml in *C. mrigala* to 3.5×10^{10}/ml in *L. calbasu* (Verma et al., 2009). Seminal plasma is an important constituent of semen that has a vital role in sperm survival, shelf-life, metabolism and motility. The cations like Na^+, K^+ and others like Cl^- in the seminal plasma establish osmotic balance, while the essential trace elements are components of many important enzymes. Hence biochemical evaluation of seminal plasma is an important criterion for assessment of milt quality. Verma et al. (2009) have listed the presence of albumin, protein, glucose, urea, cholesterol, triglycerides and other ions in the seminal plasma of Indian major carps. Alavi et al. (2009) have shown that changes in feed quality (PUFA levels) influence sperm morphology and sperm velocity of *Barbus barbus*. Research in this area of fishes has great application value for short-term extension or cryopreservation of sperm and shall constitute a title for a book of commercial importance.

Nevertheless, the amenability of piscine sperm for cadaveric preservation must be indicated. "Live, fertilazable sperm drawn from specimens that have earlier been post-mortem preserved at –20°C for periods of 30 days in the rosy barb *Puntius conchonius* (Kirankumar and Pandian, 2004), 40 days in the Buenos Aires (BT) tetra *Hemigrammus caudovittatus* (David and Pandian, 2006a) and 240 days in the Indian catfish *Heteropneustes fossilis* (Koteeswaran and Pandian, 2002) may also be used to activate or fertilize eggs of the respective species. Sex ratio of F_1 progenies sired by the cadaveric androgenics has revealed that no damage or mortality is caused to X-specific or Y-specific cadaveric sperm. The sperm count for the cadaveric sperm drawn from the 30-day post-mortem preserved specimens of BT tetra is 8.6×10^3/ml with 19% fertilizability. But it increases to 6.8×10^4/ml with 24% fertilizability, when the specimens are glycerol packed prior to preservation at –20ºC (David and Pandian, 2006). This simple method of using cadaveric sperm may be widely practicable in developing countries" (Pandian, 2011).

Interestingly, the simultaneous release of dimorphic sperm has been reported from sculpm *Hemilepidotus gilberti* (Hayakawa et al., 2002). The fish produces eusperm and parasperm. The parasperm is binuclear, due to incomplete second meiotic division and hence can not fertilize egg. It is a large (nearly 10 times larger than eusperm) erythrocyte-like cells (5–7 µm) without flagellum (see Fig. 41 from Pandian, 2010). Following fertilization of egg by an eusperm, the non-mobile parasperm moved by water form lumps and obstruct the movements of late-arriving sperm and thereby minimize sperm competition from sneakers and others. The endocrine mechanism responsible for simultaneous production of eusperm and parasperm from the same testis remains to be uncovered.

3.4 Spermatophore and spermatozuegma

Internally fertilizing males of fishes are often characterized by modifications of the genital tract specialized for packing sperm in bundles and organs involved in the delivery of these packets into the female genital tract (Mazzoldi et al., 2007). Viviparity is usually facilitated by an intromittent organ like the gonopodium (e.g., Poeciliidae), andropodium (e.g., Hemirhamphidae), pripium (e.g., Phallostelthidae), penis (e.g., Cottidae), tubular genital papillae (e.g., Athernidae) or anal fins (e.g., *Corynopoma riisei*) (Burns et al., 1995). In many oviparous fishes internal fertilization is not followed by gestation (Grier et al., 2005). Apparently, internal fertilization and viviparity in fishes seem to have evolved independently. Internal fertilization is also usually facilitated by packed spermatozoa in the form of spermatophore or spermatozeugma. But free sperm are also transferred into the female reproductive tract through the tubular gonopodium in a few viviparous fishes (e.g., *Anableps anableps, Jenynsia lineate,* Grier, 1981). Hence sperm packing is associated more with the functional morphology of intromittent organs. Incidentally, internal fertilization and sperm packing have also independently evolved in fishes. Having been evolved from their respective clades as many as 12 times (Goodwin et al., 2002), spermatophores and spermatozeugmata of varying morphologies are generated and indeed the males of viviparous piscine species seem to experiment with various designs of sperm transfer. Not surprisingly, a large number of descriptive acounts on the functional morphology and anatomy of male reproductive system and its gamete products is available (e.g., Grier and Uribe, 2005). However, publications concerning hormonal role in the formation of these gametic products are awaited.

The sperm packets covered with acellular capsule are called spermatophores (e.g., *H. setnai*). The uncapsulated ones are spermatozeugmata. In either of these packings, there are two modes of sperm packing: (i) the sperm packets are formed within the spermatogenic cysts in a species–specific pattern and are released into the efferent duct during spermiation (e.g., Poeciliidae, Goodeidae, Phallostethidae and some Hemirhamphidae and Anablephidae, Grier and Parenti, 1994; Downing and Burns, 1995). Sertoli cells of the cysts play an active role in the formation of sperm bundles within the spermatogenic cysts (Grier, 1993). In the other packing mode, loose spermatozoa are released from spermatocysts and sperm bundles are formed in the distal aspermatogenic testis (e.g., *Zenarchopterus dispar* (Grier and Collettee, 1987). The epithelial cells of the testicular tubes of the aspermatogenic part of testis are developed from hypertrophied Sertoli cells at spermiation (see Grier et al., 2005). From these epithelial cells of the efferent ducts, two types of proteinaceus secretions are recognized. In the more proximal ducts, the secretion is finely granular, whereas groups of coalescing spermatozoa are accompanied by a fibrous secretion in more distal ducts. The sperm are aligned along strings of this secretion, which is most probably instrumental in sperm clumping (Pecio et al., 2001). Hence the aspermatogenic region of the testis devoid of spermatogonia is of interest. The distal end aspermatogenic testis serves more as a storage of mature spermatozoa. This zone is narrow and is divided into chambers in some species. The area of aspermatogenic testis ranges from 7.1% in *Cheirodon calliirus* to 69.5% in *Gerphyrocharax atricauda* and to as much as 92.6% in a glandulocaudine species (Burns et al., 1995). Besides serving as the storage for sperm in its distal end, the efferent duct participates in several other functions like secretion of nutrients, steroids and pheromones, synthesis of proteins and enzymes as well as ionic regulation of seminal fluid (see Chowdhury and Joy, 2000).

3.5 Accessory glands

In various teleosts, testicular glands, blind pouches and seminal vesicles are the major accessory testicular glands (Lahnsteiner and Patzner, 2009a). Table 16 summarizes the broad distribution of these glands across the families of teleosts. Within these families, some blenniids have both testicular glands and blind pouches, while the chaenopsids have both testicular glands and seminal vesicles. Among these families, not all member species possess seminal vesicles; for example, the vesicles are present in internally fertilizing auchenipterid *Auchenipterus nuchalis* but are absent in the externally fertilizing silurid *Tatia intermedia* (Fig. 20). Similarly, the vesicles are present in the callichthyids *Corydoras* sp and *Hoplosternum littorale* but are absent in *Callichthys callichthys* (Fig. 20). In this context, Siluriformes are of special

Table 16. Accessory testicular glands present in different teleostean families (from Chowdhury and Joy, 2007; Mazzoldi et al., 2007; Knapp and Carlisle, 2011 and others).

Accessory gland	Family
Testicular gland	Dactyloscopidae, Tripterygiidae, Labriosmidae, Clinidae, Chaenopsidae some Blenniids
Steroid producing glands attached to testis but not homologous to testicular gland	Some Gobiids
Blind pouch	Some Blenniids
Seminal vesicle	Auchenipteridae, Bagridae, Callichthyidae, Clariidae, Gobiidae, Heteropneustidae, Ictaluridae, Pimelodidae, Siluridae, Batrachoididae, Chaenopsidae, Eleotridae, some Blenniids

interest. In them fertilization can be internal, or external including the 'sperm drinkers', in which the sperm pass through the females digestive tract prior to fertilization. Although the siluriformes are represented by 35 families and > 2500 species (Teugels, 1996), a comprehensive account on male reproductive anatomy and intromittent apparatus is available for internally or externally fertilizing siluriformes of 11 families alone (see Nayyar and Sundararaj, 1970; Santos et al., 2001).

Testicular glands, located ventrally to the testes, are considered as modified testicular efferent ducts (see Fig. 35—parts 2 and 6). In species characterized by semicystic spermatogenesis, spermatids do not fully mature into sperm within the testis but complete maturation within the testicular glands in seminiferous lobules. The glands contain steroidogenic cells homologous to Leydig cells and they also provide nutrition (see Knapp and Carlisle, 2011). Testicular blind pouches are tube-like evaginations of the spermatic duct. These pouches do not contain sperm (Lahnsteiner et al., 1993). But they contain 3β-hydroxysteroid dehydrogenase, and uridine disphosphoglucose dehydrogenase, suggesting a possible role in steroid glucuronide synthesis for possible use as pheromones (Knapp and Carlisle, 2011).

Seminal visicles are glandular outgrowths of the common sperm duct. They vary with regard to the number, size and shape of lobes. The number varies from a single pair in *Heteropneutes fossilis* to numerous vesicles in *Pangasius pangasius* (Joy and Singh, 1998) and also within species among male phenotypes that exhibit alternative mating strategies (Scaggiante et al., 2004). Interestingly, interspecific and intraspecific variations in size of the vesicles determine their storage capacity for seminal fluid and spermatozoa. The tubular structure of the vesicles commence as a simple protrusions from the sperm duct, develops under the influence of raising androgen levels during puberty (Cavaco et al., 2001) and becomes more complex with maturity and advancing age (see Chowdhury and Joy, 2007). Their

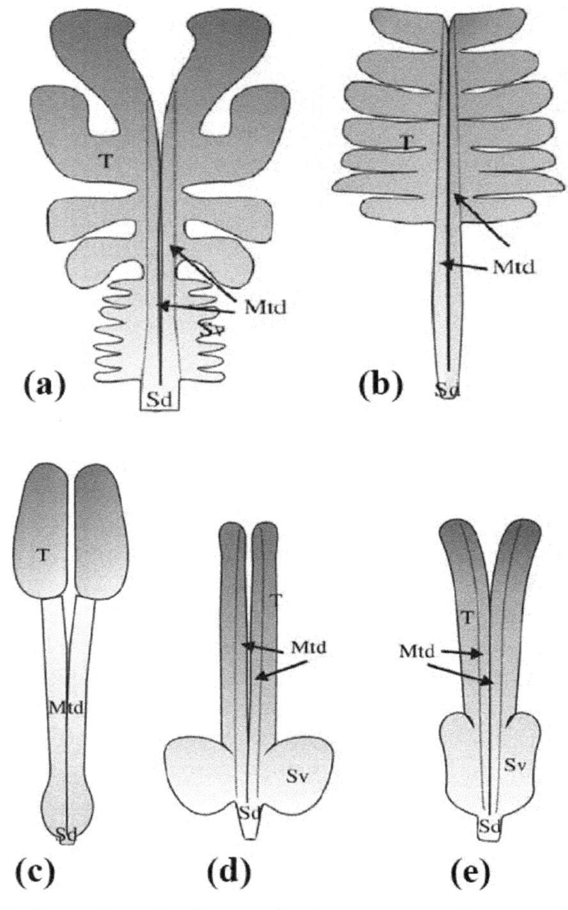

Fig. 20. Schematic illustration of male reproductive apparatus showing seminal vesicle in selected Siluriformes. (a) *Auchenipterus nuchalis*, (b) *Tatia intermedia*, (c) *Callichthys callichthys*, (d) *Corydoras* sp, (e) *Hoplosternum littorale*. Note the presence of seminal vesicle in internally fertilizing *A. nuchalis* and *Corydoras* spp, *H. littorale*, and the absence of seminal vesicle in *T. intermedia* and *C. callichthys*. Mtd = main testicular duct, Sd = sperm duct, Sv = seminal vesicle, T = testis (with kind permission of The Fishery Society of the British Isles/Wiley Blackwell; from C. Mazzoldi, V. Lorenzi and M.B. Rasotto. 2007. Variation of male reproductive apparatus in relation to fertilization modalities in the catfish families Auchenipteridae and Callichthyidae (Teleostei: Siluriformes). J Fish Biol, 70: 243–256).

columnar single layered epithelium is homologous to Sertoli cells (Loir et al., 1989) and expresses high levels of *FSHr* (Bogerd et al., 2001). The interstitial cells of the vesicles originate from the genital ridge (Fishelson et al., 1994) and are considered to be homologous to the testicular Leydig cells (Chowdhury and Joy, 2007). These steroidogenic interstitial cells contain steroid precursors as well as active enzymes of steroid hydroxylation suggesting that steroids and their glucuronides are being synthesized (see Singh and Joy, 1998). The vesicles of *Neogobius melanostonus* were recently reported to synthesize various androgens from AST *in vivo* (Jasra et al., 2007). Glucose, fructose, sucrose and hexosamines are present in concentrations varying among species. Non-motile sperm of fishes obtain energy from endogenous respiration through glycolysis, tricarboxyl acid cycle, oxidative phosphorylation and β-oxidation (Lahnsteiner et al., 1999). Proteins are the major organic constituents of the plasma of the vesicles (see Knapp and Carlisle, 2011). In *C. gariepinus* and *H. fossilis,* the vesicles secrete glucuronated steroids, which act as olfactory stimuli for females (see Chowdhury and Joy, 2007; see also p. 80-81).

The vesicles are both exocrine and endocrine secretory organs and display changes in size and secretory activity during the reproductive season. The credit for establishing the hormonal role played by the testes and seminal vesicles in regulation of cyclic reproduction in fishes must go to Indian scientists, especially those at the Banaras Hindu University. The vesicles remain small and inactive during resting and earlier pre-spawning phases, but increase during the mid spawning phase and peak during spawning season. In *H. fossilis* and *Clarias batrachus,* their size index and secretory activity are positively correlated with gonado-somatic index as well as with testicular secretory activity (Senthilkumaran and Joy, 1993; Singh and Joy, 1999; Chowdhury and Joy, 2000). Besides, steroidogenic activity of the vesicles and testes are also positively correlated suggesting that raising androgen levels promote secretory activity and their decline corresponds with quiescence of the vesicles (Chowdhury and Joy, 2007). These observations are also confirmed by unilateral castration of *Gillichthys mirabilis,* resulting in the vesicle regression (de Vlaming and Sundararaj, 1972). Contrastingly, bilateral castration induces paradoxical effect of hypertrophy and hyperactivity of the vesicles in the catfishes (Chowdhury and Joy, 2001a,b,c), which is also suppressed by anti-androgen (e.g., cyproterone acetate) (Chowdhury and Joy, 2001b).

Corydoras aeneus, C. bondi, C. erhardti, C. potaroensis and *H. littorale* are the 'sperm drinkers'. In these 'sperm drinkers', viscous mucous secretion of the vesicles embed the sperm before release (Mazzoldi et al., 2007). However, it remains to be known how the mucus embedding saves the sperm from the digestive enzymes (cf. p 118).

3.6 Corticosteroids

In fishes, the testes are known to synthesize corticosteroids including cortisol, 11-deoxycortisol (DOC), cortisterone and 11-deoxycorticosteroidione (Milla et al., 2009). A limited number of publications available on the role of corticosteroids on testicular development suggest that the responses to corticosteroids differ from species to species and ontogenetic developmental stage to stage. Cortisol is a 'stress hormone' released into the blood after activation of the hypothalamus-pituitary-adrenal axis (Bender et al., 2008). A series of publications by Consten et al. (2001a,b,c, 2002a,b) clearly demonstrates the reduction in gonado-somatic-index, impaired spermatogenesis and lower plasma T and 11-KT levels in 90–100 day-old carps, which have received a 30 d administration of cortisol through cortisol supplemented diet or intraperitoneal implants. Cortisol has only a mild effect on pituitary LH content and plasma LH levels, suggesting that the pituitary is not the primary site of inhibitory action of cortisol on testicular development. Rather it is the decrease in androgen level that contributes to the impaired maturation of testis more or less permanently, as the restoration of androgen levels in the cortisol-treated and control males do not restore the testicular development (see also Meghan et al., 2011). These findings of Consten et al. seem to confirm the earlier report that stress-induced cortisol synthesis reduces plasma sex steroid levels (e.g., Haddy and Pankhurst, 1999). With slow release of cortisol in the mature male, the brown trout *Salmo trutta* has smaller testes, reduced plasma T and reduced pituitary LH content, unlike in the carp. However, the testis size and plasma sex steroid levels are not affected by 36-d cortisol treatment in maturing *Oncorhynchus mykiss* (Carragher et al., 1989), perhaps the response of the young rainbow trout is comparable to that in juvenile carp (see Consten et al., 2002b).

On the other hand, Shankar and Kulkarni (2000) reported that the cortisol administration promotes testicular development during the early stages of spermatogenesis but inhibits it during the mature phase of spermatogenesis in *Notopterus notopterus*. The findings of Ozaki et al. (2006) seem to confirm the report of Shankar and Kulkarni (2000). In *Anguilla japonica*, Ozaki et al. (2006) found that cortisol induces DNA replication in spermatogonia, stimulates synthesis of 11-KT and potentiates its action on spermatogonial proliferation. Incidentally, there is also evidence that DOC may play a vital role in spermatogenesis (e.g., Strum et al., 2005). High levels of DOC are detected at the time of spermiation in *O. mykiss* (Milla et al., 2008).

Earlier publications have indicated the suppressive role of cortisol on E_2 synthesis in the ovarian follicles (e.g., *O. mykiss*, Carrgher and Sumpter, 1990, see also p 53). However, subsequent studies have revealed that it does not directly inhibit the synthesis of sex steroids in fishes like *Carassius auratus, Cyprinus carpio, Pagrus auratus* and *Danio rerio*. Cortisol is now known to promote hydration of the matured eggs (see Meghan et al., 2011). Corticosteroids are often elevated around the time of spawning (Milla et al., 2006), indicating the energy demand related to high metabolism at that time. They also promote *in vitro* meiotic maturation of full grown ovarian follicles by stimulating progestins (see Meghan et al., 2011). Cortisol, 11-deoxycortisol and 11-deoxycorticosterone promote ovulation *in vivo* (Milla et al., 2009).

A most unexpected role of cortisol is brought by Knapp et al. (2012). Exposure of adult female *Gambusia affinis* to 1 mg cortisol/l successfully induces 92% of the exposed females to develop masculine 'gonopodium' and copulatory behavior. In fact the frequency of copulatory behavior exceeds that by normal males. Knapp et al. (2012), however, have not given any indication regarding sex change involving transdifferentiation in the cortisol-exposed females. The masculinization index (MI), the ratio of lengths of ray 4 to 6 is used as an index of the degree of masculinization. "The cortisol-exposed females do not exhibit an MI equivalent to that of normal male" (Knapp et al., 2012). Interestingly, Turner (1947) reported the complete regeneration of gonopodium in partially castrated *G. affinis* with testis as small as 4% of the normal size. Hence the very presence of the masculinizing hormone like 11β-hydroxytestosterone in the small regenerating testis is adequate for complete regeneration of gonopodium in the cortisol-exposed female and copulating behavior are themselves not an indicator of sex change from ovary to testis. With the unisexual potency of germ cells supporting somatic cells in adult primary gonochore like *G. affinis* (see p 173), cortisol may not be able to induce sex change involving the redifferentiation of ovary into testis.

Reports available on stress caused by prolonged crowding suggest delayed ovulation and reduced egg size. For example, *Tilapia zilli* do not spawn in crowded holding tanks, but the same females spawn soon after the transfer to individual tanks. With prolonged duration of crowding the number of oocytes undergoing atresia also increases (Coward et al., 1998). This is also true of the gurnard *Chelionichthys kuma* (Clearwater and Pankhurst, 1997). On the other hand, reports on stress caused by frequent handling are not only species specific but are also contrasting responses. Adults of *O. mykiss,* subjected to repeated acute handling stress during the mature stage results in delayed ovulation and reduced eggs in females and reduced sperm counts in males (Campbell et al., 1992). In contrast, the studies by Contreras-Sanchez et al. (1998) showed that stress applied early in vitellogenesis results in smaller eggs and swim-up fry but absolute fecundity and fertilization are not affected by stress. Although an increase in maternal cortisol has effect on embryonic growth and survival of a few freshwater fishes (e.g., *Ictalurus punctatus,* Small, 2004), it has a negative effect on embryonic and larval development in many marine fishes (McCormick, 1998, 1999). High maternal cortisol levels are associated with low survival of *O. masou* (Mingist et al., 2007) and *Salmo salar* (Eriksen et al., 2006). These studies clearly suggest that the response to stress and cortisol varies between species and time, when stress is applied (see Meghan et al., 2011).

3.7 Spermiation

In captive farms, male brooders produce less and low quality sperm. In a review, Rurangawa et al. (2004) described some factors that affect sperm quality of cultured fish. In fishes, spermatozoa are stored in seminal plasma fluid and are immotile on ejaculation. They become active and motile only after release into water. In most freshwater fishes, they remain motile for less than 2 min (Kime et al., 2001). In some fishes like *Gasterosteus aculeatus* (Elofsson et al., 2003), *Anarhichys minor* (Kime and Tveiten, 2002), which are characterized by release of sticky gelatinous mass, the sperm are motile for 1-2 d within the gelatinous mass but become immotile once they come in contact with water. In captively reared *Morone saxatalis*, 3-year old male produce higher quality of sperm than those of 1- and 12-year-old fish. Broodstock nutrition and feeding greatly affect gamete quality. Polysaturated fatty acid (PUFA) enrichment of commercial diets enchances the reproductive performance of the male (e.g., *Dicentrarchus labrax*, Astuarino et al., 2001). Ascorbic acid deficiency reduces both sperm concentration and motility and consequently fertilization success (e.g., *Oncorhynchus mykiss*, Ciereszko and Dabrowski, 1995). A diet containing 4 mg E_2/kg feed suppresses spermiation after 7 mon of administration (e.g., *Acanthopagrus schlegeli*, Chang, 1995c). Stress may act by inducing changes in plasma osmolarity, which in turn, affects sperm quality. In *Morone chrysops*, 5 hr transportation of brooder male reduces seminal fluid osmolarity and hence motility at activation (Allyn et al., 2001). The cestode *Ligula intestinalis*, a common cyprinid parasite, may affect gamete production by preventing gonadal development. Infectious Pancreatic Necrosis (IPN) virus attach sperm cells of farmerd *O. mykiss* (Rodriquez et al., 1993).

Some fishes do not readily release sperm by hand stripping. For example, interatesticular sperm has to be taken out of the testes of fishes like *Clarias gariepinus;* there are others that require hormonal induction of spermiation. However, this practice increases the fluidity of the milt (e.g., *Hippoglossus hippoglossus*, Vermeirssen et al., 2000). Due to the close proximity of sperm duct and ureter or the presence of a single urogenital pore through which both milt and urine are released, contamination of sperm by urine during stripping and ejaculation is unavoidable. Since such contamination by urine prematurely induces motility, the sperm may have become immotile before fertilization can occur, especially, when contact with eggs is delayed (e.g., *Psetta maxima*, Dreanno et al., 1999). The squeezing of sperm out of testis after decapitation, or preferably collection of catheter via the urogenital papillae to obtain sperm free of blood, urine and/or feces contamination are some procedures employed to collect sperm without contamination (see Rurangawa et al., 2004).

3.8 Departures from norm

In fishes, hybridization and triploidization are the two major causes for the departure from the normal spermatogenesis. F_1 fertile hybrids are reported from 130 hybrids (Argue and Dunham, 1999). Among gonochores, spontenous triploidization occurs only in eight species: *Astyanax scabripininis, C. auratus langsdorfi, Curimata modesta, Gymnotus carapo, H. fossilis, Misgurnus anguillicaudatus, Noemacheilus barbaratulus* and *Trichomycterus davisi* (Pandian, 2011). Hence much of our knowledge on departure from normal spermatogenesis is based on fishes subjected to artificially induced hybridization and triploidization. Table 17 summarizes the available information on the recorded multiploid sperm in teleosts.

Table 17. Recorded multiploid sperm of fishes (from Pandian, 2011).

Ploidy and species	Ploidy in sperm	Euploid (%)	Relative DNA (n)
2n *Tinca tinca*	n	86	0.95
	2n	13	1.40
	4n	1	1.85
3n *T. tinca*	3n	69	1.10
	2n	34	1.50
	4n	2	1.47, 1.60
Heteropneustes fossilis		86	1.00
2n *H. fossilis*	n	94	1.00
3n *H. fossilis*	n	19	1.00
	2n	21	1.50
	3n	12	2.00
4n *H. fossilis*	n	2	1.00
	2n	90	2.00
3n *Rhodeus ocellatus ocellatus*			1.50, 1.90
3n *Oncorhynchus mykiss*			1.50
3n *Ctenopharyngodon idella*			1.40
3n *Gadus morhua*			1.60
4n *Misgurnus mizolepis*			2.00
4n *M. anguillicaudatus*			2.00
4n *Carassius auratus-Cyprinus carpio*	n 2n	40 49	1.00 2.00
2n *Squalius alburnoides*			2.00*

*clonal sperm

It is not known why hybrid males suffer gametic sterility but the female does not. The testes of *Barbus barbus* x *B. meridionalis* hybrids are irregular devoid of symmetric lobes, with no liquid milt (Phillipart and Berrebi, 1990). Intratesticular semen produced by male hybrid between *Heterobranchus longifilis* and *Clarias gariepinus* is two orders of magnitude lower (1.5–9.8 x 10^7/ml) than in parental species (2.9–4.0 x 10^9/ml) (Legendre et al., 1992). There are only smaller numbers of deformed sperm often with multiple flagellae in hybrids between *Lepomis gibbosus* x *L. cyanellus* (Dawley, 1987). Apparently, gametic sterility among male hybrid fails at different stages of spermatogenesis. However, the F_1 hybrid bester (*Huso huso* ♀ X *Acipenser ruthenus* ♂) is fully fertile but fails to spermiate spontaneously. Injection of exogenous pituitary preparation or GnRH during November–December stimulates 17, 20β-P (up to 2 ng/ml), acts as a maturation inducing steroid and results in successful spermiation in the bester (Amiri et al., 1996). Recently, 17α-hydroxyprogesterone (17α OHP) has been found to be the most potent inducer of final maturation in the triploid bester (Omoto et al., 2005).

Most triploid females are hormonally and cytologically sterile, whereas triploid males are hormonally fertile but cytologically partially or completely sterile (Pandian and Koteeswaran, 1998). Natural triploids do produce euploid and aneuploid sperm. The estimated sperm produced by natural triploid *Heteropneustes fossilis* is 69 and 31% for euploid and aneuploid sperm, respectively (Pandian and Koteeswaran, 1999). The spermatogonium size of triploid *Oncorhynchus kisutch* is 14.0 μm in diameter and 1,501 μm^3 in volume, in comparison to 11.9 μm and 964 μm^3 in diploid spermatogonium (Piferrer et al., 1994b). This may account for a 21–36% decrease in the spermatogonial number in triploid and the 25% reduction in gonado-somate index reported for *Tinca tinca* (Table 18). Hence the reduction in sperm count in triploid males is due to (i) reduction in the size of spermatogonial cells and (ii) cytological incompatibility owing to random segregation of trivalents. The reasons for the failure of fertilization success can be traced to incompatibility between the large head size of the sperm of triploid to penetrate through the micropyle of eggs of diploid and to reduced motility of sperm of triploids, as in *T. tinca* (Table 18). Interestingly, the velocity of sperm motility of tetraploid/hexaploid *Acipenser baerii* indicates motility inhibition immediately after activation, although the velocity is picked up subsequently by the 1.5 n and 2 n sperm (Psenicka et al., 2011). Notably, many diploid fishes like *Tinca tinca* and *H. fossilis* simultaneously produce n, 2n and 3n sperm. In *M. anguillicaudatus*, a cross between 2n gynogenic female and 4n male has been made by Zhang et al. (2002). As expected, triploid cells have been found in all somatic cells, such as the gills and liver. However, 17% hexaploid cells have been identified in one

Table 18. Sex ratio, sperm characteristics and their fertilizability of diploid and triploid *Tinca tinca* and *Gadus morhua:* *Relative to a haploid somatic cell (source: Linhart et al., 2006; Perruzi et al., 2009; compiled from Pandian, 2011).

Parameter	*Tinca tinca*		*Gadus morhua*	
	2n	3n	2n	3n
Mature males (%)	-	-	55	13
Gonado-somatic index (%)	1.2	0.9	-	-
Testis size (g)	2.2	1.7	-	-
Milt volume (ml)	0.6	0.1	750	500
Sperm count (n x 10^9)	2.1	0.1	12.3	9.7
Sperm head size (µm)	1.6	1.9	-	-
DNA of a sperm (n)*	1.0	1.5	1.0	1.4
DNA of a cell (n)*	1.9	2.8	2.0	2.4
Sperm motility (µm/sec)	96	70	-	-
Motile sperm (%)	93	57	124	113
Fertilization success (%)	60	16	49	43
Hatching success (%)	54	16	90	0

3n female. Among 3n males, the testes contain 14% hexaploid cells, 1.7% 9n and 3.3% 12n cells. The tetraploid metaphases in their gonads suggest the occurrence of premeiotic endomitosis, several mitotic divisions before the meiosis is commenced.

4

Gametogenesis in Unisexuals

Introduction

There are 80 unisexual biotypes among vertebrates. However, due to their amenability to rearing and experimentation (see Pandian, 2011), a volume of literature is available on eight distinct unisexual fishes (Lamatsch and Stock, 2009; Pandian, 2010, 2011). Having arisen as hybrids on multiple occasions, each unisexual fish species has explored its own mode and cycle of reproduction (Fig. 21). To avoid the onslaught of mutational meltdown and/or Muller's ratchet, the unisexual fishes have devised different escape mechanisms: (i) paternal leakage, (ii) paternal genome replacement by hybridogenesis, and (iii) paternal genome additions, which produce triploids and tetraploids. Of these, paternal genome additions occur in seven species, hybridogenesis in three species (Table 19) but paternal leakage is limited to *Poecilia formosa* and *Carassius auratus*. Hence the paternal genome addition is the preferred escape mechanism of the majority of unisexual fishes. Among the ploids, triploids occur in seven species but tetraploids are limited to three or four species. With the ubiquitous presence of triploids among the unisexuals, gametogenic function is impaired by the presence of three sets of homologous/heterologous chromosomes creating incompatibilities during gametogenesis (Pandian, 2011). However, publications describing the expression and inheritance of paternal genes or genome(s) in the unisexual progenies are limited. Even more limited are the publications that trace the cause(s) for the failure of spermatogenesis and describe the functioning of endocrine mechanism of gametogenesis. To point out the need for critical inputs in this area, this chapter has a few repetitions of some sections from earlier volumes in this series.

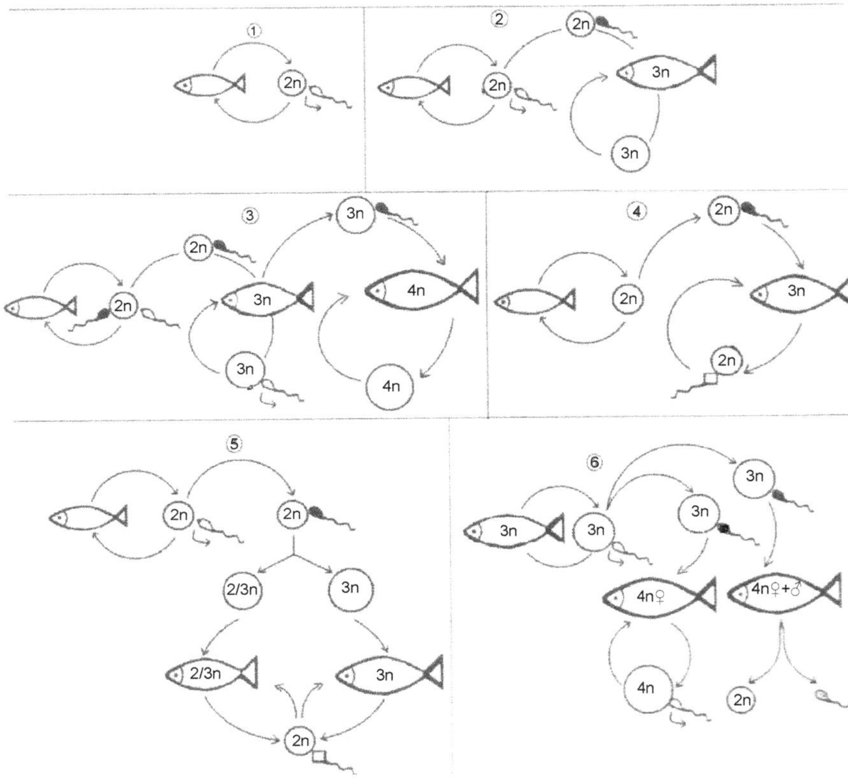

Fig. 21. Reproductive modes in unisexual fishes by 1 gynogenesis in *Menidia clarkhubbsi*, 2 gynogenesis in diploids and triploid *Poecilia formosa*, 3 gynogenesis but with genome additions to triploids and tetraploids in *Carassius auratus*, 4 gynogenesis in diploids and hybridogenesis in *Poeciliopsis monacha lucida*. Note fertilization by sperm of different species. 5 gynogenesis in diploids and hybridogenesis producing diploid-triploid mosaics and triploids in *Phoxinus eos neogaeus* and 6 gynogenesis in diploids and triploids and gynogenesis with genome addition in *Cobitis granoei taenia* (from Pandian, 2011).

4.1 Oogenesis

Essentially, all the unisexual fishes are hybrids between two or more biparentally reproducing sexual progenitors (Table 19). In general, hybridizations result in female sterility (Pandian, 2011). By producing unreduced eggs through premeiotic endomitosis, the unisexual females seem to have solved the problem of hybrid sterility. They have arisen from a particular combination of genomes, that produce diploid, triploid and tetraploid (e.g., *Cobitis granoei taenia*, Fig. 21) eggs that have not undergone recombination and reduction in ploidy. For example, cytophotometric

Table 19. Polyploidization and reproductive modes in unisexual fishes (from Pandian, 2011; updated) *Nimeitz et al. (2002).

Unisexual	Sexual pattern	Mode of reproduction	Sperm source
Menidia clarkhubbsi Athernidae Capano Bay, Texas	2n ♀ only	Gynogenesis	*M. beryllina* *M. penninsulae*
Fundulus sp Fundulidae Nova Scotia, Canada	2n ♀ only but 3n♂ rarely	Gynogenesis?	*F. heteroclitus* *F. diaphanus*
Poeciliopsis spp Poeciliidae, Mexico	2n ♀ 3n ♀	Hybridogenesis Gynogenesis	*P. lucida, P. monacha* *P. occidentalis,* *P. infans*
Poecilia formosa Poeciliidae Texas, North East Mexico	2n ♀, ♂ extremely rare, 3n ♀, ♂ rare 4n♀	Gynogenesis Gynogenesis	*P. mexicana* *P. latipinna* *P. latipunctata**
Phoxinus eos neogaeus Cyprinidae East Inlet pond, New Hampshire	2n ♀ only 3 ♀; 0.5% sterile ♂ 3n ♀ clonal; non clonal	Gynogenesis Variant in hybridogenetic direction	*P. eos, P. neogaeus*
Carassius auratus Cyprinidae *gibelio*, north China *langsdorfi*, Japan *sugu* south China	2n ♀, 50% sterile ♂ 3n ♀, 50% sterile ♂ 4n ♀, 50% sterile ♂	Gynogenesis	*Cyprinus carpio*, other cyprinids
Squalius alburnoides Cyprinidae West Spain, Iberia	2n ♀, ♂ 3n ♀ only 4n ♀, ♂	Gynogenesis Hybridogenesis Gonochorism	*Squalius pyrenaicus* *S. carolitetii*
Cobitis granoei taenia Cobitidae Moscow River	3n ♀ only 4n ♀ only 4n ♀, ♂	Gynogenesis Gynogenesis Gonochorism	*C. granoei, C. taenia*

analyses of DNA and DNA Feulgen levels in the nuclei of oogonia and primary oocytes of diploid and triploid *P. formosa* have shown that functional apomixis is the cytological basis of gynogenetic reproduction. Briefly, the first meiotic division is replaced by mitotic division, bivalents are not formed and there is no reduction of the chromosome number (Rasch et al., 1982). Interestingly, Rasch et al. (1982) have also reported that the nuclei from follicular epithelial cells and oogonia of *P. formosa* contain 1.65 pg DNA/cell, whereas those of primary oocytes have 3.65 pg/cell; this observation clearly indicates that a single cycle of chromosomal replication occurs in these cells during the period of DNA synthesis prior to the onset of meiotic prophase. Similarly, the 6C primary oocyte nuclei contain 5.05 pg/cell, which is twice the DNA content of 3C oogonia and follicular epithelial cells (2.45 pg/cell). Autoradiographic studies have confirmed that there is no additional synthesis of DNA during the course of meiotic prophase in this unisexual fish. Rasch et al. (1982) have also reported that while that primary oocyte has 3.6 pg DNA/cell, the nuclei of the oogonium and follicular epithelium contain 1.65 pg DNA/cell. Similarly, the primary oocyte nuclei of triploid have 5.1 pg DNA/cell, which is twice the DNA content of 3C oogonia and follicular epithelial cells.

As indicated elsewhere, fragmentary information available on the egg size of triploid and tetraploid gonochoric fishes provides circumstantial evidence for the genomic control on the quantum of yolk deposited and egg size in fishes. For instance, the increase in egg size of triploids is 62% only for 3n cyprinids but 89% for 3n cyprinodontids (sse p. 54-55). The fact that unisexuals occur in five families and their triploids in four of these families provide excellent opportunities to know whether the ploidy level determines the egg size. If so, does the increase in egg size with increasing ploidy level proportionate and does the proportion of increase in egg size *per se* differ from family to family? Interestingly, triploid *Squalius alburnoides* simultaneously produces haploid and triploid eggs and thus provides unique opportunity to test whether ploidies determine the egg size (Alves et al., 2004).

4.2 Paternal leakage

The possibility of 'paternal leakage' of subgenomic amount of DNA into gynogenic recipient *Poecilia formosa* has been originally suggested by Rasch and Balsano (1989) and proved by Schartl et al. (1995a). Schartl et al. have demonstrated the incorporation of subgenomic amount of DNA from the bisexual species to the gynogenic *P. formosa.* According to Rasch and Balsano (1989), the leaked paternal genes are the regulators of phenol oxidase system, which may account for the variety of patterns of melanin deposition found in *P. formosa* in its natural habitats and laboratory reared populations (Fig. 22). From their electron microscopic studies, Zhang et al. (1992b) have shown that synopsis and recombination occur between at least some homologous chromosomes in the triploid female ginbuna and the sperm donor gengoroubuna *Carassius auratus cuvieri.* Hence the scope for introduction of heterozygosity does exist in the gynogenic reproduction too. Recently, Jia et al. (2008) quantified the paternal leakage to the progenies of gynogenic unisexual *C. auratus gibelio* from China. Using microsatellite genotyping at 15 loci, they showed the incorporation of paternal genetic material in 63% of progenies. In 2004, Lamatsch et al. reported the high incidence of B chromosomes both in the laboratory reared and wild caught *P. formosa.* The credit of tracing the origin B chromosomes from the sperm-donor *P. mexicana* and their stable inheritance up to eight generations goes to Nanda et al. (2007).

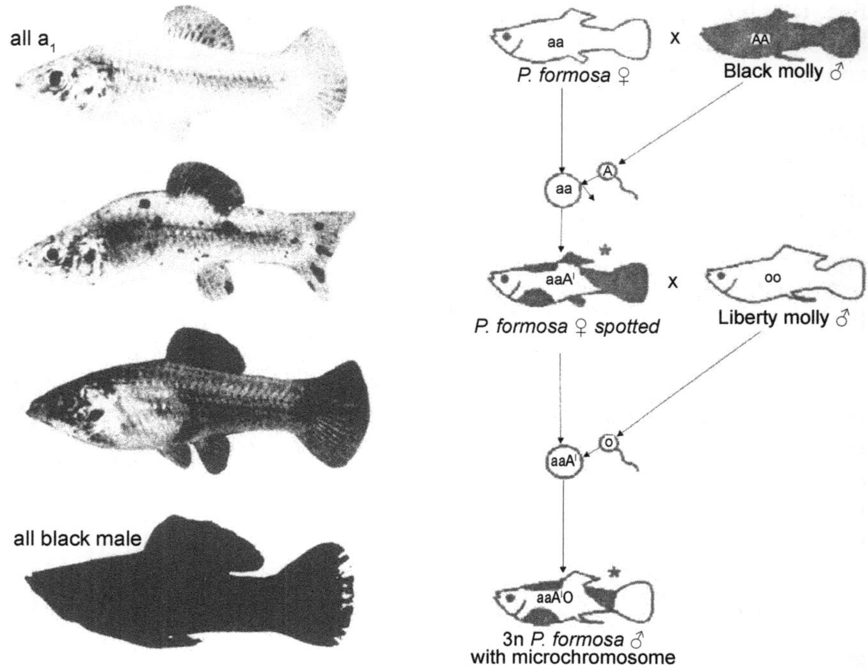

Fig. 22. Left panel: Progenies from matings of wild-type *Poecilia formosa* (top) to all-black male of *Poecilia* sp (bottom). Progenies of a_1 are (middle ones) indicating the incorporation of sub-genomic amounts of DNA from host black male (from Rasch and Balsano, 1989 printed with kind permission of the New York State Museum, Albany, New York). Right panel: Breeding complex of *P. formosa* under laboratory conditions (compiled from Dawley, 1989).

Upto now, the publication by Schartl et al. (1991) stands as a lone contribution of hormonally induced sex reversal in unisexual fish. Following masculinization by hormone treatment, *P. formosa* developed into a typical male with masculine stream-line body, stinted with xanthophore pigment and perfect gonopodium (cf. Turner, 1947, see also p 95). These developments reveal that structural genes specifying morphology of the male has remained functional, though they have not been expressed. Hence unisexuals like *P. formosa* have retained the bisexual potency for over 280,000 years (Schartl et al., 1995b). However it is not known whether the hormonal sex reversal is limited to morphological changes alone or is also extended to gonadal sex change leading to normal spermatogenesis. Due to cytological incompatibility, the males of the hybrid unisexuals like *Carassius auratus* and *Phoxinus eos neogaeus* remain sterile. The extension of the good work by Schartl et al. (1991) to other unisexuals, especially to the spermatogenesis in the sex reversed unisexuals is desired.

4.3 Hybridogenesis

The *Poeciliopsis* complex consists of three unisexual hybrid biotypes: of these, one is a hybridogenic diploid *Poeciliopsis monacha-lucida* and the other two are the gynogenic triploids: *P. 2monacha-lucida* and *P. monacha-2lucida. P. monacha-lucida* produces haploid eggs, which are fertilized by the sperm of *P. lucida* or in some cases by *P. occidentalis* or *P. latidens*. These hybrid progenies share some morphological characteristics like the dorsal fin spot in *P. monacha occidentalis* or three vertical stripes and four spots on the body of *P. monacha latidens* (Fig. 23). During oogenesis, however, the maternal *monacha* set alone is transmitted to the egg. This novel method of oogenesis through which haploid eggs are produced by hemiclonal mode of reproduction that delivers only the maternal genome of a hybrid to the functional oocyte in the unisexual fishes is called hybridogenesis (Schultz, 1989). In *P. monacha-lucida,* hybridogenesis involves the formation of a unipolar spindle before vitellogenesis. During meiosis, the maternal (*monacha*) chromosomes align unpaired on the metaphase plate, where they attach to a unipolar spindle; they are then drawn to the single pole, from where they become the sole contributor to the egg nucleus. Since the paternal chromosomes do not pair with the maternal chromosomes and are not attached to spindle fibers, no segregation or crossing-over occurs. The paternal chromosomes remain scattered in the cytoplasm and eventually get lost. The paternal chromosomes are regarded as incompatible with the

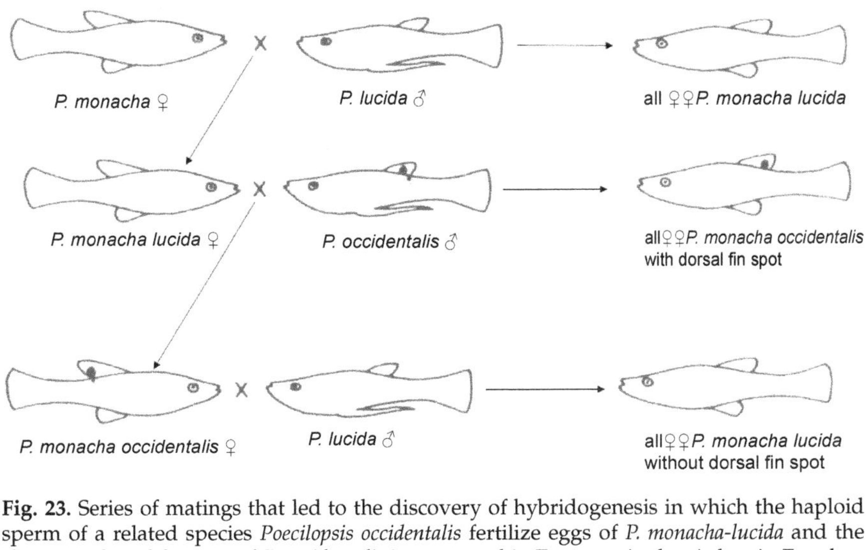

Fig. 23. Series of matings that led to the discovery of hybridogenesis in which the haploid sperm of a related species *Poecilopsis occidentalis* fertilize eggs of *P. monacha-lucida* and the character dorsal fin spot of *P. occidentalis* is expressed in F_1 progenies but is lost in F_2, when its eggs are fertilized by *P. lucida* (from Pandian, 2011).

ooplasm and hence the paternal half-spindle fails to form properly (Cimino, 1972a,b). Incidentally, similar elimination of paternal genome also occurs in gonochoric fishes: e.g., *Oncorhynchus masou* ♀ x *O. mykiss* ♂ hybrid eggs (Fujiwara et al., 1997) and triploid eggs of *Misgurnus anguillicaudatus* (Zhang et al., 1998). The causes and time of the elimination are different. In *M. anguillicaudatus,* some chromosomes in the small but full-grown oocytes get detached from the spindle, remain in the surrounding cytoplasm and ultimately these unpaired univalents get eliminated. Conversely, the paternal chromosomes of *O. mykiss* are eliminated due to cytoplasmic incompatibility in fertilized eggs (see Pandian, 2011).

Modified forms of hybridogenesis occur in the *Phoxinus* and *Squalius* complexes. The hybrids between *P.eos* and *P. neogaeus* include diploids (2n = 50), triploids (2n = 75) and mosaics (2n = 50–75). The mosaics have originated by the inclusion of the paternal genome of *P. eos* or *P. neogaeus* with the diploid hybrid genome (Dawley and Goddard, 1988). However, the paternal genome is incorporated into a few cells of some (mosaic) individuals and are discarded in the subsequent generation. The *S. alburnoides* complex is composed of diploid, triploid and tetraploid females and males with genomic composition that arose by hybridization between the sympatric bisexual *S. pyrenaicus* female and *Anaecypris hispanica*-like male ancestor (Alves et al., 2004; Robalo et al., 2006). In the northern Iberian River basins, the *pyrenaicus* genome is replaced by *S. carolitertii* genome (Carmona et al., 1997). Diploid and triploid females of the northern river basins reproduce by modified hybridogenesis, in which the maternal *carolitertii* genome is discarded during oogenesis (Carmona et al., 1997) but the triploid females of the southern river basins also reproduce by the modified hybridogenesis, where the maternal *pyrenaicus* genome is discarded in each generation followed by random segregation and recombination between the two remaining genomes. This form of modified hybridogenesis is called meiotic hybridogenesis (Alves et al., 1998). Through meiotic hybridogenesis, the eggs (carrying the maternal *pyrenaicus* genome) may be fertilized by homogametic (AA) male carrying (*anaecypris* genome) to produce all male progenies (Fig. 24).

4.4 Genome addition

In unisexual fishes, paternal genome addition not only enhances the stability and fitness but also increases its share in sympatric population. Triploids are continuosly generated at the frequency of 1.15% in *Poecilia fomosa* (Balsano et al., 1989), 3.5% in *Carassius auratus* (Umino et al., 1997) and 50% in *Phoxinus eos neogaeus* (Dawley and Goddard, 1988). The frequency of tetraploid generation ranges from 0.15% in *P. formosa* (Lampert et al., 2008) to 85–97% in *Squalius alburnoides* (Cunha et al., 2008). With only diploids

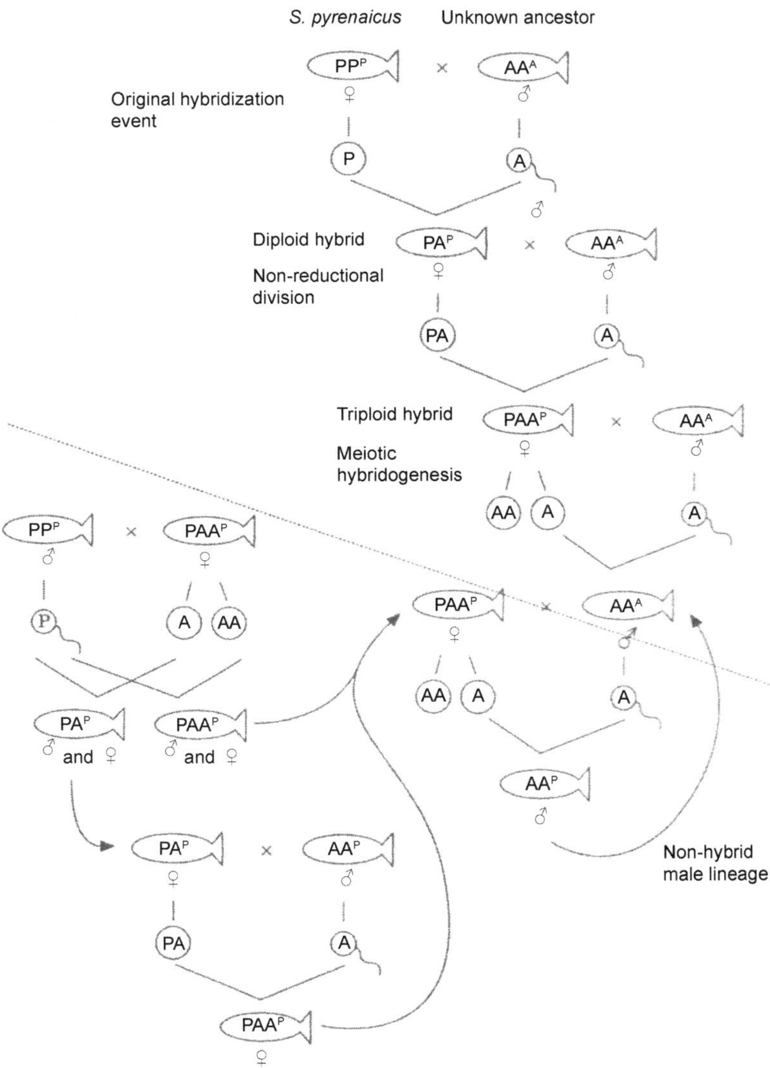

Fig. 24. *Squalius alburnoides:* Suggested reproductive pathways of all-male non-hybrid lineage within hybrid complex. Above dash line: mechanisms of the origin of the hybrid complex. Below dash line: perpetuating mechanism of the complex. P and A represents the genome of *S. pyrenaicus* and the other ancestor, respectively. The main letters indicate the nuclear genome and the prime ones the mitochondrial genome (With kind permission of The Fisheries Society of the British Isles/ Wiley Blackwell; from M.J. Alves, M.J. Collares-Pereira, T.E. Dowling and M.M. Coelho 2002. The genetics of maintenance of an all-male lineage in the *Squalius alburnoides* complex, J Fish Biol, 60: 649–662).

Table 20. Hybrid origin and reproductive strategies of unisexual fishes to escape from extinction (from Pandian, 2011; updated).

Unisexual species	Hybrid origin	Proportion in sympatric population (%)*	Reproductive strategies
Menidia clarkhubbsi Echelle et al. (1989)	*M. beryllnia* x *M. penninsulae*-like ancestor	2.8	All females; gynogenesis only; no male so far found
Fundulus sp. Dawley (1992)	*F. heteroclitus* x *F. diaphanus*	7.9	All females; gynogenesis only; no male so far found; 3n with diaphanus genome addition
Poeciliopsis spp. Schultz (1977)	*P. monacha* ♀ x *P. lucida* ♂	95.0	All females; no male so far found; (i) 2n reproduces by hybridogenesis (ii) 3n gains genome addition but reproduces by gynogenesis
Poecilia formosa Balsano et al. (1985)	*P. mexicana latipinna* ♂ x *P. latipinna* ♀	32.0	Though reproduce by gynogenesis only (i) 3n and 4n gain genome additions (ii) 2n incorporates host leaked subgenome and β chromosome to increase diversity
Phoxinus eos neogaeus Dawley and Goddard (1988) Elder and Schlosser (1995)	*P. eos* x *P. neogaeus*	59.0	Though reproduce mostly by gynogenesis (i) high incidence (50%) of fertilization by donor sperm (ii) consequent genome addition to 3n and 2n–3n mosaics (iii) Hybridogenesis ?
Carassius auratus Wu et al. (2003)	*C. auratus* ♀ x *Cyprinus carpio* ♂	13–49	Though reproduces only by gynogenesis, gains genome addition from one of 4 cyprinid species
Squalius alburnoides Ribeiro et al. (2003)	*Squalius pyrenaicus* ♀ x *Anaecypris hispanica*-like ♂ ancestor	57.0	(i) Genome addition to 3n and 4n (ii) 3n reproduces by hybridogeneis (iii) Hybrid ♂produces clonal 2n sperm and haploid sperm
Cobitis granoei taenia Vasil'yev et al. (1990)	*C. granoei* x *C. taenia*	34.0	Genome addition to 3n and 4n unisexuals

* For references see Pandian (2010); however, for data on production of unisexuals in sympatric population, references are given along with species names

in *Menidia clarkhubbsi*, its share in symapatric population is 2.8% only. But the share of unisexual fishes in the population progressively increases to 59% in *P. eos neogaeus* and to 95% *Poeciliopsis* spp., who have diploids and triploids (Table 20).

Occurrence of the male is a rarity in *P. formosa* (Lamatsch et al., 2000). The male ratio ranges from 0.5% in diploid *P. eos neogaeus* (Goddard and Dawley, 1990), to 50% in 3n and 4n *C. auratus* (Wu et al., 2003). In *S. alburnoides*, the ratio ranges from 5 to 14% among diploids, 1 to 22% among triploids and 1 to 45% among tetraploids (see Pandian, 2011). Except in *S. alburnoides*, the males in other unisexuals are sterile and the cause(s) for their sterility may be traced to chromosomal incompatibility. For instance, triploidization in *C. auratus* involves the addition of a paternal genome of a number of cyprinids, whose 2n chromosome number ranges from 48 in *Megalobrama ambycephala* to 100 in *C. carpio* (see Pandian, 2010).

Among hybrid vertebrates, the most uncommon mode of spermatogenesis is reported from *S. alburnoides* (Alves et al., 1998, 1999). The diploid males of this species are comprised of two distinct genotypes: 1. The non-hybrid genotype possessing the *S. pyrenaicus*-like mitochondria exhibits normal Mendelian meiosis and produces haploid sperm and 2. The hybrid genotype produces unreduced clonal sperm, a very rare event among vertebrates, perhaps by premeiotic endomitosis (Rasch et al., 1982). On clonal transmission of these unreduced 2n sperm to haploid eggs, triploids are generated (Alves et al., 1999). These fertile triploids, in turn, produce unreduced 3n sperm (Alves et al., 1999; Sousa-Santos et al., 2007). However, the tetraploids produce reduced 2n sperm by strictly adopting Mendelian meiosis including segregation, as confirmed by minisatellite markers (Alves et al., 1999).

5

Morphotypes and Differentiation

Introduction

In fishes, the mean gonado somatic index (GSI) is about 1.5 for males against 10.1 for females (Pandian, 2010). Energy required for normal development and functioning of testes is far less and is 6–18 times less than that required for the development of the ovary (Diana, 1983). This large difference in reproductive investment between sexes has led to (i) more restricted periods of sexual activity in females than males (see Fig. 8) and (ii) an intense selection by the female, as its fitness depends on the quality of its mating partner (e.g., Fagundes et al., 2007). As a rule, males compete for females and their fitness depends on the number of mates with whom they may breed (Oliveira et al., 2001a). For example, among oviparous fish with no parental care, maximum number of eggs are produced, when a male possesses two females, but a haremic male can effectively 'handle' as many as 12 females and fertilize all their 56,777 eggs (Pandian, 2010). Hence "the struggle between the males for possession of females.... results not in death of unsuccessful competitor but in a few or no offspring" (Darwin, 1859). With intense male competition, males of lower competitive ability may adopt Alternative Mating Strategy (AMS) [the abbreviation A*R*T does not go well with alternative *mating* tactic; hence the original definition and abbreviation of AMS of Warner and Robertson (1978) is preferred] 'to make use of a bad situation' (Taborsky, 2001) and to produce as many offspring as possible.

Within vertebrates, teleosts are the taxa that have the largest number of species displaying AMS. In 1998, Taborsky reported the occurrence of AMS in 140 species belonging to 28 families; since then the list is annually added with one or more species (Oliveira, 2006). The AMS and sexual dimorphism/ polymorphism are not synonymous, as AMS also occurs among monoandric (the term may be reserved for those among hermaphrodites) or preferably monomorphic gonochoric species, in which the same individual may behave as dominant or subordinate, according to the social situation encountered by it. The percentage of species displaying AMS ranges from about 1 (e.g., Oryziidae) to 1.7 in Blenniidae, 3.2 in Catosomidae, 6.4 in Cyprinidae, 9.2 in Salmonidae and to 11.4 in Cichlidae (Taborsky, 1998). The causes for the prevalence of AMS among these families have been traced to (i) wide differences in size at sexual maturity of males, (ii) external fertilization facilitating simultaneous spawning by 'alternatives' and (iii) paternal care rendering the 'alternative' males with high returns (Oliveira, 2006).

5.1 Theory and classification

Theoretical considerations suggest the operation of conditional strategy or proximate mechanisms in the development and sustenance of AMS (see Gross, 1996). According to the former, almost all alternative phenotypes adopt a conditional strategy and the strategy is favored by evolution, although the mean fitness of alternative strategies is unequal. The proximate mechanisms underlie the presence of many similarities among alternate phenotypes within a sex and the alternate strategy may be regulated by relatively minor differences in hormonal profile (Moore et al., 1998). Publications on endocrine role of alternative phenotypes are available but they are needed for the fixed alternative 'genotypes'. The description given below on the amazing diversity of these alternative phenotypes is likely to be more biological but shall serve to introduce them to molecular biologists. Among vertebrates, the occurrence of an array of phenotypes within male sex in fishes is astounding. However, this area of sex differentiation remains untouched by molecular biologists. These phenotypes, in a restricted sense, are like mutants, which have facilitated a greater insights into our understanding of the sex differentiation process; hence a coverage of these phenotypes by molecular biologists may be rewarding.

Taborsky (1994) gave a long list of alternative phenotypes and a little complicated classification. The more acceptable Oliveira's (2006) classification of AMS includes two major categories: (i) Fixed alternative phenotypes, adopting a specific irreversible strategy for the life time, as by jacks and hooknoses of salmonids (Gross, 1985, 1991) and sneakers and parentals of centrarchids (Neff, 2004) (ii) Flexible or plastic alternative phenotypes switch from one tactic to another during the life time. There are two patterns within the plastic alternative phenotypes (iia) transformants (the term sequential is reserved for hermaphrodites) and (iib) reversibles. With ontogenetic growth and development, the transformants switch from sneaker to satellite and satellite to territorial male (e.g., *Symphodus ocellatus Parablennuis parviconis*). The reversibles switch back and forth from one tactic to another; for instance, many monomorphic cichlids may switch from dominant to subordinate and subordinate to dominant, depending on the social status encountered by them (see also Oliveira et al., 2005b).

Fishes that display AMS seem to have explored their own patterns, resulting in wide diversity. In male fishes, the known phenotypes are: (i) territorial/parental/nest-holder/floater/dominant/breeder, collectively known as bourgeois, and (iia) cuckolder, (iib) subordinate, (iic) helper, (iid) sneaker, (iie) streaker, (iif) satellite, (iig) pirate and they are collectively called parasites (Taborsky, 1997). These male phenotypes require explanation/definition. Besides growing in size and establishing a nest/territory, the bourgeois invests time and energy on the development of secondary sexual characteristics (sscs), defence of the nest/territory aggressively and courtship behavior, perhaps at the reduced cost of developing relatively smaller testes. Morphologic investment of bourgeois males includes development, in some cases ornamented sscs including sound producing organs (Brantley and Bass, 1994) as well as fighting weapons like the hooknose in salmonids. The subordinate and cuckolder, which are 'alternatives' to dominant and parental, respectively, invest the available energy on developing larger testes. Hence the bourgeois males may readily be distinguished by the presence of one or more sscs like the supra-orbital tentacles and an enlarged distal dorsal fin (e.g., *Aidablennius sphinx*), head crests (e.g., *Salaria pavo*), glands on the anal spines (e.g., *Salaria* spp., *Parablennius* spp.), club-like glands on the tip of soft fins (e.g., *Lipophyres* spp.) (Oliveira et al., 2001a) and red-blue body color and white urinary papillae. Increased body size and the consequent dominance directly as well as indirectly through regulation of androgen levels seem to control the development of sscs and reproductive behavior (Fig. 25). Hence the corresponding gene(s) may act on behavior through hormone or on hormone through behavior (Fig. 55). In *Parablennius parvicornis,* larger and older males develop conspicuous sscs (M_+ type), whereas small and younger males do not yet express them (M-type). A fraction of the M^+ males do not establish nests but act as floaters; these floaters, instead of acting as a bourgeois, sneak fertilize or take-over a nest from a nest-holder. Indeed they act as 'parasites'. The nest-holder and floater are interchangeable ethotypes (Oliveira et al., 2001b). The presence of floater males is also reported among cichlids, for example *Oreochromis mossambicus, Pseudocrenilabrus philander, Aulonocranus dewindti* (see Oliveira and Almada, 1998). The presence of such ethotypes, which can interchange from bourgeois to parasitic floater/'sneaker' status depending on the availability of space for construction of a nest, may question the veracity of the designation of 'parasites' by Taborsky (1997) to all the 'alternative' phenotypes.

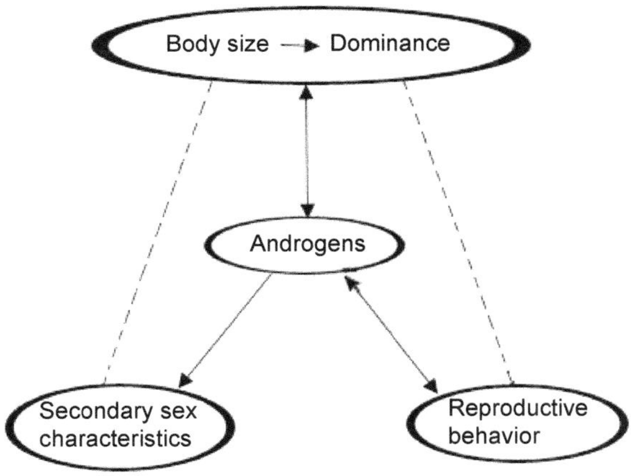

Fig. 25. Suggested role of androgens as mediators of social modulations causing expression of traits characteristic of alternative morphotypes. Implied pathways of gene action on behavior through androgens or androgens through behavior (compiled from Cardwell et al., 1996 and Oliveira and Almada, 1998).

Sneakers are relatively small males; they dart and steal spawning with the territorial female, who is waiting to spawn with the territorial male (Warner and Robertson, 1978). According to Goncalves et al. (2003), they are parasitic males that rely on inconspicuous approaches to fertilize eggs during spawning events between bourgeois males and females (e.g., *Lepomis macrochirus, Zacco temmincki, Symphodus* spp., see Pandian, 2010). They occur in and around the nests/harems but are not attached to any specific nest/territory/harem (Oliveira et al., 2001b). Adopting the female's nuptial body color and behavior, they may deceive the bourgeois, although the bourgeois of *Salaria pavo* can discriminate large sneaker from female (Goncalves et al., 2003). Female mimicry by the sneakers has been reported from more than 30 species (Taborsky, 1998). The sneakers possess as much as 3.5–9.0 (Bass, 1992) times higher GSI (Oliveira et al., 2001c) and their high quality sperm that contain more ATP (Vladic and Jarvi, 2001; Burness et al., 2005; Lacatello et al., 2007), swim faster (Burness et al., 2004; Vaz Serrano et al., 2006; Lacatello et al., 2007) and display greater motility (Rudofsen et al., 2006); for example, ATP contents are 0.6 nmol/ml for the dominant male, but as much as 3.2 nmol/ml for the subordinate male *Salmo salar* (Vladic et al., 2010); the swimming speed at 1 min after activation is 24 µm/sec for the sperm of sneaker, against 20 µm/sec for those of territorial and satellite, and 13 µm/sec for the pirate male of *Telmatochromis vitattus* (Fitzpatrick et al., 2007). In *Symphodus ocellatus*, the GSI decreases from 10 in sneakers (2–4 g size) to 8-5 in satellites (4–6 g) and to 2 in territorials (8–11 g) (Taborsky, 1998). Increased sperm production by sneaker males involves development of storage organs like seminal vesicles. Streakers join the territorial male and female at the climax of spawning like the spawning rush (Warner and Robertson, 1978). Pirates overtake temporarily the territory and steal spawning with the territorial female awaiting the territorial male for spawning (e.g., *T. vittatus,* Ota and Kohda, 2006). In a Taborskian (1997) sense, these phenotypes are all parasitic, as they exploit the investment of bourgeois males.

In cooperatively breeding cichlid groups like those of *Neolamprologus pulcher* and *N. multifasciatus,* helpers are social subordinates assisting breeders by caring for the progeny (Taborsky, 2001). In *N. pulcher,* a substrate breeding cichlid, the typical brood cycle runs to 2–4 mon, but the previous cohorts of helpers remain in the natal nest for up to a year resulting in several size cohorts of helpers usually co-existing in a family group and assisting in defence of the territory, digging out shelters and direct brood-caring. It has the most advanced social system of any fish known to date (Duftner et al., 2007). Breeders and helpers of both sexes, including mature and immature, related and unrelated individuals of the cichlid share in cleaning and defending the brood and territory (Bender et al., 2008). Non-reproducing individuals of both sexes and all size classes may also live in large groups without holding a territory or 'aggregation'. However, on introduction into a territory with a mate, the adult commences breeding within 3–7 d. Notably, many helpers may also share paternity with the breeding male (Dierkes et al., 1999, 2005) and have similar levels of T and KT (Oliveira et al., 2003). Evidently, there is a potential for conflict between breeders and helpers. Breeders aggressively threaten helpers and helpers run the risk of being expelled from the group (Skubic et al., 2004). Such helpers attempt to appease the breeders by helping and displaying more submissive behavior. Thus a delicate balance between helpers and breeders is maintained because of the benefits of group living outweigh the cost of competition among the group members (Bergmuller and Taborsky, 2005; Bergmuller et al., 2005; Heg et al., 2005). Bender et al. (2006) manipulated the potential for competition between breeder males and helpers by combining either small or large breeder males with adult male helpers in experimental families. They monitored the KT, T and cortisol (C) levels in the breeders and helpers. Similar androgen levels in equal sized breeder and helper confirmed that helpers readily participate in reproduction. The C level is significantly lower in helpers that are more submissive towards the breeder, suggesting a relationship between C and submissive behavior of the helper. Incidentally, the contrasting results on C level in subordinates of *Astatotilapia burtoni* reported by Parikh et al. (2006) call for further studies on this aspect. KT and T hormones that are involved in dominance and reproduction of *N. pulcher* breeders, are lower in helpers that are more submissive than in similar sized helpers that are less submissive. As the submissive helpers are more of symbionts than parasites, they can not be designated as 'parasites' as has been considered by Taborsky (1997).

Among the dimorphic phenotypes, parasitic sneakers are readily distinguishable. Within polymorphic phenotypes the term 'satellite' has been used very indiscriminately and they are described as parasites in some species but as 'symbionts' in others. For example, Katano (1983, 1992) describes the cannibalic male and 'female' satellites, who intensively compete against pair-spawning *Zacco temmincki*. In bluegill sunfish *Lepomis macrochirus*, the parentals mature at the age of seven years. Cuckolders are parasitic males that mature precociously and steal fertilizations in the nests of parental through two tactics: 2–3 year-old 'junior sneaker' that darts into the nest, while the female is spawning; 'satellites', preferably called 'senior sneakers' are older cuckolders aged 4–5 yr that express female color and behavior, and deceive parentals. Both junior and senior cuckolders die before they reach the size of mature parental and there is no evidence that they ever become parentals themselves (Gross, 1982). Hence the 'sneaker' and 'satellite' described by Gross (1982) may be designated as junior and senior parasitic sneakers. In *Porichthys notatus* too, the term cuckolder has also been wrongly used to connote sneaker male by Lee and Bass (2006). In a series of publications, Oliveira et al. (2001a,c) and Goncalves et al. (2005) have described the sneaker of *Parablennius parvicornis* to assume female's nuptial color and courting behavior to deceive the nest-holder; the satellite of *Salaria pavo* gets attached to a specific nest, which it defends. Hence the term satellite, may be restricted to a relatively larger male, who is attached to a specific nest (Santos, 1985; Bentivegna and Benedetto, 1989) and defends it. In various species, satellites are tolerated, and are allowed to share paternity. Dominant satellites may obtain about 30% share of paternity (Taborsky, 1994). Published examples for tolerance of satellites by bourgeois males are available for 22 species belonging to Cyprinidae, Cyprinodontidae, Cichlidae, Embiotocidae and others. In the presence of the satellite male, the nest holding male has a higher condition factor, an index of good health; such nest-holders receive more female visitors and more spawnings, clearly indicating that the nest-holders are benefitted by the presence of satellites (Oliveira et al., 2002). Evidently, the satellites are symbionts and cannot be designated as parasites. These examples and explanations on the symbiotic role of helper and satellite designate these phenotypic males, who adopt AMS as 'alternatives' rather than as 'parasites'.

5.2 Plastic reversibles

In these reversibles, the same individual acts as a dominant breeder, subordinate or helper/depending on its social status (see Fig. 26). These phenotypes are common among cichlids (e.g., *Astatotilapia burtoni*,

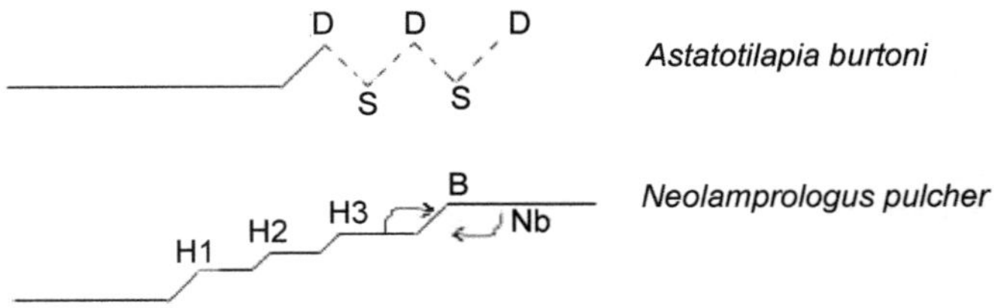

Fig. 26. Schematic illustration of plastic reversible alternative phenotypes in fishes. The same individual male reversibly changes from dominant (D) to subordinate (S) status in *A. burtoni* and from cohort helpers (H1, H2, H3) to breeder (B) in *N. pulcher*, depending upon the social situation, which they encounter. Note the helper may reversibly ascent to breeder status; similarly the breeder, on loss of territory, may reversibly become a non-breeder (Nb) like the cohort helper.

Neolamprologus pulcher) and salmonids. In *N. pulcher* too, a breeder may not breed, when he is not holding a territory or in aggregation. And a helper may also share paternity. Hence *N. pulcher* may be considered as a plastic reversible (see Oliveira, 2006). Males of *A. burtoni* express two distinct reversible phenotypic tactics: (i) larger, more brightly colored, more aggressive and reproductively competent dominant T male and (ii) smaller, camouflage-colored S male (Fig. 26). T male has higher levels of GnRH-R1 mRNA and larger testes containing mature sperm, as against S males with regressed testes containing mostly immature sperm (White et al., 2002). The two key elements that influence the status of a reversible phenotypes are the circulating hormone level and the social context. Interestingly, White et al. (2002) have designed experiments to induce changes in social status of T and S males of *A. burtoni*. When a T male is housed with a larger T male, the smaller T male descends to S status within 3 w, but on placing a relatively larger S male with a smaller one, the larger S male quickly ascends to T status. Monitoring the endocrine profile of descending T male for 24 hr following the induction, Parikh et al. (2006) have found significant reductions in pituitary level mRNA of GnRH-R (receptor) 1 and plasma level of testosterone. They have also reported a significant increase in cortisol, which calls for further study. In laboratory reared *Oncorhynchus mykiss*, Cardwell et al. (1996) have measured higher plasma levels of T and 17, 20β, P in dominant T males than those in 'S' males. Steroid level increases in a S male, when it ascends to T status, after T male is experimentally removed. Since steroid hormones respond to social interactions and have receptors extensively expressed in the social behavioral network, Oliveira (2009) proposed that steroids play a key role in the hormonal modulation of social plasticity. However, the hormone-behavior relationship is bidirectional (Cardwell et al., 1996, see also Fig. 25).

Monitoring the social status influencing the GnRH1 neurons at the apex of BPG axis, Maruska and Fernald (2011), Maruska et al. (2011) traced the linear changes in the BPG axis in an ascending S male *A. burtoni*. The lower but measurable levels of LH and FSH in stable S males suggest that the BPG axis is not totally arrested in these socially suppressed individuals. GnRH acts as a secretagogue for both LH and FSH from the pituitary of fishes, as in mammals. In an ascending male, LH and FSH are elevated rapidly as the mRNA levels within the pituitary as well as at the functional protein level in the circulation. Briefly, the S male ascends to dominant status within five days. Using their and other relevant information, Maruska and Fernald (2011) proposed a chronologically sequenced pathway in the BPG axis, through which a socially suppressed S male ascends to reproductively competent dominant status. At 0.5 hr, both LHβ and FHβ are up-regulated in the pituitary; in turn, LH, FSH and KT are elevated in the blood stream and FSHr and others in the testis. After 6 hr, Arα, Arβ and fErα levels are also increased in the testis. When the fish ascends to dominance, GnRH-R1 is increased in the pituitary and acquires eye bar, bright coloration and dominant behavior. Maruska and Fernald (2011) reached the following conclusions: 1. The socially suppressed S males are not reproductively incompetent (cf. Darwin, 1859), but rather sustain some level of activity at every level of the BPG axis, that results in the production of sperm at the density of 0.12 sperm/μm^2 (cf. White et al., 2002), against 0.23 sperm/μm^2 in a dominant male. 2. The rapid changes in mRNA levels suggest that both transcriptional and post-transcriptional mechanisms like the changes in mRNA stability or small RNA-mediated regulation (e.g., micro RNAs) may play a role in social ascend. 3. The quick changes in mRNA levels in the testes indicate that there are other pathways that bypass the BPG linear axis namely GnRH→LH/FSH pathway.

5.3 Plastic transformants

Within the plastic alternative phenotypic transformants, there are differences in the number of phenotypes through which ontogenetic progression occurs; for example, sneaker → territorial, as in *Symphodus roissali*, *Porichthys notatus*, sneaker → satellite → territorial, as in *S. ocellatus* or sneaker → satellite → territorial → pirate, as *Telematochromis vittatus* (Fig. 27). In the pygmy swordtail *Xiphophorus nigrensis*, Zimmerer and Kallman (1989) recognized smaller (25.5 mm body length) with yellow caudal fin, large (37.5 mm) yellow/blue colored dominant and two intermediate satellites I and II as well. The genetically controlled polymorphism in the swordtail is apparently maintained by the variations of many variables like growth rate, size and age at sexual maturity and so on. There are indications that the satellites may ultimately progress to become dominants (Ryan et al., 1998), although

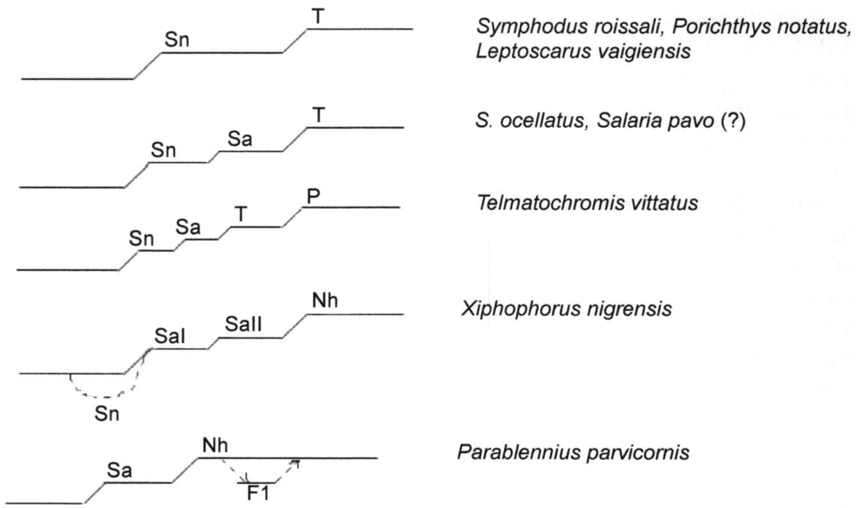

Fig. 27. Schematic illustration of flexible transforming alternative phenotypes in fishes. Note the ontogenetic progression and transformation of a phenotype like the sneaker (Sn) → territorial (T) or Sn → satellite (Sa) → T or Sn → Sa → T → pirate (P). Note that Sa may become floater (Fl) and revert to nest-holder (Nh), as in *P. parvicornis*. In *X. nigrensis* ontogenetic growth may also transform Sn → satellite I (SaI) → satellite II (SaII) → Nh.

no experimental evidence is yet available on whether the sneakers shall also progress to become satellites and subsequently to dominants. Neither is there an analysis to confirm that the so called 'P' locus is sex-linked.

In *P. notatus*, the territorial (Type I) males (15 cm body size, 2 years old) generate a long-duration vocalization at frequencies that range from 90 to 100 Hz with harmonics of up to 400 Hz (a hum up to 15 min) that attracts females to their nests (see Sismeros et al., 2009). The sneakers (Type I males of 6 cm size, 1 yr old) and females do not hum but grunt for shorter duration of less than a sec (Bass, 1993). Within the large and diverse family of Gobiidae, the occurrence of gobies displaying AMS with a sound communication system has been documented in many species. For example, the larger territorial males of the burrowing grass goby *Zosterisessor ophiocephalus* with elongated fins are territorial and emit sounds during courtship. The smaller sneaker males do not have elongated fins. Both T and S males produce grunt lasting for 300 msec made up of pulses repeated at low rate. Pulse duration, number and repetition do not differ between the males but frequency and sound amplitude of dominant does (Malavasi et al., 2003). To minimize the negative effects of sperm competition and sperm loss, the grass goby adopts a sperm trail technique. The trail consists of a band of mucin, in which sperm are embedded (cf. p 91). The mucin producing seminal vesicles are highly developed in T males but the vesicles are used as merely a storage organ by the S male (Mazzoldi et al., 2000).

Among the transformants, the change from sneaker to parental phenotype seems to be under the control of social status in the black goby *Gobius niger*. Under the exclusion of male-male competition, the sneaker males have been kept with and without females. When kept with females, the sneakers change tactics, court females, begin to spawn, and perform parental care. Thus sneaker males are able to reallocate energy in sscs like the elongation of fourth ray of dorsal fin, nuptial coloration of body and anal papilla as well as mucin secretion in the seminal vesicles by reducing testis size (Immler et al., 2004).

Thanks to the excellent series of investigations on the peacock and rock pool blennies by the team led by Prof R.F. Oliveira, it has become possible to understand the role played by androgens in the dimorphic phenotypes. Their studies permit the following generalizations: (i) the bourgeois/nest-holding males have significantly higher levels of KT than the alternatives/sneakers (Table 21), (ii) KT levels are higher during

Table 21. Comparison of sex steroid levels between bourgeois and alternative males of gonochoric fishes. B = bourgeois, A = alternatives, KT = 11-Ketotestosterone, T = Testosterone, M = Monomorphic, D = Dimorphic, P = Polymorphic.

Family/Species	KT	T	References
Batrachoididae			
Halobatrachus didactylus	B > A	B = A	Modesta and Canario (2003a)
Porichthys notatus (D)	B > A	B < A	Brantley et al. (1993b); Remage-Healey and Bass (2007)
Blenniidae			
Salaria pavo (D)	B > A	B > A	Oliveira et al. (2001b)
Parablennius s parvicornis (D)	B > A	B = A	Oliveira et al. (2001c)
Centrarchidae			
Lepomis macrochirus (D)	B > A	B = A	Kindler et al. (1989)
L. megalotis (D)	B > A	B = A	Knapp (2004)
Cichlidae			
Neolamprologus pulcher (M)	B = A	B = A	Oliveira et al. (2003)
Oreochromis mossambicus	B > A	B > A	Oliveira and Almada (1998a)
Sarotherodon galilaeus	B = A	B = A	Ros et al. (2003)
Labridae			
Symphodus melops (P)	B > A	B < A	Uglem et al. (2000, 2002)
Poeciliidae			
Poecilia velifera	B = A	B = A	Oliveira et al. (unpublished)
Salmonidae			
Salmo salar (M)	B > A	B < A	Mayer et al. (1990)
S. trutta	B = A	B > A	Cardwell et al. (1996)
Salvelinus alpinus (M)	B > A	B > A	Elofsson et al. (2000)
S. fontinalis	B = A	B = A	Cardwell et al. (1996)

courting/nest establishment phase than during parental phase, (iii) KT plays a more important role in male reproductive behavior like the courtship, nest construction and territoriality (see Oliveira and Goncalves, 2008) and (iv) in differentiation of sscs (Oliveira et al., 2005b; Oliveira, 2006). For instance, KT treated sneakers of *Salaria pavo* differentiate sscs, and their female-like courtship behavior and display of nuptial coloration are inhibited as well (Oliveira et al., 2001d). The satellite males of *Parablennius parvicornis,* on KT or MT administration, develops sscs like the longer and broader genital papilla, and anal gland, but no testicular gland, which is a part of testis (see Fig. 35, p 144). The administration has also no effect on the GSI (Oliveira et al., 2001d,e). KT levels decrease, when the nest-holders enter parental phase. Kindler et al. (1991a,b) found no effect of KT on the parental behavior of *Lepomis macrochirus.* In *Gasterosteus aculeatus* too, the administration of KT has no effect on intact or castrated nest-holding males (Pall et al., 2002).

At this juncture, it has become necessary to take a side step to have an overall idea on the effects of castration and hormonal therapy in gonochorics and AMS displaying fishes. The reported observations on the effects of castration differ for a few species. For example, on administration of 11-Ketoandrostenedione (11-KA) the castrated *Gasterosteus aculeatus* displays courtship behavior (Borg, 1987) but not to display it (Pall et al., 2002). Secondly, the following descriptions are made on the basis of information (Table 22) provided by Oliveira and Goncalves (2008) alone. With these limitations, it is still possible to note the following: (1) The nest-holding males, on castration, lose their male characteristics like nest-building, nuptial coloration and courting behavior in a dozen species including *Trichogaster trichopterus, G. aculeatus, Lepomis gibbosus* and *L. megalotis,* clearly indicating the role of testicular hormones on these characteristics. (2) However, some of these characteristics except spermiation are restored on administration of MT or T in seven of the nine treated species. The castrated male *O. mykiss,* which does not respond to the administration of 11-KA, displays courtship behavior on administration of 17, 20βP. In *T. trichopterus* female matures 25% of its vitellogenic oocytes on viewing the completion of nest-building by the male (Degani, 1993). (3) Interestingly, males of secondary gonochores like *T. trichoptorus, G. aculeatus* and *Macropodus opercularis,* on MT and T therapy, have spermiated, respectively. (4) Remarkably, the diandric protogynic *Thalossoma bifasciatum* following castration or ovariectomy (Table 12) fails to spermiate or spawn but the ovariectomized female develops male's body coloration on KT supplementation (Semsar and Godwin, 2003). (5) However, the reports on spermiation by the normal females of *Carassius auratus auratus* and gynogenic females of *C. auratus langsdorfi,* on supplementation of KT, requires further investigation (not shown in Table 22). In fact females commence to display masculine courting behavior on administration KT

Table 22. Effect of castration and hormonal manipulations on reproductive behavior of gonochoric fishes. + = increase, – = decrease, 0 = no effect, ND = not described (compiled from Oliveira and Goncalves, 2008 and Prof RF Oliveira).

Species/gonodectomy	Hormone	Nest building	Nuptial coloration	Courtship behavior	Spermiation/ spawning
Trichogaster trichopterus	–	–	–	–	–
Gasterosteus aculeatus	–	–	–	–	ND
Macropodus opercularis	–	0	ND	ND	–
Pseudocrenilabrus multicolor	–	–	–	–	ND
Lepomis gibbosus	–	–	–	ND	ND
L. megalotis	–	–	–	ND	ND
Oncorhynchus mykiss	–	ND	ND	–	ND
Morone americana	–	ND	ND	–	ND
Sarotherodon melanotheron	–	0	–	ND	ND
Hemichromis bimaculatus	–	ND	0	0	ND
Oreochromis upembae	–	0	ND	ND	ND
Bathygobius soporator	–	ND	ND	0	ND
Thalassoma bifasciatum	–	ND	ND	–	–
T. trichopterus	MT	+	+	+	+
P. multicolor	T	+	+	+	ND
L. gibbosus	MT	+	+	ND	ND
L. megulotis	MT	+	+	ND	ND
G. aculeatus	MT	+	+	ND	ND
G. aculeatus	11-KA	+	+	+	+
S. melanotheron	MT	ND	+	+	ND
M. americana	T or MT	ND	ND	+	ND
M. opercularis	T	0	ND	ND	+
O. mykiss	11–KA	ND	ND	0	ND
O. mykiss	17,20βp	ND	ND	+	ND

in *C. auratus auratus*, and 17, 20βP in *O. mykiss*. Many other fishes are also known to display the courtship behavior, when supplementated with PGF2α (Oliveira, 2006 see also Table 12).

In a unique publication, Goncalves et al. (2008) have reported the hormone-mediated transition from sneaker to nest-holder status in the plastic transformant peacock blenny *S. parva*. Due to the direct effect of estrogen and/or decrease in androgen availability, the transitional males pass through a non-reproductive phase with relatively underdeveloped testes. The transition involves morphological (e.g., development of sscs), anatomical and behavioral transformation as mediated by androgens. For instance, the GSI decreases from about 3.0 in sneaker to 0.4 in the transitional but subsequently increases to about 1.0 in the nest-holder. Aromatase activity (AA) is about 10 times higher in the testes and fairly high in hypothalamus of the transitional suggesting that AA suppresses the testicular development during the transition. Subsequently, the plasma levels of T and KT are progressively increased. More studies are required to understand the role of specific androgens in completion of the transition process.

5.4 Fixed sex linked genotypes

In these phenotypes, preferably known as genotypes, the traits responsible for morphotypic differences are fixed for life time and are sex-linked. The morphotypes such as the hooknose and jack among salmonids have genetic basis. The jacks display AMS parasitizing the hooknoses. Precociously fast growing jacks return to the natal streams to reproduce at least one year before the hooknose males of their age class (Fig. 28). Jacks are characterized by small body size; for example, the size at maturity of *Oncorhynchus kisutch* is 34 cm fork length for the jacks against 52 cm for the hooknose (Gross, 1991). Using body size as a cue, the larger juveniles irreversibly adopt the jack's tactics and smaller ones the hooknose's tactics. Juvenile body size is determined by many variables, of which genotypes and egg size are important. With increasing body size of the fry, the fraction of juveniles

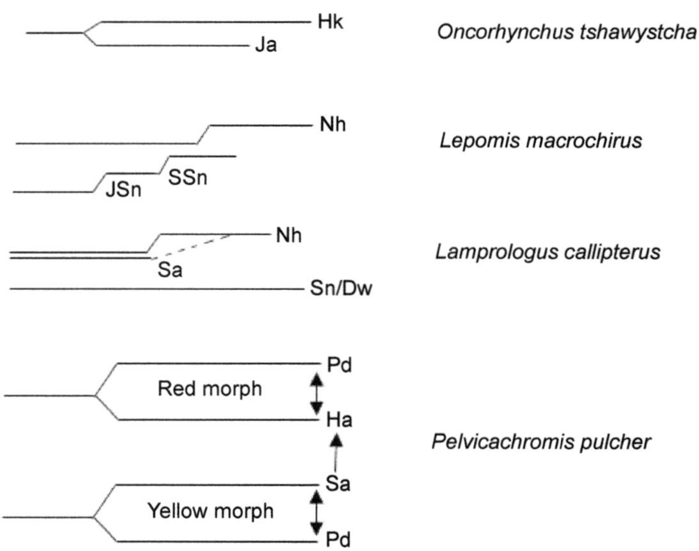

Fig. 28. Schematic illustration of fixed, sex linked alternative phenotypes in fishes. Note the precocious maturity of sneakers in *O. tshawystcha*, *L. macrochirus* and *L. callipterus*. Also note the shorter life spawn of JSn and SSn in *L. macrochirus* but a longer one of Sn/Dw in *L. callipterus*. Sn = Sneaker, JSn = Junior sneaker, SSn = Senior sneaker, Sn/Dw = Dwarf sneaker, Sa = Satellite, Hk = Hooknose, Ja = Jack, Nh = Nest-holder, Pd = Paired, Ha = Haremic.

opting to become jacks are proportionately and linearly increased. With many other environmental factors controlling the fry size, the percentage of jacks in a population varies widely. For example, the percentage of jacks in *O. tshawystcha* ranges from 10 to over 90 among populations (Heath et al., 1991). Iwamoto et al. (1984) have reported sire effect on jacking in *O. kisutch,* i.e., jacks produce significantly higher proportion of jacks than those sired by the hooknose. However, whether the jacking is a sex linked trait or not has remained unknown. From cross breeding experiments involving the normal male (XY) and sex reversed male (XX) as well as sires nested with dam and dams nested with sire, Heath et al. (2002) have reported that (a) XY males sire 31 jack and 69% hooknose progenies and (b) the sire-based heritability is high, against the negligible dam-based heritability. These results suggest (i) the presence of sex-linked component to jacking but with some autosomal contributions, (ii) the polygenic trait of jacking and (iii) the threshold model of heritability of jacking, where the threshold being partially environmentally controlled. Incidentally, selective harvesting of large fish provides an immediate increase in the probability that jacks, relative to hooknose males, survive to breed. The one-year earlier returning jacks have also an edge over the hooknoses in being selected as brooders in hatcheries. With growing interest in salmon aquaculture, the chances for the precociously fast growing jacks becoming fathers of more offspring are greater (Gross, 1991).

In the Atlantic salmon *Salmo salar,* the non-migratory males precociously attain sexual maturity as parr (1–3 yr) and become sneakers or preferably called 'jacks' that are 10-times smaller than their migratory bourgeois preferably called 'hooknoses' (3–7 yr). To quantify the molecular changes that cause this striking difference among the jacks and hooknoses, Aubin-Horth et al. (2005) surveyed 2,917 genes, of which 432 (15%) genes are differentially expressed in the brains of these male phenotypes. These genes are involved in a wide range of processes including growth, reproduction and neural plasticity. According to the life-history theory, trade-offs between growth and reproduction are expected to result in differential allocation of resources in different phenotypes. The number of up-regulated genes associated with feeding (10 genes), growth including somatolactin (3 genes) and reproduction (10 genes) are significantly higher in jacks than in hooknoses [feeding (one gene); growth (nil) and reproduction (one gene)]. On the other hand, as many as 48, 6 and 7 genes associated with protein synthesis, neurodegeneration and transcription regulation, respectively are up-regulated in hooknoses, against 5, 0 and 3 in jacks (see Table 1 of Aubin-Horth et al., 2005). Surprisingly, the cluster analysis has shown that brain gene expression profiles of jacks are more similar to immature females than hooknoses.

In the sunfish *Lepomis macrochirus,* both junior and senior cuckolders die before they reach the size of mature parental (Fig. 28) and there is no evidence that they ever become parental themselves (Gross, 1982). Reporting an enhanced performance of cuckolder fry, Neff (2004) demonstrated that alevins sired cuckolders (prior to exogenous feeding, 5.39 mm) are significantly longer (21%) than those (5.15 mm) sired by parentals and also have larger eye area. Against the predation of *Hydra canadensis,* known to predate 20% of sunfish fry in natural habitats, the alevins of cuckolder have over three times higher survivorship than those of parental. The characteristics like the larger egg size, longer alevin, predator avoiding ability and precocious maturity are apparently genetic and are sex-linked.

Rearing *Lamprologus callipterus* in the laboratory, Taborsky (2001) recognized "three genotypes: (ia) large haremic bourgeois (ib) 'nest male' that performs reproductive parasitism in earlier stages of life but finally reproduces as bourgeois and (ii) dwarf males that halt growth after sexual maturity and remain parasitic throughout their reproductive life" (Fig. 28). Hence there is evidence that "the dwarf male tactic is indeed genetically determined" (Taborsky, 2001). Hence *L. callipterus* is dimorphic with bourgeois and dwarf sneaker, as is hooknoses and jacks in salmonids and parentals and cuckolders in centrachids. It is likely that the trait for the dimorphism in *L. callipterus* is also sex-linked, though experimental evidence is awaited.

A more complicated but interesting genotype is found in the Nigerian cichlid *Pelvicachromis pulcher* (Martin and Taborsky, 1997). The cichlid is small, sexually dimorphic with color polymorphism. Females have an intensely colored purple belly. In males, there are red and yellow colored morphs that are fixed for life (Fig. 28). None of the sons sired by yellow morph males ever belong to red morph and so are the sons of yellow morphs. Apparently, body color is genetically determined, most probably by gene(s) linked to Y chromosome. The sex ratio ranges from 0.33♂ : 0.67♀ in the upper wet region of the Sombreiro River to 0.5 ♂ : 0.5♀ in the lower dry region. Red morphs are more aggressive and about 50% of them are haremic and hold two or more females in each of their harems. The other 50% are pair forming males. Individual red morph can change between haremic to pair forming status but can not change into a yellow morph. The yellow morphs develop into pair-forming or become parasitic 'alternative', when the partner or territory is lost, although they can change between paired and 'alternative' status. With red and yellow morphs, *P. pulcher* males are also dimorphic, as the other fixed sex linked genotypes are. However, within each genotype of *P. pulcher*, there are plastic reversibles.

5.5 Gonado somatic index

In general the sneaker males have higher GSI than that of the bourgeois; for example, the reported GSI values are 4.7 for the sneaker but just 0.9 for the bourgeois of *Parablennius parvicornis* (Oliveiro et al., 2001b). However, the values reported for the GSI of sneaker males range from 4.7 in *P. parvicornis* to 10.0 in *Symphodus ocellatus* (Taborsky, 1998). The cause for such wide variations in the higher GSI among sneaker males has remained unknown. Oliveira's (2006) analysis provides the first cue to understand the wide variations in the GSI among sneaker males. Besides GTH, androgens also participate in spermatogenesis; KT stimulates maturation stages of spermatogenesis. But T provides the negative feedback to control KT-dependent spermatogenesis. Hence a balance between KT and T is required for the control of spermatogenesis. This suggests that the KT (divided by) T must be more informative. Androgen level, i.e., the ratio between androgen level in the bourgeois over that of sneakers, as reported in Fig. 8.4 of Oliveira (2006), is expected to reduce the negative control of T on spermatogenesis, resulting in larger testes. From the data available for eight gonochoric species displaying AMS on KT-T ratio as well as GSI, Oliveira (2006) have inferred that the high GSI reported among the sneakers is associated with relatively lower T level rendering the sneaker males to have larger testes but without the expression of sscs and courting behavior, which are typical characteristics of bourgeois males. Accordingly, KT levels in bourgeois males are about eight times higher than in sneaker males *Lepomis macrochirus,* which display genetically fixed and sex linked AMS. This difference in KT levels between sneakers and bourgeois is about five times in the plastic transformants like *Porichthys notatus* and *P. parvicornis.* But the difference is reduced to only one time in the plastic reversibles like *Oreochromis mossambicus* and *Sarotherodon galilaeus.* Conversely, the relative levels of T is just one time, two and a half times, and three and a half times higher in the fixed, transformant and reversible categories respectively (Oliveira, 2006). Consequently, the KT-T ratio is about 8, 2 and 0.3 for the fixed, transformant and reversible categories. Clearly, the difference in values reported for the GSI for the sneaker males of fishes displaying fixed AMS is far wider than that of the plastic reversibles.

5.6 Sneakers and paternity

Parasitic fertilization by sneaker males has necessitated an estimation on their share of paternity, which may be an yardstick for their evolutionary fitness. Different methods like (i) behavioral observation (e.g., Gross, 1982), (ii) protein electrophoresis (e.g., Phillipp and Gross, 1994), (iii) microsatellite markers (e.g., Jones and Avise, 2001) and (iv) microsatellite multiplexing (e.g., Neff et al., 2000) are employed to detect paternity of the progeny of fishes that display AMS and that are characterized by polyandric or polygynic mating system. In several nest-tending species that are polyandric as well as polygynic, an average of 14% of embryos in a nest are not sired by the present nest-attendant; successful nest-holding males have received eggs from an average of 3.5 mothers per nest (DeWoody and Avise, 2001). In the nests of *Pomatoschistus minutus,* the sneaker males fertilize about 50% eggs (Jones et al., 2001). Among the colonial nests of *Lepomis macrochirus,* sneaking cuckolders fertilize 21.3% of the eggs (Neff, 2001), a value not very different from that (27%) estimated from observation of reproductive behavior (Gross, 1982) and 25% arrived from protein electrophoretic analysis (Phillipp and Gross, 1994). From available data on both male ratio and level of paternity by the sneaker males of fishes displaying AMS, it is apparent that the sneaker's paternity is likely to increase with increasing male ratio (Fig. 29). The causes for the wide scatter of data can be traced to the differences in various mating procedures adopted by the fishes. For instance, the territorial male of *Rhodeus ocellatus* ejaculates following a sequence of courtship timing with the female spawning her eggs into mussel's respiratory chamber through inhalant siphon. Despite the high male ratios (0.55–0.77), the paternity of sneakers is limited to 40%. This limitation is due to the inappropriate timing of sperm ejaculation (Kanoh, 2000). A mating pair of *Zacco temmincki* is followed by more than 40 cannibalistic sneakers,

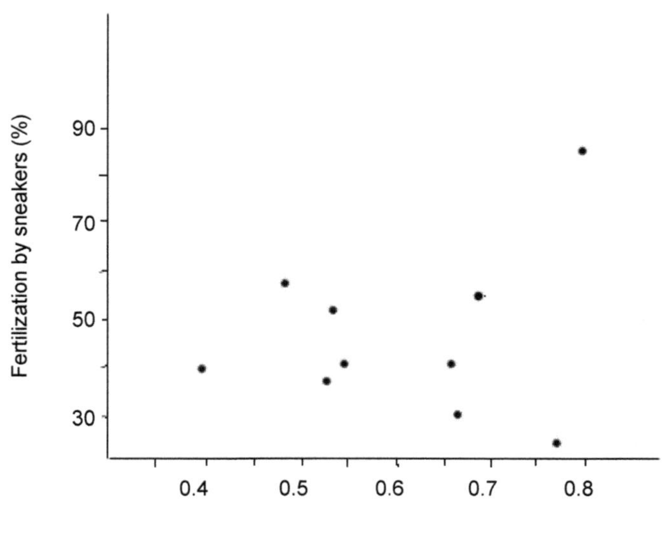

Fig. 29. Fertilization frequency of sneakers as function of male ratio in some fishes displaying alternative morphotypes.

who prevent the fertilized eggs from being buried and reduce the number of progeny (Katano, 1992). More information on this aspect is required before any generalization can be made.

5.7 Aggressive behavior

The display of aggression is a ubiquitous trait of animals. It occurs in the context of competition for limited resources including mate and territory, and is an adaptation that may lead an individual to its success and evolutionary fitness (Filby et al., 2010). With a view to unravel the genetic and neurophysiological basis of aggression in fish, Filby et al. (2010) selected dominant and subordinate males of zebrafish *Danio rerio* as a model, and 40 genes, that are known to play a role in aggression in mammals; the selected genes are spread over seven of eight interrelated neurological pathways (Fig. 30). The greatest differences between dominant and subordinate zebrafish occur in eight genes in telencephalon, 17 in hypothalamus and the other eight in hindbrain (Fig. 31). Of these, only five namely *avpl, tph1b, slc6a3, gnrh3* and *sstr1* show the same expression pattern in more than one brain region. Table 23 selectively lists the number of times some of these genes are overexpressed in dominants than in subordinates.

The product of *avpl* is arginine vasotocin (AVT) and its mammalian homologue arginine is vaso pressin (AVP). These products in the HNS pathway (see Fig. 30) have a well established role in aggressive behavior and social status across vertebrate taxa (Cardwell et al., 2008). Expectedly, the expression level of *avpl* in hypothalamus is 4.3 times higher in dominant male zebrafish than in subordinate (Filby et al., 2010). This observation is consistent with a microarray analysis with 3.6 K unique sequences of the whole brain in *Astatotilapia burtoni* (Filipenko et al., 2001). In the highly territorial polygynous damselfish *Stegastes leucostictus,* AVT facilitates aggression in males (Santangelo and Bass, 2006). Dominant males have higher levels of AVT mRNA in the preoptic regions than subordinate males (Greenwood et al., 2008). Other pharmacological experiments by Filby et al. (2010) have shown that even within a single species, AVT can have opposite effects between behavioral phenotypes (see also Semsar et al., 2001).

In the dopamine pathway, *th* encodes the tyrosine hydoxylase for biosynthesis of dopamine and *slc6a3* for the dopamine transporter (DAT), which ultimately terminates dopaminergic activity. These genes *th1b* and *slc6a3* are two to three times overexpressed in dominant male zebrafish. Similarly *sst1* and its receptor *sstr1* are also overexpressed in dominant males. In the HPG axis, the elevated expression of *ar* in the dominant male suggests the enhanced responsiveness of its brain to sex steroids, which have stimulatory roles in aggression/dominance. The 2.4 times difference reported for the overexpression of *gnrh3* in dominant male zebrafish by Filby

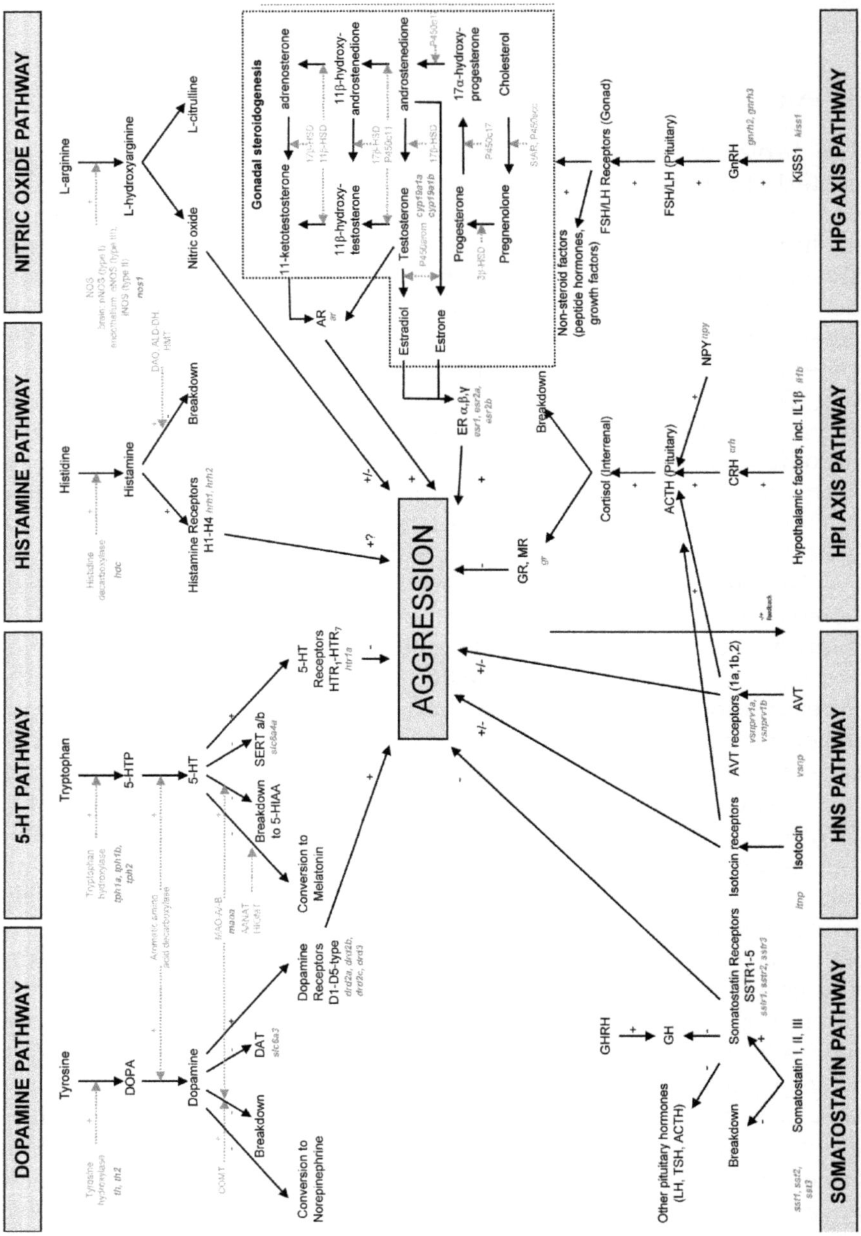

DOPAMINE PATHWAY
5-HT PATHWAY
HISTAMINE PATHWAY
NITRIC OXIDE PATHWAY
Tyrosine
DOPA
Dopamine
Breakdown
DAT
Conversion to Norepinephrine
Dopamine Receptors D1-D5-type
Tryptophan
5-HTP
5-HT
SERT a/b
Breakdown to 5-HIAA
Conversion to Melatonin
5-HT Receptors
Histidine
Histamine
Histamine Receptors H1-H4
Breakdown
L-arginine
L-hydroxyarginine
L-citrulline
Nitric oxide
Gonadal steroidogenesis
11-ketotestosterone
adrenosterone
11β-hydroxy-testosterone
11β-hydroxy-androstenedione
Testosterone
androstenedione
Progesterone
17α-hydroxy-progesterone
Pregnenolone
Cholesterol
Estradiol
Estrone
AR
ER α,β,γ
AGGRESSION
GR, MR
Breakdown
Non-steroid factors (peptide hormones, growth factors)
FSH/LH Receptors (Gonad)
FSH/LH (Pituitary)
GnRH
KiSS1
Cortisol (Interrenal)
ACTH (Pituitary)
CRH
NPY
Hypothalamic factors, incl. IL1β
GHRH
GH
Other pituitary hormones (LH, TSH, ACTH)
Somatostatin Receptors SSTR1-5
Breakdown
Somatostatin I, II, III
Isotocin receptors
Isotocin
AVT receptors (1a,1b,2)
AVT
SOMATOSTATIN PATHWAY
HNS PATHWAY
HPI AXIS PATHWAY
HPG AXIS PATHWAY

et al. (2010) is also comparable to those between dominant and subordinate *A. burtoni* (White et al., 2002). Conversely, corticosteroids, which remain elevated for long time in subordinates, suppress aggression via elevated brain 5-HT. Both *crh* and *npy* influence aggressiveness of fishes (Carpenter et al., 2009). The elevated expression of *crh* and *npy* in the telencephalon of subordinate male zebrafish may represent an enhanced stress response of the subordinate (see Filby et al., 2010). Briefly, the described differential expressions in the brain regions related to social rank representing seven functional pathways suggest a multifactorial control of aggressive behavior of zebrafish. The commonality of the pathways and identified genes in zebrafish suggests the high degree of conservation in the control of aggression between fishes and mammals (Filby et al., 2010).

Despite the fact that *avpl* expression shows the greatest difference between dominant and subordinate male zebrafish, divergent effects of AVT on aggression have been reported among teleosts with examples of AVT increasing, decreasing or having no effect on aggressive behavior (see Oliveira and Goncalves, 2008). However, AVT seems to play a role in facilitating sneaker males to mature early and commence reproductive activity. For example, the number and size of AVT cells are correlated with sexual morphotypes of *Salaria pavo*. Female-like sneakers and females produce about two times more AVT mRNA per cell than the nesting males (Grober et al., 2002). In *Parablennius parvicornis* too, the number and size of AVT cells in the forebrain develop initially to allow reproduction in young non-nesting males (Miranda et al., 2003). Very few studies have been conducted on the direct effect of GABA on aggression. As in AVT, the available information on the effect of GABA on reproductive behavior remains inconclusive. Five-day intraperitoneal implants of T or progesterone

---

Fig. 30. Simplified illustration of regulatory neurological pathways of aggression common to mammals and fishes (highlighted in red color). 5-HIAA = 5-hydroindoleacetic acid, 5-HT = 5-hydroxytryptamine (serotonin), 5-HTP = 5-hydroxytryptophan, AANAT = arylalkylamine N-acetyltransferase, ACTH = adrenocorticotropic hormone, ALD-DH = acetaldehyde dehydrogenase, AR = androgen receptor, AVT = arginine vasotocin, COMT = catechol-O-methyl transferase, CRH = corticotropin releasing hormone, DAO = diamine oxidase, DAT = dopamine transporter, DOPA = dihydroxyphenylalanine, ER = estrogen receptor, FSH = follicle-stimulating hormone, GH = growth hormone, GHRH = growth hormone releasing hormone, GnRH = gonadotropin releasing hormone, GR = glucocorticoid receptor, HIOMT = hydroxyindole-O-methyltransferase, HMT = histamine-N-methyltransferase, HNS = hypothalamo-neurohypophysial system, HPI = hypothalamo-pituitary-interrenal, HPG = hypothalamo pituitary-gonadal, IL1β = interleukin 1β, LH = luteinizing hormone, MAO = monoamine oxidase, MR = mineralocorticoid receptor, NOS = nitric oxide synthase, NPY = neuropeptide Y, SERT, 5-HT transporter, TSH = thyrotropin-stimulating hormone. Plus (+) and negative (–) symbols indicate the proposed action of a gene/neurotransmitter/pathway on aggression (i.e., stimulatory or inhibitory, respectively) (from Filby et al., 2010).

Color image of this figure appears in the color plate section at the end of the book.

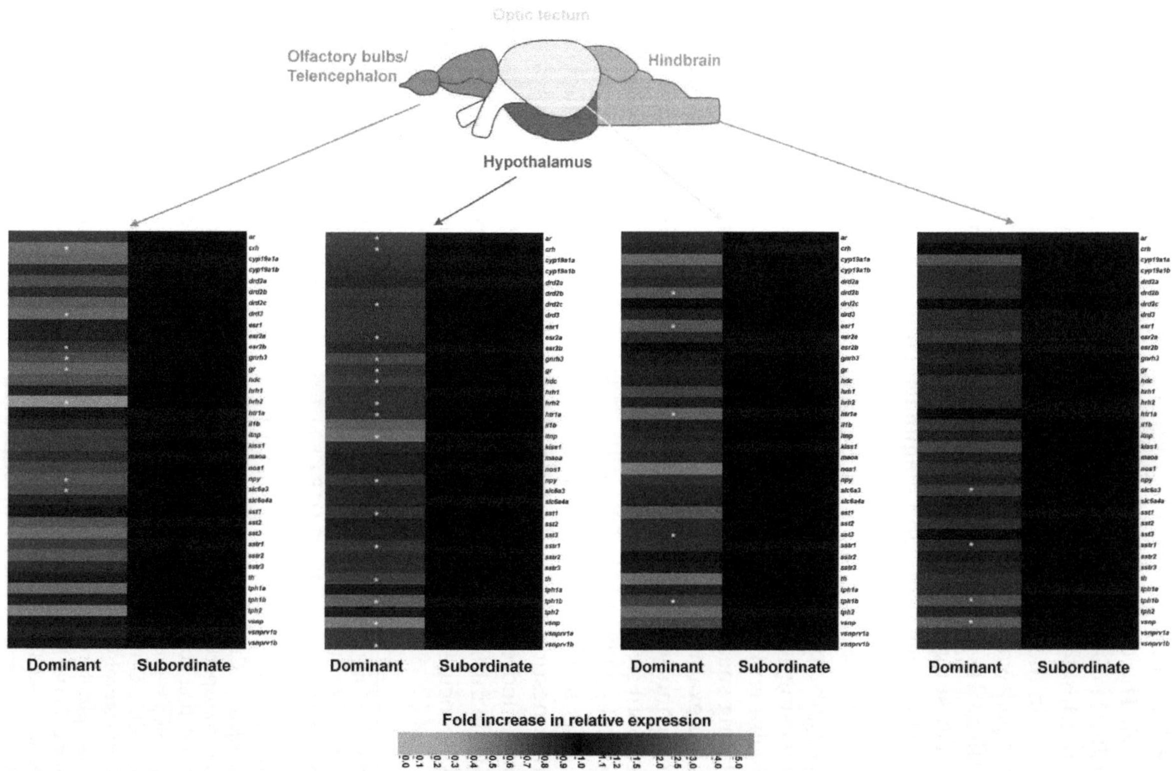

Fig. 31. Regions of the brain associated with differences in aggressiveness in *Danio rerio*. Heat maps show the consistently expressed genes in four regions of the brain in dominant and subordinate males. Genes that are differentially expressed, are denoted with an asterix (from Filby et al., 2010).

Color image of this figure appears in the color plate section at the end of the book.

Table 23. Genes associated with aggressiveness and selected pathways in dominant (high aggression) and subordinate (low aggression) zebrafish (compiled from Filby et al., 2010).

Gene	Description	Pathway	Fold increase
avpl	arginine vasropressin-like	HNS	4.31*
th1b	tyrosine hydorxylase 1b	Dopamine	2.98*
slc6a3	neurotransmitter transporter dopamine member 3	Dopamine	2.25*
gnrh3	gonadotropin releasing hormone	HPG	2.16*
ar	androgen receptor	HPG	1.96*
sst1	somatotropin 1	Somatostatin	1.56*
sstr1	somatotropin receptor	Somatosatin	1.67*
npy	neuropeptide Y	HP	0.68**
crh	corticotropin releasing hormone	HPI	0.56**

*in dominant male; **fold increase in submissive males

decreases glutamic acid decarboxylase (GAD), an enzyme responsible for GABA in telencephalon of male goldfish. But experimental lesions in the ventral region of telencephalon, an area expressing GAD mRNA disrupt sexual behavior. Despite the availability of a few publications, the negative role of serotonin on aggression, however, seems to be conclusive. Serotonergic activity is inversely associated with social status (e.g., *Salvelinus alpinus, A. burtoni*). In *Cichalsoma meeki,* the treatment with p-chlorophenylalanine, a serotonin synthesis inhibitor, increases mirror-elicited aggression. Dietary supplementation of L-tryptophan, the precursor of serotonin, reduces aggression in *Oncorhynchus mykiss* (see Oliveira and Goncalves, 2008).

The two neuropeptides that have been most extensively studied with regard to behavioral plasticity are GnRH and AVT. Increase in the number or size of GnRH neurons reflects the differences in the potential to deliver GnRH to the pituitary and thereby stimulate steroidogenesis. Among the teleost species displaying AMS, the GnRH neuron size or its number is divergent in association with relative size of gonad. Pre-optic AVT neurons also exhibit male dimorphism. In *Porichthys notatus,* GnRH-ir neurons are larger in territorial males (Type I) than in (Type II) sneaker males and females. But the number of neurons per gram body weight is significantly higher in sneakers than territorial males, as are the AVT cells per gram body weight (Foran and Bass, 1999).

Ros et al. (2004a) showed that KT increases aggressive behavior in *Parablennius parvicornis.* In the yellowtail *Seriola quinqueradiata,* the J-posture larvae became more aggressive and metamorphose into dominants. Labeled dominant larva retains dominance at least for one week. The onset of aggressive behavior coincides with significant increase in tissue cortisol level (Sakakura and Tsukamoto, 1999). The presence of higher levels of KT in the bourgeois (Table 21) or a treatment with KT increases metabolism, especially during the night (Ros et al., 2004b). *Oreochromis mossambicus* male exposed to its mirror image shows increased metabolism due to escalated aggressive behavior (Oliveira et al., 2005a). The mean routine oxygen uptake of male *Betta splendens* is 0.36 mg/h. During a fight, it increases by five times to 1.8 mg/h. There is no difference in O_2 uptake between the winner and loser, but the winner's O_2 uptake is higher during the night after the fight (Castro et al., 2006).

As males have higher androgen levels than subordinates, gonadal steroids have been considered as major modulators of aggressive behavior (Oliveira and Goncalves, 2008). However, successful castration in a dozen fish species has either decreased aggressiveness or evoked no response (Table 24). A reason for the failure of castration to evoke no effect in

Table 24. Effects of castration and/or hormonal treatment on aggressive behavior of male phenotypes of fishes. + = increase, – = decrease, 0 = no effect, B = breeder ♂, N = nesting ♂, Su = subordinate ♂, T = territorial ♂, Pn = pre-nesting ♂, NH = nest-holding ♂, Sn = sneaker ♂ (compiled from Oliveira and Goncalves, 2008, with permission by Science Publishers and Prof RF Oliveira).

Species	Effect	Treatment	Effect
Xiphophorus maculatus	–		
Bathygobius soporator (B)	–		
Pseudocranilabrus multicolor	–		
Gasterosteus aculeatus (N)	–		
Astatotilapia burtoni (Su ♂)	–	T, KT or DHT	0
A. burtoni (T)	+		
G. aculeatus (Pn)	0		
Lepomis megalotis (T)	0		
L. gibbosus (T)	0		
Trichogaster trichopterus (B)	0		
Betta splendens (B)*	0		
Macropodus opercularis (B)*	0	Tenanthate	+
Apteronotus albifrons		T, KT or DHT	0
Salario pavo (Sn)		KT	0
Porichthys notatus (Sn)		KT	0
G. aculeatus (N)	–	T-propionate	0
X. helleri	–	T-propionate	+
Acquildens pulcher		T or T-propionate	+
Parablennius parvicornis (Nh)		KT	+
P. parvicornis (Sn)		KT or MT	–
Pterophyllum scalare		LH	+
L. gibbosus		LH	+

*Incomplete castration or immediate testicular regeneration
+Goncalves and Oliveira (2011)

some fishes was shown by Cavaco et al. (1996, 1997). In *Clarias gariepinus,* castration and/or surgical removal of seminal vesicles results in drastic decrease in plasma KT levels but with no change in the levels of 11β-hydroxyandrostenedione (OHA), the precursor of KT. Hepatodectomy and *in vitro* experiments have conclusively shown that the enzymes, 11β-hydroxysteroid dehydrogenase (11β-HSD) and 17β-hydroxysteroid dehydrogenase (17β-HSD) required for the conversion of OHA are present in the liver and liver is the site of KT synthesis. Extra testicular synthesis of KT is also known in *Gasterosteus aculeatus,* in which androstenetrione (OA) of testicular origin is converted to KT by 17β-HSD activity residing in blood cells (Mayer et al., 1990b). It is likely that in fishes like *Lepomis megalotis, L. gibbosus* (Centrarchidae), *Trichogaster trichopterus, Betta splendens* and *Macropodus opercularis* (Anabantidae), the liver is the site of KT synthesis and expectedly the castration evokes no response in them. But liver may not be the site of KT synthesis in *Xiphophorous maculatus* (Poeciliidae), *Bathygobius saporator* (Gobiidae) and *Pseudocranilabrus multicolor* (Cichlidae), in which castration reduces the level of aggression.

The aggressiveness is restored by tenanthate supplementation in the castrated *M. opercularis.* The castration evokes different levels of aggression in breeding and non-breeding/nested males. It aggravates the level of aggression in breeding (but castrated) male *A. burtoni* but evokes no response in castrated subordinate male. This is also true of intact males; exogenous KT administration into intact sneaker males of *P. parvicornis, S. pavo* and *P. notatus* has negative or no effect on aggression (Table 24). Taken together the stimulation of aggressive behavior in these fishes seems to be not under testicular control. However, administration of LH stimulates aggressive behavior in males of both *Pterophyllum scalare* and *Lepomis gibbosus,* which suggests that some other neuroendocrine factor in association with the hypophysis-pituitary axis seems to stimulate aggressiveness in males. Peripheral administration of E_2 has an inhibitory effect on male aggression in fishes (Bell, 2001; Clotfelter and Rodriquez, 2006; Goncalves et al., 2007). According to Goncalves and Oliveira (2011), a meta-analysis by Hirschenhauser and Oliveira (2006) confirmed a positive effect of exogenous androgen administration on male aggressiveness in teleosts. Nevertheless, futher research is urgently required on this relevant subject. Serotonin is a neurotransmitter that regulates social interactions among animals. In fishes like *Salvelinus alpinus* (Winberg et al., 1992), *Oncorhynchus mykiss* (Winberg et al., 1993) and *Haplochromis burtoni* (Winberg et al., 1997), serotonergic activity is lower in aggressive dominants than in subordinates.

5.8 The challenge hypothesis

In an excellent review on social modulation of androgen levels and responsiveness of male teleost fish, Oliveira et al. (2002) tested the validity of the challenge hypothesis with reference to fishes, especially those which display AMS. Agonistic interactions and male-male competition usually result in stimulation of androgen synthesis and the levels of androgens may reflect the stability of the social environment encountered by the animal at that time. Having found experimental evidences among birds, Wingfield et al. (1984 a,b, 1990) proposed what is presently known as the challenge hypothesis. Using the hypothesis, a number of predictions can be generated. The first prediction is that during the territory establishment phase, the androgen levels are expected to be higher than those in the non-territorial/ subordinate/sneaker males. On testing the validity of the prediction, Oliveira et al. (2002) found that in many gonochoric fishes, for which data are available, the levels of KT and T are higher in the bourgeois of almost all the fishes that display AMS (Table 21). The second prediction expects the androgen levels to be higher during the commencement of breeding/ establishing territory phase than during the subsequent parental phase. Table 25 shows that KT and T levels are higher during the commencement (e.g., *Syngnathus* spp.)/territory establishment (e.g., *Porichthys notatus*) phase in all the gonochoric fishes, for which relevant data are available.

The third prediction expects the androgen pattern during the breeding season to vary among species with different mating and brood care systems. For example, the degree of mating competition is likely to be very low among monogamous species in comparison to polygynous species that encounter an intense male-male competition. To verify the divergence between the results reported for different mating and parenting types, the data available on the effects of mating system from that of the parenting

Table 25. Effect of mating and parental phases on the levels of KT and T in selected fishes. ND = Not described.

Species	KT		T		Reference
	Mating	Parental	Mating	Parental	
Syngnathus acus	3.6	0.9	15.2	6.4	Mayer et al. (1993)
S. typhle	2.4	0.9	3.2	2.0	Mayer et al. (1993)
Lepomis macrochirus	55.0	14.0	21.0	7.0	Kindler et al. (1989)
Parablennius parvicornis	6.0	2.1	15.0	8.1	Oliveira et al. (2001b)
Liphophyres pholis	2.5	1.5	9.0	1.4	Oliveira and Canario (unpublished)
Sarotherodon melanotheron	ND	ND	22.0	1.9	Kishida and Specker (2000)
Porichthys notatus	11.5	8.1	0.1	0.0	Knapp et al. (1999)

system of fishes have been separated by Oliveira et al. (2002). As expected, the androgen responsiveness does vary widely; it varies from about 7.5 in paternal brooders like *Sarotherodon melanotheron* (Kishida and Specker, 2000) to about 5.0 in paternal brooders with satellite helpers like *Parablennius parvicornis* (Oliveira et al., 2001b), to about 6.0 in biparental brooders like *Ictalurus nebulosus* (Rosenblum et al., 1987), to about 3.0 in maternal brooders like *Salmo trutta* (Cardwell et al., 1996), to about 2.0 in species like *Stizostedion vitreum* (Malison et al., 1994) with no parental care. Similarly, the responsiveness also varies from about 1.6 in monogamous species like *I. punctatus* (Rosenblum et al., 1987) to 1.3 in polygynous species like *Gasterosteus aculeatus* (Mayer et al., 1990a) to 1.0 in polyandrous species like *Syngnathus typle* (Mayer et al., 1993) and to about 0.5 in *P. parvicornis* (Oliveira et al., 2001b). Figure 32 represents similar analysis by Oliveira (2009) on the androgen responsiveness to different mating system. With its validity and applicability to fishes, the challenge hypothesis "may emerge as a general principle in vertebrate endocrinology" (Oliveira et al., 2002).

With a visionary approach, Oliveira et al. (2002) have beautifully conceptualized how the mediating hormones in the social context and diverse expression of male traits have opened the gate for the origin and evolution of AMS. In vertebrates including fish, social modulation of androgens is a widespread phenomenon. Then hormones playing a mediating role between the outcomes of social interactions and expression of male traits open an avenue for the evolution of a number of alternative

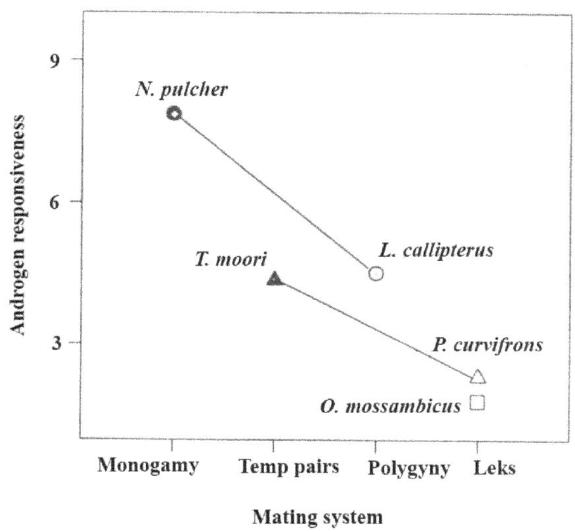

Fig. 32. Androgen responsiveness as function of mating systems of selected cichilds, *Neolamprologus pulcher, Lamprologus callipterus, Tilapia moori, Pseudosimochromis curvifrons, Oreochromis mossambicus* (from Oliveira, 2009).

life history patterns, as it has happened in fishes. "The social status of a fish may affect a number of life history traits like sexual maturation, the adoption of alternate male tactics.... All these possibilities may allow compensatory responses to evolve so that subordinate fishes to minimize the disadvantages of their status by adopting the life history pattern that makes the best of their situation.... The same compensatory principle has also been proposed both for AMS and for socially controlled mechanisms of sex change" (Oliveira et al., 2002).

6

Sex Changers

Introduction

Among vertebrates taxa, sex change in fishes is a unique, fascinating but an intriguing phenomenon. It is a complex integrative, multilevel phenomenon (Wu et al., 2010) involving multiple aspects of multiple endocrine axes (Larson, 2011) and requires coordinated interplay of signals to regulate (Wu et al., 2010) transdifferentiation of different tissues of the reproductive system. The available publications hint that only the tip of iceberg has been touched in our understanding of these complex social and neurobiological control mechanisms of sex change (Larson, 2011).

6.1 Secondary gonochores

In many cichlids, like *Amphilophus citrenellus, Archocentrus nigrofasciatus, Cichlasoma portelegrense, Etroplus maculatus, Hemichromis bimaculatus, Pseudotropheus lombardoi* and *Sarotherodon melanotheron*, sexual ambiguity and potential sex change have been suspected (e.g., Polder, 1971); the aberrant or regular (e.g., *P. livingstoni*) presence of underdeveloped (e.g., *P. lombardoi*), previtelligenic (e.g., *S. melanotheron*) or degenerated (e.g., *C. portelegrense*) oocytes scattered within the testis has been reported. As per the definition of Pandian (2012), these fishes are now shown as secondary gonochores (Oldfield, 2005, 2007). In fact Naish and Ribbink (1990) showed that in many of these cichlids, in which Peters (1975) suggested sex change, the changes are limited to morphological secondary sexual features alone. Nevertheless, some of these cichlids may have to be reinvestigated, especially, *P. livingstoni,* in which oocytes are scattered among spermatogenic tissue attached to the walls of the tubuli of all tested testes, and in *C. portelegrense* and *Tramitichromis intermedius*, in which established social groups may provide more reliable information (Oldfield, 2005).

Reports on sex change/reversal in secondary gonochoric mature adults are limited to a poeciliid *Xiphophorus helleri*, cichlids *Crenicara punctulata* and *Matriaclima* cf. *livingstoni* and an anabantid *Macropodus opercularis*. Providing evidence for its secondary gonochorism in *X. helleri*, Essenberg (1926) indicated sex change in one of the two females observed by him. Lodi (1979) confirmed that three hermaphroditic individuals in the normally secondary gonochoric *X. helleri* switch to males. Similar but socially mediated one time unidirectional sex change/reversal is reported from cichlids: *C. punctulata* and *M. livingstoni* (Table 26). On the other hand, *M. opercularis* undergoes social-mediated bidirectional sex change more than once. In fact, a single individual has been induced to undergo sex reversal as many as six times but producing eggs or sperm upto the fourth time sex change only (R Koteeswaran, pers comm). Sex change is completed within 30 d, irrespective of the direction of sex change. A sex changed female progressively increases her clutch size with successive spawnings and achieves complete compensation by the fourth spawning (see Pandian, 2011). Briefly, the cichlids and poeciliid change sex once in a single direction from female to male but the paradise fish changes sex more than once in either direction.

Table 26. Social mediated sex change/reversal in adult gonochoric fishes. J = Juvenile, ♀-sm = sub-male possessing male's body color and aggressive behavior.

Social components	Reported observations
Crenicara punctulata (Carruth, 2000)	
(i) In natural territory ♂ + ♀♀♀♀	On removal of territorial male, the dominant α female changes sex to male
(ii) Group rearing ♂ + ♀sm♀sm♀♀♀	2-year old females differentiate into sub-males and display male's body color and aggressive behavior. On removal of male, one of the sub-males reverts to male
(iii) J reared in isolation	Passes through female phase to become an adult male at the age of 7–10 mon
Matriaclima cf. *livingstoni* (Stauffer and Ruffing, 2008)	
(i) Six tanks each housing ♀♀♀	One female from a tank reverts to male and fertilized eggs of one female, "resulting in three large broods"
(ii) Two tanks each housing ♀♀♀♀♀♀	One female from a tank reverts to male, as "indicated by a female holding eggs"
Macropodus opercularis (R. Koteeswaran pers com)	
(i) J reared alone	Matures into male
(ii) ♂♂ reared individually	Reverts to female
(iii) Reared in group of ten females or more than 19 females	2–3 females revert to males

At this point it has become necessary to take a look into the endocrine sex reversal in zebrafish *Danio rerio*. The exposure to 5 ng EE_2/l in a flow through system for four months from egg until sexual maturity results in the production of all-female adults. Reared in clean water for eight months, eight out of 25 EE_2-treated adult females revert to males, indicating that estrogenic (protogynic) sex reversal is possible in zebrafish (Larsen et al., 2009). However, none of the 17β-trenbalone (TB)-treated adult males, on being reared in clean water for six months, reverts to female suggesting that androgenic (protandric) sex reversal is not possible in zebrafish (Larsen and Baatrup, 2010). Being a secondary gonochore, the zebrafish initially develops ovary-like gonads leading to aromatase production and feminization (Fenske and Segner, 2004). However, about 20 d after hatching, the ovary along with primordial germ cells (PGCs) suffer apoptosis (Uchida et al., 2002). With the loss of ovary and *aromatase*, the presumptive males develop into males. But the loss of PGCs along with the ovary, the TB-treated male zebrafish is unable to regenerate the ovary (see Larsen et al., 2009). In this context, it is also interesting to note the ovariectomized secondary gonochoric *Carassius auratus* female reverts to male on supplementation of KT (Kobayashi et al., 1991) but fails to regenerate the ovary, as indicated by no spawning (Table 12) (Kobayashi and Stacey, 1993). Also the castrated *Trichogaster trichopterus* (MT), *Macropodus opercularis* (T) and *Gasterosteus aculeatus* (11-KA) (see Table 22) regenerate testis, as indicated by spermiation, following the administration of one or other steroid.

Notably, ovariectomy fails to regenerate the ovary in the presumably seconday gonochore *Betta splendens* (Table 27). In two other experiments also, the ovariectomized *B. splendens* regenerate testis only from the remaining oviduct, and ovarian wall (?). In another secondary gonochore

Table 27. Ovariectomy and regeneration of testis in secondary gonochoric fishes.

Species and Reference	Reported observations
	Betta splendens
Lowe and Larkin (1975)	42% surviving ovariectomized females regenerate functional testis from the cut end of the remaining oviduct or ovarian wall(?). Crossed with normal female, the regenerated male produces F_1 all female progenies or at sex ratio of 0.43♀:0.57♂
Kaiser and Schmidt (1951)*	Ovariectomized female regenerates testis. Crossed with normal female, the regenerated male sires two broods and the second time with 102 all-female progenies
Noble and Kumpf (1937)*	Treated with E_2, the ovariectomized female fails to regenerate ovary
	Carassius auratus
Kobayashi et al. (1991)	Treated with KT, the ovariectomized female regenerates testes

*cited in Atz (1964)

C. auratus (see Pandian, 2012) too, the ovariectomized female reverts to a male but on KT-implantation. Hence the adults of the secondary gonochores belonging to the families Poeciliidae (*X. helleri*), Cichlidae (*C. punctulata, M. livingstoni*), undergo social-mediated sex change once from female to male, i.e., they are capable of protogynic sex change but not protandric sex change. Following ovariectomy, *B. splendens* and *C. auratus*, of course on the administration of KT, regenerate testis and undergo sex reversal from female to male. Castration in *Trichogaster trichopterus*, *Gasteroestus aculeatus* and *Macropodus opercularis* (Table 22) also regenerate the testis and undergo sex reversal from female to male. Briefly, the secondary gonochores are amenable to change sex from female to male alone either on surgical removal of ovary or testis, as in *B. splendens*, *C. auratus*, *T. trichopterus* and *M. opercularis* or administration of androgens, as in *D. rerio* or when socially-mediated as in a poeciliid and cichliids.

Apart from this, the paradise fish *M. opercularis* undergoes sex change more than once in either direction. Whereas the primary gonochores directly develop from undifferentiated gonad to ovary or testis, the juveniles of the tertiary gonochores like *Anguilla anguilla* (Beulbens et al., 1997) commence with bisexual potential gonad, which then develops directly into either ovary or testis. It is not known whether the paradise fish is a tertiary gonochore that has retained the bisexual potency during the mature adult stage also. But still it is difficult to comprehend, how *M. opercularis* is amenable for social-mediated sex change in either direction, but from male to female only,when castrated.

6.2 Structural diversity

Based on the presence or absence of a non-functional interphase during ontogenetic pathway of sex differentiation, Yamamoto (1969) recognized two categories of sex differentiation namely primary and secondary gonochores. Intersexuality is the result of interference between the opposing tendencies; consequently, neither of them can express decisively and remain non-functional (Atz, 1964). The primary gonochores commence with an undifferentiated gonad and the differentiation proceeds according to genetic sex determination. Hence dimorphic differentiation becomes recognizable prior to hatching (e.g., *Orizias latipes*) or after hatching (e.g., *Oreochromis niloticus*). But the secondary gonochores commence with a differentiated or intersexual gonad; their genotypic and phenotypic sexes differentiate in parallel lines for a short (e.g., *Danio rerio*) or longer (e.g., *Pterogymnus laniarius*) duration but only one sex, as genetically or environmentally determined, becomes functional (Atz, 1964). On the basis of commonality regarding commencement of sexual development with differentiated or interphase gonads (Fig. 33), hermaphrodites are considered to have

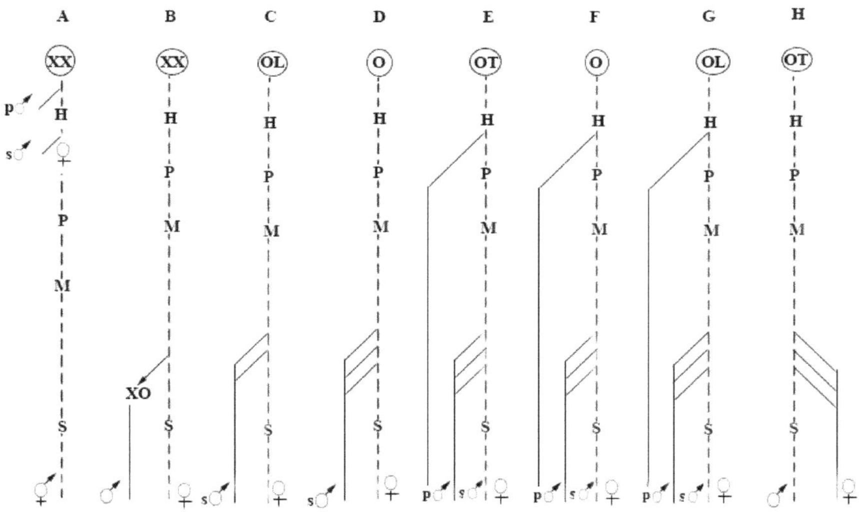

Fig. 33. Ontogenetic pathways of sex differentiation in selected hermaphrodites. Simultaneous hermaphrodites: A = *Kryptolebias marmoratus*, B = *Serranus baldwini*, Monoandrics: C = *Cromileptes altivelis*, D = *Tigrigobius multifasciatum*, Diandrics : E = *Parapercis snyderi*, F = *Suezichthys gracilis*, G = *Pagrus pagrus*, Protandric : H = *Diplodus vulgaris*. XX = Somatic egg, O = Egg, OT = Ovotestis, OL = Ovarian lumen, H = Hatching, P = Puberty, M = Maturation, S = Senescence?, p♂ = Primary male, s♂ = Secondary male, XO = Male from sex changed hermaphrodite (from Pandian, 2012, corrected).

originated and evolved from secondary gonochores (see Pandian, 2012). Considering that hermaphrodites have evolved independently in more than 10 lineages (Warner, 1978) and that a substantial evolutionary period may have separated each of these lineages (Frisch, 2004), it may not be difficult to comprehend a wide range of structural diversity in the reproductive system and possible differences in the neuroendocrine mechanisms controlling sex change. In fact Ross (1990) proposed five different sex change mechanisms based on proximate causes and socio-ecological attributes.

Hermaphroditism is defined as the expression of both male and female reproductive function in a single individual either simultaneously or sequentially (Sadovy de Mitcheson and Liu, 2008). In fishes three patterns of functional hermaphroditism have been recognized (Pandian, 2010): simultaneous, sequential and serial (Fig. 34). Simultaneous hermaphrodites function as male and female at the same (e.g., *Kryptolebias marmoratus*) or within a short span of time (e.g., *Serranus tortugarum*). They do not undergo natural sex change. But, the sequential and serial hermaphrodites undergo natural sex change; the sequentials change sex only once in a single direction during their life time but the serials do it more than once in either direction. The sequentials are further divided into (i) female to male sex changing protogynics and (ii) male to female sex changing protandrics.

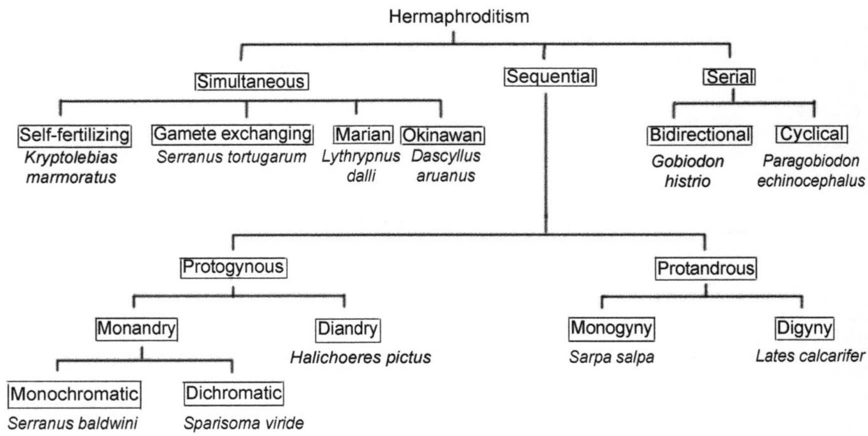

Fig. 34. Patterns of functional hermaphroditism in fishes (from Pandian 2011, modified).

In the monandric protogynics, secondary males arise indirectly from sex changing females but in the diandrics, primary males also arise directly, in addition to the secondary males arising from sex changing females (see Pandian, 2012). A similar dichotomy has been recognized in the protandric species, and the corresponding terms are monogyny and digyny. The monandrics are further divided into monochromatics with all mature (IP) males displaying the same dull body color as the females and dichromatics, in which the (TP) males change their body color to bright terminal one (see Pandian, 2010, 2011).

Among simultaneous hermaphrodites, the simultaneous presence of testicular and ovarian structures are recognizable in *Kryptolebias marmoratus* (see Orlando et al., 2006). Similarly separate ovarian and testicular structures with their respective ducts to the outside have been described (Fig. 35). The sequentials have either a pair of ovotestes or gonads. Sadovy and Shapiro (1987) have classified them into two groups. (i) Delimited type of ovotestis consisting of distinct heterologous zones of testicular and ovarian lobes, which are separated by a connective tissue (Fig. 35) is present among protogynous species (e.g., many sparids, *Pagrus pagrus*) and protandrous species (e.g., *Acanthopagrus schlegeli*). In the ovotestis, the ovary is located dorsally and the testis latero-ventrally. The undelimited group is further divided into two types. In the undelimited type 1, the ovarian and testicular zones are in contact with each other but never intermixed (e.g., *Hypoplectrus puella*) and in the type 2, the ovarian and testicular tissues are intermingled during the course of sex change (e.g., *Epinephelus guttatus*). Recently, a rather different gonad type has been described by Takamoto et al. (2003). In the balastid *Sufflamen chrysopterus,* the female to male sex change is accompanied by the diminishing anterior ovarian component, and enlarging and maturation of the posterior latent testis (Fig. 35) Among the Marian hermaphrodites, ovarian and testicular tissues are present in the same gonad, but the amount of allocation, for the testicular tissues varies (1 to 90%) from individual to individual, for example *Lythrypnus nesiotes* (Fig. 36).

Typically, the sparids and serranids possessing delimited ovotestis commence with the same. On differentiation into ovaries in protogynics like *Pagrus pagrus* functional females are produced. Concomitant with sex change, the ovaries degenerate, functional testes are derived from the testicular zones of the ovotestes. The reverse is true of protandrics like *A. schlegeli.* During sex differentiation, the PGCs and their derivatives OSCs or SSCs are drawn from the gonadal primordium into respective heterologous zones of the ovotestes. Conversely, the labrids, gobiids and parapercids possessing undelimited gonad types, commence with bisexual gonads; on differentiation into ovaries, functional females are developed (Fig. 35). Simultaneous with sex change, the ovaries degenerate and functional testes are derived from the former ovarian wall and secondary males are produced. Though the presence of PGCs, OSCs and SSCs has been reported in many sex changing fishes (see Pandian, 2012), their exact location on the former ovarian wall of the gonad is not yet reported; they may remain attached to the tubuli of the former ovarian wall (cf. *Cichlasoma portelegrense*). The sex changing hermaphrodites have retained bisexual potency of the germ cells (Brusle-Sigard et al., 1992; Yao et al., 2007) and germ cells supporting somatic cells, as well. However, it is not known whether the preformed oogonial/spermatogonial stem cells in the sex changing hermaphrodites await a specific signal from neuroendocrine system, and/ or increasing and decreasing E_2 levels to stimulate mitotic proliferation and oogenesis and spermatogenesis, respectively.

Despite these diversities in reproductive structures, Pandian (2012) has shown that sex differentiation genes and sex steroid receptor genes are not only common to gonochores and hermaphrodites but are also homologous, and thereby confirm that the hermaphrodites have originated from gonochores and have retained the same genetic mechanism of gonadal differentiation. However, the following must be indicated: 1. The surgical removal of the testicular zone from the ovotestis of 1+ yr-old *A. schlegeli* has successfully induced precocious but complete sex reversal, comparable to the natural sex change in 2+ and 3-year-old porgy (Wu et al., 2008a). As of now, successful gonadectomy in protogynic has been achieved only in the bluehead wrasse *Thalassoma bifasciatum* (Godwin et al., 1996; Semsar and Godwin, 2003, 2004). In contrast to castration in *A. schlegeli* inducing gonadal and behavioral sex change, the available information on ovariectomy in *T. bifasciatum* suggests that the sex change is limited to behavior alone. The sex change in *T. bifasciatum* is limited to changes in nuptial coloration and

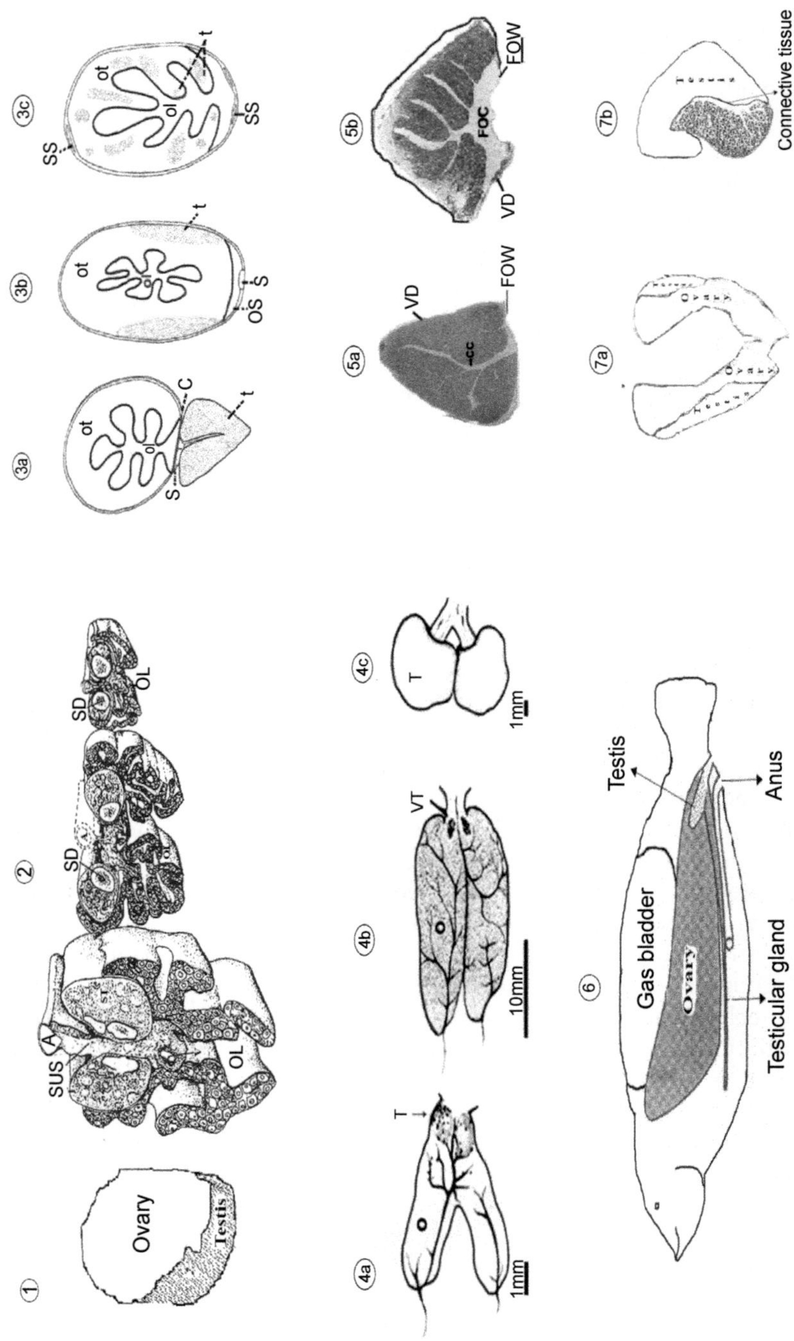
1
Ovary
Testis
2
SUS
A
OL
SD
SD
OL
3a
ot
ol
S
C
t
3b
ot
ol
OS
S
t
3c
ot
ol
SS
SS
t
4a
T
O
1mm
4b
VT
O
10mm
4c
T
1mm
5a
VD
cc
FOW
5b
FOC
VD
FOW
6
Gas bladder
Ovary
Testis
Anus
Testicular gland
7a
Ovary
Testis
7b
Testis
Connective tissue

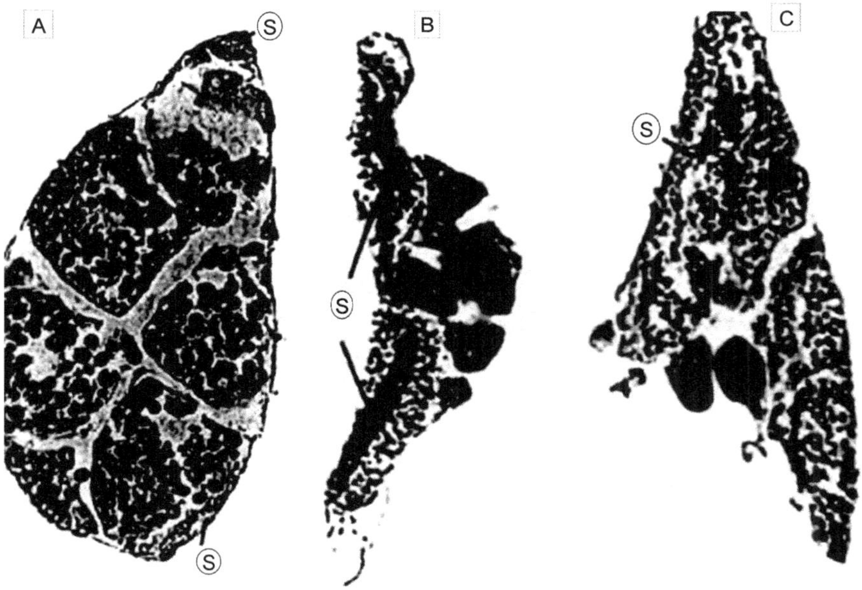

Fig. 36. Transverse section through the gonad of *Lythrypnus nesiotes*. (A) allocation to male tissue = 1%, (B) allocation to male tissue = 74%, and (C) allocation to male tissue = 90% (form St Mary, 1998).

courtship display but no spermiation, even after the hormonal therapy of the ovariectomized female with the potent androgen KT. Clearly, this raises the question, as was suspected by Frisch (2004), whether some non-classical hormone (Kime, 1993) and/or other than BPG/HPG axes are responsible for the male behavior in the ovariectomized bluehead females (see p 155).

Fig. 35. Structural diversity in gonads of selected hermaphrodites. 1. Cross section through the gonad of a serranid simultaneous hermaphrodite. 2. Transverse section through the gonads of a potential simultaneous hermaphrodite *Siderea grisea*, as schematically as viewed from the left side. OL = ovarian lamella, T = testis, ST = spermatogonial tissue, SD = sperm duct (from Fishelson, 1992). 3. Changes in the configuration of ovarian and testicular tissues of sequential hermaphrodites. (a) Delimited type, e.g., *Pagrus pagrus*, (b) Undelimited type 1, e.g., *Hypoplectrus puella*, (c) Undelimited type 2, e.g., *Epinephelus guttatus* C = connective tissue, ol = ovarian lumen, os = ovarian sinus, ot = ovarian tissue, s = sperm duct, ss = sperm sinus, t = testicular tissue (from Sadovy and Shapiro, 1987). 4. Gonads of *Sufflamen chrysopterus* (a) Gonads of juvenile with female coloration, (b) a breeding female, (c) a breeding male. O = ovarian part, VT = undeveloped testicular tissue, T = testicular tissue (from Takamoto et al., 2003). 5. Transverse section of testis of (a) *Lethrinus rubrioperculatus* (from Ebisawa, 1997) and (b) *Thalassoma lutescrens* (from Shibuno et al., 1994). 6. Schematic illustration to show the locations of ovary, testis and testicular gland in *Trimma okinawae* (redrawn from Nakamura et al., 2005; Kobayashi et al., 2005). 7. (a) Schematic illustration show the latero-ventral position of testes and dorsal position of ovaries in a typical sparid. (b) A section across the ovotestis to suggest the delicate surgery required for ovariectomy or castration.

6.3 Sexual phenotypes

The change from (Initial Phase) IP male to (Terminal Phase) TP male among the diandric sequentials is limited only to the reproductive role but the change from female to IP or TP male and the reverse involves extensive restructuring and redevelopment of the existing reproductive system (Cardwell and Liley, 1991a,b, Morrey et al., 2002). In protogynics, an undifferentiated diandric juvenile with bisexual gonad may pass through one of the pathways of sexual differentiation and differentiate directly into an IP male or a functional female (Fig. 37, e.g., *Thalassoma bifasciatum,* Munday et al., 2006). Primary or IP males resemble the females in body coloration and appearance but the secondary males have a bright body color during the terminal phase (TP). Shapiro and Rasotta (1993) reported the instances of primary males of *T. bifasciatum* changing their

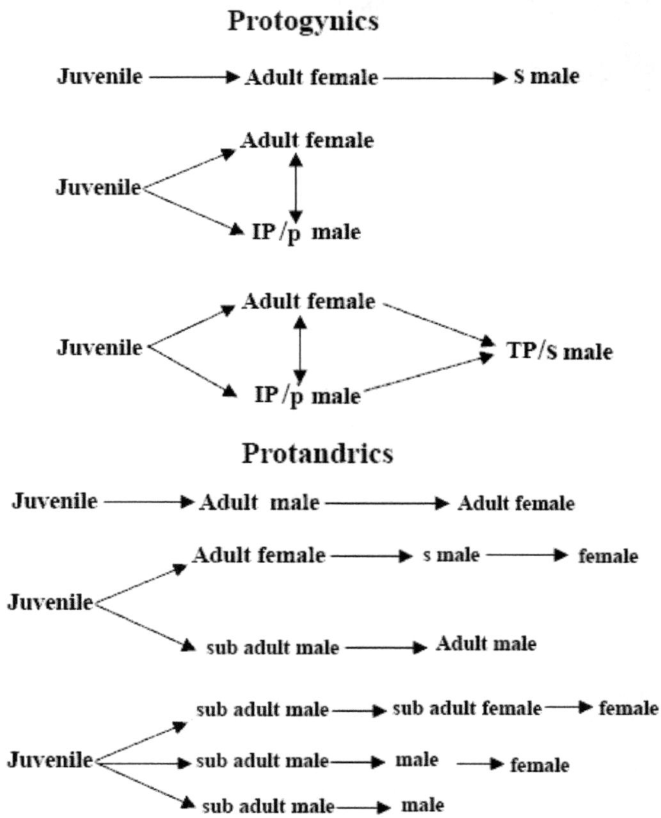

Fig. 37. Plasticity in sexual phenotypic differentiation among sex changing hermaphrodites. IP/p = initial/primary male, TP/s = secondary/terminal male.

(intial phase, IP) color into TP males and the female too re-differentiate into a 'secondary male'. Munday et al. (2006) showed that the juveniles of *T. bifasciatum* may choose to differentiate into primary males or females, depending on the female population size in a location. From their long time field and aquarium observations, Kuwamura et al. (2007) showed that a primary male of *Halichoeres trimaculatus* changes to female and an adult female, even after a few spawnings, changes into primary male, i.e., *H. trimaculatus* is a bidirectional serial hermaphrodite. By E_2-treatment, Kojima et al. (2008) are the first to show sex reversal in the IP male to female of *H. trimaculatus*. Following the treatment, spermatogenic germ cells in the testis undergo rapid degeneration with concomitant appearance of female germ cells. The E_2-treatment suppresses the expression of genes involved is steroidogenesis in the testis such as 11β hydroxylase and *Dmrt1*, and reduces androgen receptor transcripts (see Pandian, 2012). Administration of androgen inhibitor (AnI) to *H. trimaculatus* transforms IP males into TP males and females to sex reverse to become TP males (Kobayashi et al., 2010c). Similarly the administration of (AnI) (exemestane) for 10 wk in an IP male wrasse completely blocks estrogen synthesis, suggesting the obligate requirement of estrogen for normal spermatogenesis (Kobayashi et al., 2011).

In majority of protandrics, a juvenile differentiates into an adult male, which subsequently changes to a female phenotype. However, some clownfishes may directly differentiate into a female, then transdifferentitate into a secondary male and may even transform again into a female phenotype (e.g., *Amphiprion sebae*). A juvenile *A. clarkii* may choose one of the three alternative pathways to differentiate directly into a female, or into a male and then to a female. Rarely, an adult male may permanently remain as a male. Briefly all phenotypic changes involve change in sex, i.e., the changes in phenotypes are intersexual among hermaphrodites.

Zic family members encode zing finger transcription factors. Genome of some teleosts encodes seven *zic* genes. *zic2* is one of the five highly conserved among vertebrate groups. These conserved genes play a critical role in up-regulating proliferation factors and delaying differentiation during the development of neurulation. McCaffrey et al. (2011) examined the relative abundance of *zinc2* mRNA expression among the sexual phenotypes of *T. bifasciatum*. The *zic2* mRNA expression is abundant in all the examined seven areas of the brain including hypothalamus. However, they are more abundant in the POA of TP males than in IP males and females (Fig. 38), suggesting their potential role in remarkable sexual plasticity.

"The available observations clearly reveal that (i) the morphotypes, such as primary male (IP), functional female and TP male are only sexual phenotypes and (ii) these phenotypes retain the bipotential gonadal germ cells and germ cells supporting somatic cells, even after sexual maturation,

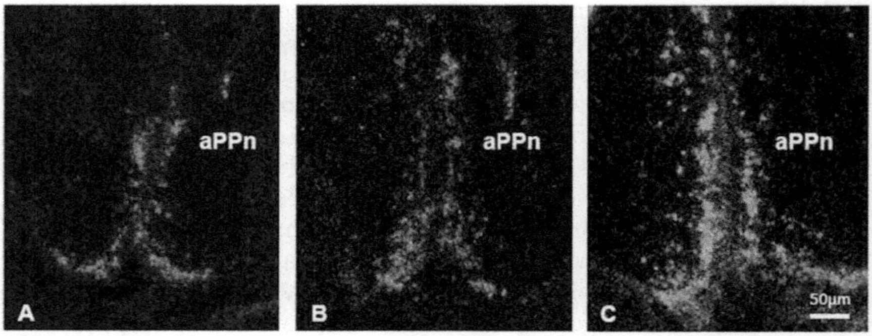

Fig. 38. Expression of *zic2* mRNA in the anterior parvocellular preoptic nucleus (aPPn) of the hypothalamus in diandric protogynous bluehead wrasse *Thalassoma bifasciatum*. Expression across female (A), initial phase (IP) male (B), and terminal phase (TP) male (C) (from McCaffrey et al., 2011).

and spermiation or spawning: Most remarkably, these sexual phenotypes among hermaphrodites can change from primary males to females and females to primary IP males and secondary TP males. Hence these sex changes are radically different from those among gonochores, in which the phenotypic morphotypes change at the best from sneaker male to satellite male, satellite male to territorial male and/or territorial male to pirate male—all these changes occur within a single sex namely male. Thus the changes in morphotypes are *intrasexual* among gonochores but are *intersexual* among hermaphrodites. This generalization may implicate that the flexibility/plasticity of sex determination and differentiation processes are several times greater among hermaphrodites than among gonochores" (Pandian, 2012).

6.4 Simultaneous hermaphrodites

These hermaphrodites are known not to undergo natural sex change; however, a couple of them merit consideration. Recent findings have shown the amazing features of sex differentiation in the simultaneous hermaphroditic cyprinodontid rivulus *Kryptolebias lebias*, the only vertebrate known to self-fertilize. A reconstructed figure shows the sexually dimorphic trends for the mitotic proliferation of PGCs (germ cells) in the female-turned hermaphrodite and male rivulus (Fig. 39). Sakakura and Noakes (2000) have shown that the rivulus are indeed females (<60 dph) and the frequency of hermaphrodites increases with advancing age. The sex change from female to hermaphrodite can be prevented by immersion of the juveniles in 1 ppb ethynylestradiol during the first 28 d. Hence sex steroids, their receptors, and aromatase play an important role in sex differentiation. The rivulus has

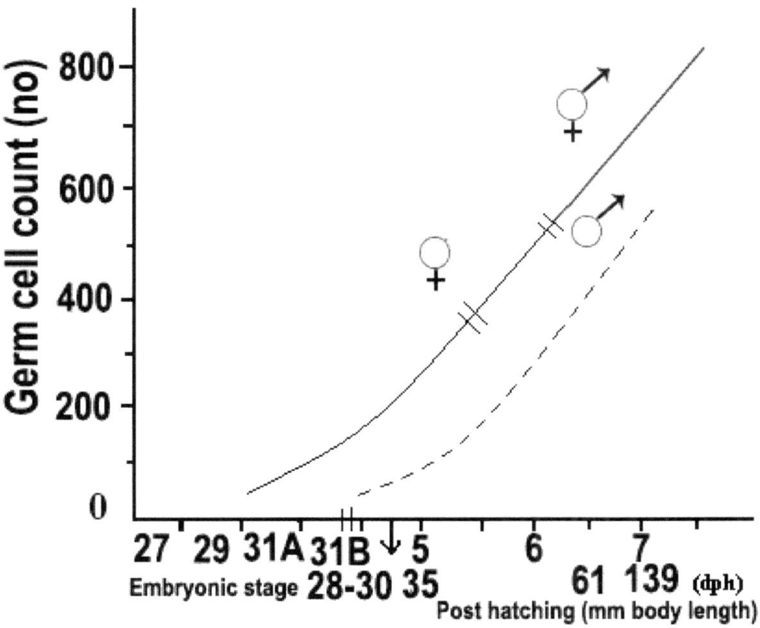

Fig. 39. Mitotic germ cell proliferation in embryos/hatchlings of *Kryptolebias marmoratus* incubated at 19°C and 26°C leading to differentiation of males (- - - -), females (—) and hermaphrodites (**—**), respectively. The numbers 27–31A, 31B represent embryonic stages, and a body size of 4.8 mm on 30 day post hatching (figure redrawn by compiling data from Harrington, 1975).

erα, *erβ*, and *aromβ* genes that are more strongly expressed in the liver, brain and gonad of hermaphrodites than those in males (Orlando et al., 2006).

When eggs of the rivulus are incubated at <19°C, 5% primary males appear. However, continuous exposure of post-hatchlings to <20°C for 3–6 mon fails to produce primary males (Harrington, 1967). Apparently, sex is determined prior to hatching. Nevertheless, high temperature (28°C) maintained in the laboratory (Harrington, 1971) or prevailing in its natural habitats like in the Belize (Tatarenkov et al., 2009) induces the production of secondary males. From the 60th day, the post-hatchling begins to add a male reproductive system in its natural course of changing sex from female to hermaphrodite. The lower temperature (<19°C), which induces the production of primary males prior to hatching, fails to do it on exposure of the post-hatchling from 90–120 d but the exposure to 28°C does produce secondary males. Hence, it is difficult to comprehend the plasticity in sex differentiation in the rivulus.

With the presence of hermaphrodite or primary and secondary males, the rivulus is considered as a diandric protogyny (Cole and Noakes, 1997).

Being a self-fertilizing hermaphrodite, the rivulus is perhaps the only fish to provide a homozygous lineage, and its potential use as a unique animal model has evoked much interest (e.g., Park et al., 1994). The first publication on its reproductive endocrinology was by Minamimoto et al. (2006). In hermaphroditic rivulus, the plasma E_2 level is higher (2453 pg/ml) than that of KT (346 pg/ml). Yet these steroid levels are nearly 10-times lower than those in bisexually reproducing cyrinodontid killifish *Fundulus heteroclitus* (Bradford and Taylor, 1987; Cochran, 1987). In primary male rivulus, the plasma levels of E_2 (6104 pg/ml) and KT (1180 pg/ml) are nearly 2–3 times higher than those of hermaphrodites. Using T or OHP (17α-hydroxyprogesterone) as precursors, the ovarian follicles of hermaphrodite and testis pieces of primary male simultaneously produce T, E_2, KT and 17α, 20β-P (progestin), while the early yolk stage oocytes produce 17α, 20β-P and E_2 alone. Values for GSI are 1.9 and 3.6 for the hermaphrodites and primary males. In the former, many stages of oocyte and spermatocytes are simultaneously present (Soto et al., 1992). The secreted steroids regulate each stage of oogenesis and spermatogenesis specifically, without any effect of other steroid hormones. Like the rivulus, *Serranus subligarius,* a gamete exchanging simultaneous hermaphrodite also secretes E_2, KT, 17α, 20β-P and 20βs during spawning (Cheek et al., 2000). Hence there seems some commonality among these simultaneous hermaphrodites.

In the case of *K. marmoratus, Serranus baldwini* and *S. fasciatus,* a publication by Zarkower (2006) is relevant. In the nematode *Caenorhabditis elegans,* there are two natural sexes namely XX hermaphrodites and XO males. Hermaphrodites are somatic females but produce progenies either by self fertilization or by mating with males. Matings generate broods of 50% XX and 50% XO progenies. Selfing progenies are mostly XX but one out of 500 is a XO male, which is due to non-disjunction of X chromosomes. Whether a similar non-disjunction of X chromosome is induced by temperature below 19°C in incubated eggs and 28°C in post-hatchlings resulting in the production of primary and secondary males of *K. marmoratus* remains to be tested. The same holds true of gamete exchanging older *S. baldwini* and *S. fasciatus,* who commence life as gamete exchanging hermaphrodites, lose the female function, become functional males and hold harems consisting one to seven hermaphrodites (see Pandian, 2010). It is also to be discovered whether the XX female *Coris julis* lose X due to non-disjunction during sex change to become XO males (see Pandian, 2012).

6.5 Sequentials and serials

The ability to change sex seems to have independently evolved in at least 23 families including 350 species (Frisch, 2004). Unlike the simultaneous hermaphrodites, the sequentials and serials are natural sex changers, although fractions of female populations of protogynic *Pagrus pagrus* (Sparidae, Kokokris et al., 1999), *Epinephelus itajara* (Serranidae, Bullock et al., 1992) and male population of protandrous *Diplodus vulgaris* (Sparidae, Pajuelo et al., 2006) do not change sex during their entire life time. It will be interesting and informative to know whether the testicular zone of these fractions of female protogynics has lost the PGCs and/or SSCs and whether the ovarian zone of the fraction of males in *D. vulgaris* has lost the PGCs and/or OSCs. Being a strange phenomenon, sex change has evoked much interest. The remarkable discovery of sex change induced by social factors has accelerated the research output (Sadovy de Mitcheson and Liu, 2008). The seminal publication on serial sex change by Kuwamura et al. (1994) has indeed catalyzed the numerous of publications in this area. Earlier publications on endocrine sex differentiation are limited to gonadectomy and histological studies. With availability of radioimmunoassay (RIA) in 1980s and subsequently the enzyme-linked immunosorbentassay (ELISA) techniques, rapid and precise determination of steroid concentration in biological samples has accelerated more accurate studies on the endocrine role in sex differentiation and sex change. At a given point of time, the sex changing transitionals in natural populations comprise 10.2% (Pandian, 2010) but the relatively longer durations required for 50% sex changers in a population (Table 36) provide ample scope to have snapshots of the ongoing sex change in sequentials. Relevant information on reproductive biology and genetic mechanism of sex differentiation in sex changing fishes is recently comprehensively summarized by Pandian (2010, 2011, 2012). The forthcoming account in neuroendocrine mechanisms of sex change is thankfully based on Frisch (2004); Godwin (2009); Larson (2011) and other relevant literature.

In fishes sex change has attracted attention for three reasons: (i) Among vertebrates, fishes are the only taxa that displays sex change and provides a unique opportunity for the academic understanding of the very plastic and sometimes, elusive process of sex differentiation, (ii) The sex changing sparids, serranids, scarids and lethrinids sustain valuable fisheries both in temperate and tropical seas (Chopelet et al., 2009) and some of them like the groupers, seabass and porgies are cultured as food fishes and clown fishes as ornamentals, and (iii) It is important to develop policies to minimize fishing the reproductively very fragile, sex changing, commercially important fishes. In the process of developing an evolutionarily stable life history strategy (ESS) model, Sattar et al. (2008) have noted the reduced age and size at maturity (see p 48), and increased female to male sex ratios among sex changing fishes. Increasing fishing pressure on the long living, large serranids provide a good example for the changing sex ratio; for every male, there are as many as 15 females in the black grouper *Mycteroperca bonaci* in Florida, 30 in Cuba and 77 in Mexico. In the gag *M. microlepis,* the number of females available per male is increased from six during 1977–80 to 50 during 1991–93 and the scamp *M. phenax* from five during 1977–82 to 17 in 1995. In the brown marbled grouper *E. fuscoguttatus*, there have been eight females per male showing intense fishing pressure at the Great Barrier Reef, as in the Gulf of Mexico (see Pandian, 2010). In female-biased groupers, females exhibit a relatively low reproductive success (Chopelet et al., 2009). Among the gonochores too, the size-at-sexual maturity of female red porgy *Pagrus pagrus* is decreased from 27 cm in 1972–74 to 18 cm in 1991–94, these smaller porgies may have hardly spawned once or twice, when they have reached the legally permissible catch size (Harris and McGovern, 1997).

Gonadal steroids: Considerable efforts have been made to induce artificial sex reversal in sequentials and serials. Following the exogenous administration of androgens in protogynics, the level of estrogen decreases with a concomitant increase in androgens (Nakamura et al., 1989). Incidentally, the grouper *E. suillus* reverts back to female after four month MT administration (Tan-Fermin, 1992). Non-aromatizable androgens are more potent inducers of sex reversal than aromatizable androgens like T or MT. In the goby *Coryphopterus nicholsii*, KT or 11-ketoandrosterone, the precursor of KT, induces gonadal sex reversal but MT implant fails to do it (Kroon et al., 2005). When placed with a TP male, a socially non-permissive environment, female *Thalossoma bifasciatum* implanted with KT completes the entire sequence of morphological and behavioral sex reversal in about two weeks, while the T-implanted female fails to do it. Hence KT is the predominant steroid associated with sex change in protogynics. It induces permanent sex reversal in *Sparisoma viride* (Cardwell and Liley, 1991a,b) and *C. nicholsii* (Kroon and Liley, 2000).

In protandrics, the opposite occurs (Godwin and Thomas, 1993). Estradiol 17β is the dominant estrogen in most (but not all, see Matty, 1985; Kime, 1993) protandrics. In sex reversing *Amphiprion melanopus*, elevated E_2 levels are not apparent until the oogonial proliferation (Godwin and Thomas, 1993). This suggests that E_2 is not necessary to initiate sex change/reversal in this fish. However, administration of E_2 successfully induces permanent sex reversal in protandrics like *Sparus aurata* (Condeca and Canario, 1995) and *Acanthopagrus schlegeli* (Lee et al., 2000). Hence the phenotypic sex of an individual may be a function of the ratio of E_2 to KT in the gonad (Frisch, 2004). In serials, the estrogen level increases, as the individual reverts to female (e.g., *Trimma okinawae*, Sunobe et al., 2005) and decreases, when it switches over to male. Remarkably, the androgen level remains unchanged during sex reversal (Lorenzi et al., 2008). Clearly, estrogen seems to play a more important role in sex changing/reversing serials.

In diandric protogynic *T. duperrey*, KT levels are higher in terminal phase (TP) males than in initial phase (IP) males (Nakamura et al., 1989). Using the radio-label technique, Hourigan et al. (1991) found that the Leydig cells are better developed in terms of both number and size in TP male *T. duperrey* than in IP male and the TP male is more efficient sperm producer than the IP male. However, the IP male with lower levels of T, the androgen that controls the KT-dependent spermatogenesis through the feedback pathway, is a prolific sperm producer. In many diandrics, GSI of primary (IP) males is more than three-times greater than that of secondary (TP) males (Table 28). Not surprisingly, *T. bifasciatum* IP males with nearly 10-times greater

Table 28. Gonado somatic index of some labrids and scarids. PM= Primary male, SM=Secondary male (from Pandian, 2010, as compiled from Warner and Robertson, 1978; Robertson and Warner, 1978).

Species	Gonado somatic index	
	PM	SM
	Labridae	
Bodianus rufus		0.09
Halichoeres radians		0.13
Clepticus parrae		0.12
Mean		0.11
	Scaridae	
Scarus vetula		0.11
S.taeniopterus		0.05
Mean		0.08
	Labridae	
Halichoeres bivittatus	0.35	0.59
H. garnoti	0.08	0.15
H. maculipinna	0.16	0.16
H. pictus	0.30	0.13
Thalassoma bifasciatum	2.71	0.22
Mean	0.72	0.30
	Scaridae	
Cryptotomus roseus	1.15	0.41
Sparisoma rubipinne	1.57	0.12
S. chryopternum	0.16	0.07
S. viride	0.33	0.06
S. radians	3.35	0.76
S. aurofrenatum	0.67	0.11
S. atomarium	0.27	0.20
Mean	1.07	0.25
Mean of means	**0.90**	**0.28**

GSI than TP males spermiate six-times more sperm during pair spawning than TP males (Shapiro et al., 1994).

Estrogen signaling through aromatase activity has now emerged as a key regulator of sex change. In the rice field eel *Monopterus albus,* on initiation of protogynic sex change p450 arom decreases but p450 11-β (p450-11β-hydroxylase) mRNA levels increase in the transitional and males (Liu et al., 2009). Aromatase immunoradiactivity is higher in the ovaries of *T. duperray* than in TP male. It declines with female to male sex reversal in *E. merra* (Bhandari et al., 2005) and *Lythrypnus dalli* (Black et al., 2005a). In *Corypterus nicholsii,* Kroon and Liley (2000) found that the aromatase inhibitor, fadrozole is as effective as KT in inducing female to male sex reversal. In *E. merra* fadrozole treatment depletes estrogen but increases KT level leading to complete sex reversal of females to males and spermiation by sex reversed males (Bhandari et al., 2004). Fadrozole also induces female to male reversal in *Gobiodon erythrospilus* (Kroon et al., 2005) and *Halichoeres trimaculatus* (Higa et al., 2003), in which the typical coloration is changed along with sex reversal. Clearly, the aromatase activity is also the initiator of sex change. Support for this suggestion comes from Blusle's (1987) report that oocyte atresia occurs long before spermatogonial proliferation in MT-treated sex reversing *Coris julis.* Sex reversal is suppressed in *H. trimaculatus* on co-administration of E_2 with aromatase inhitor (AI) or KT (Higa et al., 2003). The investigation by Wong et al. (2006) on *Sparus aurata* suggests that the differential expression of *cyp9a1a* is associated with sex reversal and gonadotropin signals, especially LH may serve as a major player in regulation of *cyp9a1a* during the process of sex reversal. The changing steroid profiles are a result of sex change rather than its cause. These studies have brought evidence that the steroids are not primarily important in initiation of sex change but the signal may be of neuroendocrine in origin. Aromatase mRNA has been detected in undifferentiated gonadal tissue of gonochoric rainbow trout *Oncorhynchus mykiss,* indicating that gonadal cells have estrogenic potential well before sex differentiation (Guiguen et al., 1998, 1999). But the protandric hermaphroditic *A. melanopus* undergoes oogonial proliferation much before E_2 levels are elevated.

Neuropeptide hormones and neurotransmitors: The peptides of GHs and GnRH are known to control the observed changes in steroid hormones in sex changing/reversing hermaphrodites. GnRH alone is capable of inducing sex reversal in the wrasse *T. bifasciatum* (Koulish and Kramer, 1989). For example, treatment with sGnRHa either alone or in combination with dopamine induces **precocious** sex reversal from female to male in *M. albus;* within 3–4 mon following the treatment, the sex reversed male spermiates (Tao et al., 1993). GnRH is capable of it in the wrasse only in combination with dopamine antagonist (Kramer et al., 1993). Notably, increase in the number of immune-reactive (*ir*) neurons in preoptic area (POA) is a correlate of sex reversal in *T. bifasciatum* (Grober and Bass, 1991). In the monoandric *Labrus berggylta* also, the number of AVT-*ir* neurons correlate with body size (Elofsson et al., 1999). However, the presence of greater number of AVT-*ir* neurons in the protandric male *A. melanopus* than in the female suggests that GnRH is associated more with the male function rather than sex reversal/ change in hermaphrodites (Table 29).

Table 29. Reported observations on the number and size of AVT neurons and their mRNA expression in selected sex changing hermaphroditic fishes.

Species and References	Reported observations
I Protogynics	
Thalassoma duperrey Grober (1998)	Dominant TP male has more and larger AVT-ir neurons than IP male and female
T. bifasciatum Godwin (2009) Semsar and Godwin (2003) Semsar et al. (2001)	Ovariectomy or castration has no effect on AVT mRNA expression or AVT-ir neuron size of magnocellular POA (mPOA). KT implantation in ovariectomized female results in no difference in AVT mRNA expression but induces TP male body coloration, courtship behavior and establishment of a temporary spawning territory. AVT mRNA abundance is typical of TP male and can be developed and maintained independent of gonadal inputs
Labrus berggylta Elofsson et al. (1999)	Males have higher number of GnRH-ir cells within POA than females. Both size and number of GnRH-ir cells are positively correlated with gonadal development in male
II Protandrics	
Amphiprion ocellaris Iwata et al. (2008)	Dominant female and second ranked male have less number of AVT-ir neurons in mPOA than in the sexual maturation-arrested juveniles
III Serials	
Lythrypnus dalli Reavis and Grober (1999)	Female → male (137/μm^2) sex changer has more number of AVT-ir neurons in mPOA (135/μm^2) than that of a female
Trimma okinawae Grober and Sunobe (1996)	Male has less number of AVT-ir neurons (30/μm^2) than male → female (45/μm^2) and female → male sex changers. Neuron size is larger in females than in males

Neuropeptides have a behavioral role in sex reversing fishes. In the serial goby, *T. okinawae,* the level of AVT is usually lower during the male phase than during the female phase; when the goby reverses sex in either direction, the AVT level also rapidly reversibly changes synchronous with the gender phase (Grober and Sunobe, 1996). Similarly, the levels of AVT as well as AVT mRNA are usually higher in the male than in the female *T. bifasciatum* (Godwin et al., 2000, 2003). Amazingly, following the removal of T male, the ovariectomized female *T. bifasciatum* not only displays territorial and other behavior of a male but also shows increased AVT mRNA level (Semsar and Godwin, 2004). Apparently, AVT stimulates territorial aggression and dominance in males as well as in sex changing/reversing females (Table 29). Serotonin is another neurotransmitter that regulates social interactions among animals. Experimental and field studies of Perrault et al. (2003) seem to suggest no effect on the activity of *T. bifasciatum* that received fluoxetine, a selective serotonin re-uptake inhibitor. However, serotonogeric activity is lower in aggressive dominants than in the subordinates in *Salvelinus alpinus* (Winberg et al., 1992), *Oncorhynchus mykiss* (Winberg et al., 1993) and *Haplochromis burtoni* (Winberg et al., 1997). Incidentally, high ranked *Dascyllus aruanas* consumes greater proportion of copepods and animal prey than low ranked ones depending more on phytoplankton (Forrester, 1991).

Monoamines: In gonochoric fishes, monoamines or biogenic amines are the most important neurotransmitters to control secretion of GnRH and GTH, which in turn control secretion of gonadal hormones. For example, intraperitoneal administration of DA (dopamine) and its agonist apomorphine (PO) decrease serum GTH levels, whereas intraperitoneal injections of DA antagonists pimozide and metoclopramide have the opposite effect (Peter et al., 1986). The injection of norepinephrine into the third cranial ventricles increases GTH release (Chang and Peter, 1983). Hence these amines regulate secretion of hypothalamic and pituitary hormones that may have a role in the pathway mediating conversion of social cue into a neuroendocrine event. Surprisingly, no one had studied the role of monoamines on induction of sex change, until Larson et al. (2003a,b) took up an investigation in the saddleback wrasse *T. duperrey*. *T. duperrey* is diandric and haremic; in its social organization, females have a larger home range and do not always mate with the same male.

Larson et al. (2003b) undertook three experiments to elucidate the role of monoamines in initiation and completion of gonadal sex reversal in *T. duperrey*. Experiments 1 and 2 tested whether increasing NE (norepinephrine) or blocking DA can trigger the initiation or completion of gonadal sex reversal under nonpermissive social conditions. Experiment 3 tested whether adrenergic blockade or increase in DA or 5-HT can block sex reversal under nonpermissive social conditions (Ross et al., 1983). From their studies on changes in monoaminergic activity from different regions of brain in relation to natural sex change in *T. duperrey*, Larson et al. arrived at the following conclusion: "The most striking changes occur in the POA (preoptic area) and the RN (Raphe nucleus). In the POA, a vast reorganization of all monoaminergic systems occurs on day three, the day when courting is first observed. There is an increase in noradrenergic activity concomitant with drastic decrease in both serotonergic and dopaminergic activity" (Larson, 2011). Perhaps male removal induces a decrease in serotonergic activity and an increase in adrenergic activity, which in association with the decrease in dopaminergic activity triggers gonadal sex reversal. "There is a drastic decrease in the serotonergic activity in the RN from day one to five that may be important both in the acquisition of dominance and the transition from female to male sexual behaviors" (Larson, 2011). There are also other studies, which support the findings of Larson et al. Increase in 5-HT accompanied by 5-HT re-uptake inhibitor fluoxetione decreases aggression in TP males of *T. bifasciatum* (Perrault et al., 2003). Fluoxetine-treated TP males of *T. bifasciatum* decrease AVT expression in the POA (Semsar et al., 2004). Hence 5-HT and AVT are also important in sex changing fishes. "The rapid change in expression of the monoamines during sex change/reversal is an important step in the complex reorganization of the entire HPG axis during this process involving not only monoamines but also steroids and various peptides" (Larson, 2011).

6.6 Model fishes

Considering *Epinephelus coioides* as a model, Yao et al. (2007) and Lu et al. (2010) attempted to elucidate the molecular mechanism of genetic differentiation. PGCs are located along the germinal epithelium; they differentiate into oogonia or spermatogonia, as development progresses. Oogonia are recognized by their enlarged size and spermatogonia by their intense staining. Assuming *sox3* as an ancestral of *sry*, Yao et al. (2007) prepared an anti-*Ecsox3* polyclonal antibody and tested the antibody for its specificity. *Ecsox3* is expressed and localized in the PGCs; hence the *Ecsox3* expression is a key marker for the time-point, at which the PGCs enter into oogonia or spermatogonia. At the entry of PGCs into oogonia, the *Ecsox3*-positive speckles appear in the PGCs. Similarly, *Ecsox3* transcription begins to appear in some cysts of male germ cells marking the initiation of sex change/reversal. *sox9* is expressed and localized specifically in Sertoli cells of the gonad. The events in the histological changes that occur during the transition from ovary to testis are briefly summarized in Table 30. A short description is given below on sex change/reversal in selected model fish groups that are economically or academically important.

Groupers: "The subfamily Epinephelinae, commonly known as groupers, includes 159 species in 15 genera; the genus *Epinephelus* alone comprises of 98 species. Groupers are widely distributed throughout the tropical and subtropical waters. Being a favorite marine food-fish, they are commercially important. They are amenable to culture, grow fast and resist disease. Hence there is a growing interest for aquaculture of groupers. Being protogynous hermaphrodites, the males are, however, not readily available" (see p151–152) (see Pandian, 2012). Not surprisingly, considerable efforts are being made to understand the mechanism that ensures sex reversal from

Table 30. Chronological sequence of events that take place in MT-treated sex reversing protogynous *Epinephelus coioides* (compiled from Yao et al., 2007; Lu et al., 2010) and naturally sex changing protandrous *Amphiprion ocellaris* (compiled from Madhu et al., 2010).

Sex reversal or change (week)	Reported observations
	Epinephelus coioides
First	Oocytes at different stages still exist in the gonads. *sox9* mRNA expression significantly increases
Second	*sox9* mRNA expression decreases
Third	Oocyte number largely decreases. Spermatogonia and spermatocytes appear and increase gradually in number. *Ecsox3* positive signals begin to appear in Sertoli cells
Fourth	The gonad is basically composed of testicular tissue. *sox9* expression increases again and the level is maintained subsequently
Fifth	Testicular tissue is well organized in lobules and cyst of male germ cells at all developmental stages are distributed within the lobules
Sixth	Numerous free spermatozoa are congregated into the lobules. Consequently, milt can be squeezed in about 95% of the sex reversed males
	Amphiprion ocellaris
Second	The first sign of sex reversal with reduction in the number of spermatocytes becomes perceptible
Eight	Spermatocyte number is further reduced
Twelve	Oogonia and oocytes in perinuclear stage become apparent
Sixteen	The number of oocytes increases
Twenty five	Testicular zone disappears. Gonad looks like maturing ovary
Sixty four	Spawning occurs

female to male (see Pandian, 2012). Incidentally, efforts are also underway to 'test-tube culture' of sperm of the groupers (see p 78), many investigations have identified the effective androgen, its minimum dose and mode of its administration. MT seems to be the most preferred androgen (Table 31) and the mode of administration ranges from dietary administration (e.g., *E. akaara*, Wang et al., 2004) to injection (e.g., *E. suillus*, Tan-Fermin et al., 1994) and to implantation, which itself ranges from simple silastic (e.g., *E. septemfasciatus*, Tanaka et al., 1999) to specialized EVAc release (e.g., *E. marginatus*, Sartar et al., 2006).

The reported observations on the effect of steroid administration on sex reversal in *Epinephelus* spp. lead to the following conclusions: 1. Most strikingly, the juvenile, immature and mature stages of the tested groupers are readily amenable to reverse their either undifferentiated gonad (e.g., *E. malabaricus*) or differentiated ovary (e.g., *E. septemfasciatus*) to develop into testis and produce fertilizable sperm. Precocious sex change or direct differentiation to testis is reported for *E. marginatus*, *E. suillus*, *E. malabaricus* and *E. merra*. It is advantageous to aquaculture; for example, *E. marginatus*,

Table 31. Effects of MT administration on sex reversal in selected groupers (*Epinephelus* spp.).

Species and Reference	Reported observations
E. akaara Tanaka et al. (1990)	**Precocious** males appear at the age 14 mon, but sex change mostly occurs at the age of 26 mon
E. akaara Li et al. (2006)	Implantation of MT or MDTH suppresses *p450 aromA* expression
E. akaara Wang et al. (2004)	Dietary supplementation of 2 mg MT/kg fish for 6 wk in 2-year old females induces sex reversal and spermiation
E. septemfasciatus Lee and Kim (1996)	Dietary supplementation of 0.5 mg MT/kg for 30 mon induces sex reversal but production of spermatids only
E. septemfasciatus Tanaka et al. (1999)	Implantation of silastic capsule delivering 4 mg MT/kg for 3 mon in 4-yr old female induces sex reversal and spermiation
E. coioides Yao et al. (2007)	Dietary supplementation of 2 MT/kg for 6 wk induces sex reversal and production of spermatozoa
E. coioides Lee et al. (1998)	Implantation (0.5 mg MT/kg) followed by second one of the same dosage after 4 mon induces permanent sex reversal
E. coioides Yeh et al. (2003a)	Implantation of 1 mg MT + androgen mixture (T, MT, KT)/kg induces sex reversal and spermiation. MT + androgen mixture is more effective than T, TP or KT
E. tauvina Mathew (2011)	Supplemented accumulation of 55 mg MT/kg or daily administration of diet containing 3 mg MT/kg for 3 mon fails to induce sex reversal
E. tauvina Chen et al. (1977)	Accumulated dose of 80–145 mg MT/kg or daily administration of 0.5–1.0 mg MT/kg for 5 mon induces sex reversal and spermiation in 3-year old grouper
E. fario Kuo et al. (1988)	Accumulated dose of 70–160 mg MT/kg or 145 mg MT/kg by daily administration of 5 mg MT/kg for five mon induces sex reversal and spermiation in 2-year old grouper
E. marginatus Sartar et al. (2006)	Females mature at the age of 5 yr and change sex from the age of 9+ to 16+. Controlled EVAc release of 5 mg MT/implant (≈ 11 mg MT/kg) plus hCG for 12 wk in 1+ year old **juveniles** induces **precocious** sex reversal and production of spermatozoa. Accumulation of 33 mg MT/kg via EVAc release is equivalent to accumulation of 85–113 mg MT/kg via dietary supplementation
E. suillus Tan-Fermin et al. (1994)	Females mature at the age 2+. Two year old **juveniles** (1.2–1.5 kg) undergo **precocious** sex reversal to males and produce spermatozoa, following biweekly injections of 1 mg MT/kg with net accumulation of 5 mg MT/kg
E. malabaricus Murata et al. (2010)	A 10-year old female changes to male. But supplementation of 50 mg MT/kg diet for 6 mon from 144 dph induces **precocious** sex reversal to male. 30–80 dph is the labile period suitable to induce sex reversal
E. tukula Yeh et al. (2003b)	Females attain sexual maturity at the body size of 16–24 kg (95 cm) and grow to 90 kg (150 cm). Implantation of 1 mg androgen mixture of T, KT and TP/kg fish in 90 cm females induces **precocious** sex reversal and spermiation
E. merra Kobayashi, Y et al. (2010b)	FSH (50 or 500 mg/kg fish)-treated females reverse to males in 3 wk treatment
E. merra Bhandari et al. (2003)	Implantation delivering 10 mg AI (fadrozole)/kg reduces E_2 level, degenerates oocytes and induces **precocious** sex reversal in 20 cm female at the threshold of sexual maturity. AI-treated male fertilizes more eggs than the wild male

which undergoes natural sex change at the age between 9+ and 16+, can be induced to sex reverse at the age 1+ (Sartar et al., 2006). Similarly, sex can also be reversed at the age of 30–80 dph (days post hatching) in *E. malabaricus,* which undergoes natural sex change at the age of 10+ (Murata et al., 2010). 2. Almost all the groupers tested so far are readily amenable to sex reversal on induction with MT, an aromatizable androgen. 3. A critical minimum accumulation MT is required to initiate and complete sex reversal. The accumulation of 80–145 mg/kg *E. tauvina* or daily dietary administration of 1 mg MT/kg for 5 mon successfully induces sex reversal (Chen et al., 1977). Understandably, the accumulation of 55 mg MT/kg or daily dietary administration of 3 mg MT/kg for 12 wk fails to induce sex reversal in *E. tauvina* (Mathew, 2011). 4. Delivery through silastic or EVAc implantation requires significantly less MT than that administered through dietary supplementation. For example, the accumulation of 33 mg MT/kg via EVAc release is equal to that of 80–113 mg MT/kg through dietary supplementation (Sartar et al., 2006). Incidentally, Hishida (1965) showed that the required estrone dose, when administered intraperitoneally or intramuscularly, can be reduced to a tenth of that required through diet. 5. MT is a potent androgen but a mixture expecially with KT is more effective. 6. The others like FSH or AnI (androgen inhibitor, fadrozole) deplete E_2 level and thereby ensure initiation of sex reversal.

Clownfishes: Among pomacentrids, there are 30 species of ornamental clownfishes belonging to two genera: *Amphiprion* and *Premnas*. The clownfishes are small (5–10 cm in length), attractively colored (Allen et al., 2010) (Fig. 40), adaptable to captivity, hardy, and display entertaining behavior. Because of their suitability for home aquaria, they are in great demand in the international market. They are characterized by protandrous sex change. Reared in groups of two, pair formation takes place sooner or later with larger one directly differentiating into female and smaller one into male. However, no experiment has been performed to know whether a female, when housed with a larger male, shall redifferentiate into a male, or the smaller male, when kept with a still smaller female, shall redifferentiate into a female. The possibility does exist, as the anemones have ovotestes (see below, Madhu et al., 2010). The clownfishes are monogamous but a few like *A. akallopisos* (Fricke and Fricke, 1977) and *A. clarkii* (Hattori and Yanagisawa, 1991) hold social units. Known also as anemonefishes, many clownfishes obligately require an association with a specific anemone (e.g., *A. frenatus* with *Stichodactyla haddoni*), some of them may associate with one or other of a few anemones (e.g., *A. ocellaris* with *Heteractis magnifica* or *S. haddoni*) and a few like *A. akallopisos* may not require the association of anemone. The credit for (i) developing and perfecting rearing and hatchery techniques (e.g., Ajithkumar et al., 2012) for more than a dozen

Fig. 40. Left upper: *Amphiprion sebae* in its host *Stichodactyla haddoni*, Right upper: *Premnas biaculeatus* in its host *S. haddoni*, Left lower: *Amphiprion percula* in its host *S. haddoni*, Right lower: *Amphiprion ocellaris* in its artificial sea anemone (curtsey, TT Ajithkumar and T Balasubramanian).

Color image of this figure appears in the color plate section at the end of the book.

clownfishes and (ii) some in estuarine water (e.g., *A. ocellaris*, Ajithkumar and Balasubramanian, 2009, *A. sabae*, Ajithkumar et al., 2010, *A. akallopisos*, Dhaneesh et al., 2011, *A. clarkii*, Ghosh et al., 2012) and (iii) attempt to rear clownfishes in association with artificial anemone must go to the Centre for Advanced Studies in Marine Biology, Annamalai University, India.

In a social unit of clownfish, there is a larger dominant 'harem mistress' with functional ovaries including remnants of degenerated testicular tissues, a large male with not only functional testes but also latent ovarian cells ('ovotestis'), one or more subadult males and juveniles, in which sexual maturation remains arrested (Fricke and Fricke, 1977; Casadevall et al., 2009; Madhu et al., 2010). In *A. akallopisos*, the presence of distinct hierarchy among the members of a social unit has been reported (Fricke and Fricke, 1977). The social rank is modulated by the AVT-ir *A. ocellaris* (Iwata et al., 2008).

The protandric sex reversal has been described for *A. akallopisos* and *A. bicinctus* (Fricke and Fricke, 1977), *A. melanopus* (Ross, 1978), *A. frenatus*, *A. clarkii*, *A. polymnus*, *A. perideraion* and *A. sandracinus* (Moyer and

Nakazona, 1978) and *A. percula* (Madhu and Madhu, 2005). With body growth and the attendant color changes, the following pathways for sex change in *A. clarkii* have been reported: 1. Subadult ♂ → subadult ♀ → adult ♀, i.e., females that do not pass through a functional male phase, and it occurs also in *A. ocellaris* (Madhu et al., 2010), 2. Subadult ♂ → adult ♂ →adult ♀, and 3. Subadult ♀ → adult ♂ → adult ♂, i.e., males that permanently remain as males (Hattori and Yanagisawa, 1991). The different transitional stages through which sex is reversed from male to female are briefly described in Table 30. On removal of the female, the male *A. akallopisos* reverses to female in 63 d in the laboratory but in only 26 d in the field experiments. An undifferentiated juvenile *A. clarkii* may also establish a social unit and directly become a functional female within 42 d (Hattori and Yanakisawa, 1991). The experiments by Madhu et al. (2010) suggest that complete sex reversal of a male to female in the absence of a female requires a period of 15 ± 3 mon, while that of a juvenile to female 22 ± 2 mon. Table 32 lists the durations required by some clownfishes to initiate and complete sex differentiation into male and female from date of pair formation. The process of ovarian differentiation seems to require a relatively shorter duration in the laboratory after the removal of the female or when juveniles alone are held together (e.g., *A. percula*).

Table 32. The duration required for socially mediated sex differentiation in protandric *Amphiprion* spp. after the pair formation (data kindly provided by T.T. Ajithkumar and T. Balasubramanian).

Species	Duration required for sex differentiation (mon)
A. ocellaris	8–9
A. percula	8–9
A. nigripes	9–10
A. clarkii	11–12
A. sebae	11–12

Gobies: Gobidae is a most speciose family (≈ 1,500 species, see Pandian, 2010) comprising of many protogynous gobies like *Coryphopterus nicholsii* and serials like *Trimma okinawae* (Table 33). They are small and hardy, and serve as good experimental fishes. Not surprisingly, there is growing interest in them. Understandably, the whole body content of KT is two-times higher in a male than a female *Lythrypnus dalli* (Kroon and Liley, 2000). KT implant induces male-like genital papillae in female *L. dalli* (Carlisle et al., 2000) and female to male (protogynic) sex change in *C. nicholsii* (Kroon and Liley, 2000). In the wrasse *Pseudolabrus siebaldi,* activation of 17β-HSD, responsible for production of T and subsequent biosynthesis of KT, is the first step in inducing morphological and functional change from the ovary to testis. KT level is also positively correlated with the amount of male tissue in gonads

Table 33. Effects of hormone manipulation and male removal in sex changing protogynous gobies and a labrid.

Species and Reference	Reported observations
Lythrypnus dalli Carlisle et al. (2000)	KT implant elevates urinary KT output and induces male-like genital papillae in female
L. dalli Black et al. (2005a)	Aramatase activity (AA) levels of brain (bAA) and gonad (gAA) are estimated. Male removal induces 200% increase in aggressiveness of large female. bAA level is lowest in male, lower in sex reversing female and highest in female. Aggression is inversely related to bAA but not to gAA. A possible socially-induced neurochemical mechanism is identified
L. dalli Black et al. (2005b)	Within 15 d following male removal, sex reversing large female fertilizes eggs, as males. This female has gonads, genital papillae and accessory gonadal structures with morphology in between that of males and females. KT level is positively correlated with male tissue in gonad
Coryphopterus nicholsii Kroon and Liley (2000)	T and KT levels of whole body are two-times higher in males than in females. KT implant depletes E_2. KT implant or fadrozole also induces sex reversal
Gobiodon erythrospilus Kroon et al. (2005)	Aramatase activity determines androgen/estrogen ratio. In single females, E_2 inhibits natural sex change from female to male. In single males, E_2 induces male to female sex reversal
Trimma okinawae Kobayashi, Y, et al. (2009)	Gonads with ovarian and testicular tissues but only one gonad remains active at a time. Changes in gonadotropin receptor (*GtHR*) gene expression is limited to active gonad changes at the onset of sex change. To be regressed gonad does not respond to exogenous gonadotropin
Pseudolabrus siebaldi Ohta et al. (2003)	Activation of 17β-hydroxysteroid dehydrogenase (HSD), which produces T as an intermediate in KT biosynthesis, is the first step in inducing morphological and functional changes from ovary to testis

of *L. dalli* (Black et al., 2005a) (see also Fig. 42). These findings clearly suggest that KT plays a major role in protogynic sex change.

However, the implantation of KT is concomitantly followed by depletion of E_2 level in *C. nicholsii* (Kroon and Liley, 2000). In a serial *Gobiodon erythrospilus* (?), E_2 induces male to female (androgynic) sex reversal in single males but inhibits sex change in single females (Kroon et al., 2005). That aromatase inhibitor (AI) induces sex reversal in protogynic *C. nicholsii* (Kroon and Liley, 2000) and *Epinephelus merra* (Bhandari et al., 2004) but blocks natural sex change to female in protandric *Acanthopagrus schlegeli* (Lee et al., 2001) clearly shows that a major role is played by estrogen in sex change. Then it is likely that the androgen/estrogen ratio, determined by aromatase activity (AA), is responsible to induce sex change. Brain AA is the lowest in the male, low in sex reversing female but the highest in the female *L. dalli,* suggesting a role for the socially induced neurochemical mechanism (Black et al., 2005a). Indeed, Kobayashi et al. (2009) showed that the changes in the gene expression level of gonadotropin hormone receptor (GtHR) initiate sex change in the serial *T. okinawae.*

Porgy: The black porgy *Acanthopagrus schlegeli* is an euryhaline protandrous hermaphrodite. Typical of a sparid, the porgy has a delimited ovotestis with a connective tissue separating the heterosexual zones (Lee et al., 2001). This hermaphrodite with a bisexual gonad is a functional male during the first two years of life and spermiates during the spawning season from January to March. About 30–50% of the 3 yr old males undergo natural sex change to females (Chang C.F et al., 1997). The 3 yr-old females (≈ 650 g, Liu et al., 2010) are larger than the 2 year-old males (310 g) and have 'the size advantage' of being more fecund. The need for female brooder porgy has been the major driving force of Taiwanese scientists, who have contributed more than two dozen publications to enrich our understanding of the endocrine mechanism that induces sex reversal. Their findings have led to a greater insight into hormonal mechanism/induction that leads to natural/artificial sex change/reversal in the porgy (Table 34). Briefly, administration of 4 mg E_2/kg feed from September to March in the next calendar year in 2+ year old male induces complete and non-reversible sex change. The elevated exogenous E_2 as a feed back stimulates GTHII expression, which in turn stimulates higher gonadal aromatase activity leading to sex change. Incidentally, Chinese scientists, who cloned the *p450 arom* of the porgy, showed a relatively higher transcription of *p450 arom* in 2+ yr old porgy than in the 1 yr-old male and 5 year-old female porgy, which breeds annually until the age of 5+ (914 g) (Liu et al., 2010). However, the role of monoamines, which are known to initiate sex change in protogynics (Larson, 2011), is yet to be tested in a protandrics like the porgy.

Nevertheless, the Taiwanese group has provided a greater insight on the role of genes in sex change in the porgy. Amazingly, but understandably, the surgical removal of testicular zone from a 1 year-old porgy successfully induces precocious sex reversal, comparable to the natural sex change in 2+ and 3-year-old porgy. Vitellogenic oocytes begin to appear in the ovary following the removal of testicular tissues from the bisexual gonad, clearly indicating the inhibitory role of testicular tissue on further development of the ovarian tissue in the porgy (Wu et al., 2008a). On removal of a testis, the transcripts of sf-1 (steroidogenic factor 1), Wnt4 (Wingless type MMTV integration site family 4), *StAR* and *cyp19a1a* increase but with simultaneous decrease in *dax-1* (dosage-sensitive sex reversal adrenal hypoblasia congenital critical region on the X chromosome, gene 1) (Wu et al., 2010). Incidentally, exogenous E_2 is known to stimulate increased transcripts of *cyp19a1a* and *Wnt4* and gonadal aromatase activity (Wu et al., 2008b). Hence the low expression of *dax-1* may be important in activation of ovarian tissue in the porgy (Wu et al., 2008a).

Table 34. Effects of hormonal treatment on sex change in *Acanthopagrus schlegeli.*

Treatment and Reference	Observed results
4 mg E_2/kg feed for 7 mon to 1 year old male (Chang et al., 1995a,b)	Inhibits spermiation; induces reversible sex change in 10% of treated porgy with gonad containing primary oocytes
4 mg E_2/kg feed for 4 ½ mon to 2 year old male (Chang and Lin, B-Y, 1998)	Also induces reversible sex change in 38% of treated porgy with gonad containing vitellogenic oocytes. Gonadal aromatic activity is high
4 mg E_2/kg feed for 7 mon to 2 year old male (Lee et al., 2000)	Administered E_2 stimulates increased aromatase activity in the gonad and brain. E_2 also stimulates plasma level GTH II. The observed parallel pattern between plasma GTH II and gonadal aromatase activity (but not between plasma E_2 level and gonadal aromatase activity) indicates that this relation is important in controlled sex change
4 mg E_2/kg feed from September to March in the next year to 2+ year old male (Lee et al., 2001)	*era* is less expressed in sex changing bisexual gonad than in the vitellogenic ovary. Levels of plasma GTH II and gonadal aromatase are high in sex changing porgy. Elevated E_2 feedback stimulates GTH II expression, GTH II induces higher gonadal aromatase activity; E_2 stimulates *era*. Administration of the 4 mg E_2/kg feed from September to March to 2+ yr old male induces non-reversible sex change. The selected E_2 dose, duration of labile period and age of male are appropriate to induce non-reversible, complete sex change
Implantation of T for one year in 2 year old male (Lee et al., 2001)	Does not block natural sex change; apparently, T is converted into E_2, which induces the known course of events leading to natural sex change
10 mg fadrozole 10 mg AST/kg feed from April 1999 to January 2000 to 2 year old male (Lee et al., 2002)	Completely inhibits the natural sex change, clearly showing the important role of E_2 in natural sex change

6.7 Social mediation

To investigate the process of socially mediated sex change in sequentials and serials, two different experimental approaches have been made: 1. Removal of the haremic/dominant male (e.g., *Labroides dimidiatus*) or female from a harem or social unit (e.g., *Amphiprion akallopisos*) or a monogamous pair (e.g., clownfishes) and 2. Isolated housing of one or more, equal and unequal sized individuals of the same sex (Table 35). The former induces immediate and dramatic changes in behavior and in some species morphological (e.g., body color) modifications (e.g., haremic *L. dimidiatus,* Nagashima et al., 2000) and (e.g., non-haremic *Lythrypnus dalli,* Black et al., 2011) in the dominant female, which ultimately changes its gonadal sex to a male. Similar observations from the latter are awaited. There are also publications on the changes in neuroendocrines and/or endocrines in the former during

Table 35. Socially-mediated sex change in selected fishes. J = juvenile, I = immature, M = mature, S = small, L = large.

Species, Reference, Social components	Reported observations
	Protogynics
Centropyge fisheri, C. acanthops C. flavissimus, Hioki and Suzuki (1996) 2 ♂ ♂	One male changes to female
C. ferrugata Kuwamura et al. (2002, 2011) L ♂ + S ♂	Smaller widowed male, on pairing with larger widowed male, differentiates into female
Labroides dimidiatus Kuwamura et al. (2003, 2011) L ♂ + S ♂	Smaller widowed male, on pairing with larger widowed male, differentiates into female
Halichoeres trimaculatus Kuwamura et al. (2007) Multiple IP ♂♂ of 4.8 to 10.4 cm	Five IP males change sex to females, of which the largest one reverts to IP male
Thalassoma bifasciatum Munday et al. (2006) J reared in isolation J reared in group of three	One J differentiates into a female Larger one J differentiates into primary (IP) male, while the other two into females
T. bifasciatum Morrey et al. (2002) L♀ + L TP♂ or S TP♂ L♀ or S♀ + S or L IP♂	Irrespective of the size of TP male and female, TP male inhibits sex change in female, which has a high GSI IP males fails to inhibit the initiation of sex change in female, which has a low GSI
Parapercis snyderi Kobayashi, K et al. (1993a,b) Single ♀ Multiple ♀♀♀	No sex change Largest female changes to male; gonads of some other females too become transitional
	Protandrics
Amphiprion spp. Ajithkumar and Balasubramanian (pers comm) J + J	Among the paired juveniles of *A. sebae, A. clarkii, A. ocellaris, A. percula, A. nigripes,* the larger ones differentiate directely into females
A. sabea, Ajithkumar and Balasubramanian (pers comm) L ♂+ s ♀	Larger male transforms into female and smaller female into male

Table 35. contd....

Table 35. contd....

Sex changing serials	
Epinephelus akaara Tanaka et al. (1990) Multiple ♀♀♀ (> 30 cm)	 Larger female with mature oocytes changes sex to male
Cirrhitichthys aureus Kobayashi, K and Suzuki (1992) 9.4 cm ♀ vs 7.5 cm ♀ 9.4 cm ♂ vs 8.9 cm ♂	 Larger female changes to male Larger male changes to female
Apolemichthys trimaculatus Moyer (1990) ♂♂	 Smaller male begins to change sex
Gobiodon erythrospilus Hobbs et al. (2004) Single juveniles Paired with M 2.2 cm J paired with 5 cm M ♀ 2.2 cm J paired with 5 cm M ♂	 50% Js attain sexual maturity at 3.5 cm 50% Js attain sexual maturity at 2.5 cm J differentiates into male J differentiates into female
Gobiodon histrio Munday et al. (1998) 2 ♀♀ 2 ♂♂	 Larger female changes to male One male changes to female
Paragobiodon echinocephalus Kuwamura et al. (1994) Multiple J	Monogamous mating system. Larger one initially changes to male but male and female become equal in size subsequently About 75% J differentiate into females but 25% of them into male
Trimma okinawae, Gobiodon spp. (Kuwamura and Nakashima, 1998)	Polygynous mating system. Large one changes to male

the transitional period (e.g., Kobayashi et al., 2009; Black et al., 2011), but again they are awaited for the latter. Clearly, research on isolated holdings of individuals of the same sex remain a virgin field of research. Incidentally, these investigations also provide information on whether or not the body size has an advantage in sex change over the smaller ones.

Firstly, pairing an undifferentiated juvenile (2.25 cm) with a larger (> 5 cm) adult results in socially-mediated precocious sexual maturation at the size of 4 cm instead of 5 cm, when the serial *Gobiodon erythrospilus* is reared singly (Hobbs et al., 2004). Secondly, when kept singly in isolation, juveniles differentiate always into females in *Thalassoma bifasciatum* (Munday et al., 2006) but almost always into males in *Cephalopholis boenak* (Liu and Sadovy, 2004). On the one hand, juveniles of protogynic *T. bifasciatum,* when held in groups of three each, perhaps the larger one differentiates into male, while the other two into females. On the other hand, smaller juveniles of the serial *G. erythrospilus,* when paired with larger mature male or female, differentiate into female or male. When a pair of juvenile *Amphiprion* spp. is kept together, the larger one is directly differentiated into female (Ajithkumar and Balasubramainan, pers com, see Table 32, 35). Clearly, pairing a juvenile with another juvenile or mature adult induces precocious sexual maturity and the juvenile hermaphrodite with its bisexual potency (see Pandian, 2012) yields to either direction of socially mediated sex change in both protogynics and serials. Amazingly, the protogynics, which are considered to undergo female to male sex change only once in their life time (see p. 141), are shown to undergo a social mediated sex change to females, when their male pairs of the genus *Centropyge* spp. are held in isolation (Hioki and Suzuki, 1996). Kuwamura et al. (2007) reported that on growing from 9.6 cm to 13.6 cm during a period of about 10 mon *Halichoeres trimaculatus* IP male also changes to functional female. When multiple IP males (4.8–10.4 cm) were kept together in an aquarium, five IP males changed to females, of which the largest one reverted to an IP male. Kuwamura et al. (2011) have also shown that when two widowed males of *L. dimidiatus* are kept together in the field, the smaller widowed male changes to female. The same is also true of *C. ferrugata*. For the first time (Ajithkumar and Balasubramanian, pers comm) have also shown that the same is also true of protandrics like *A. sebae*. Whereas the first sex differentiation requires a long period of 11–12 mon (Table 32), the second one occurs within 115 ± 15 d (cf. Table 37). These observations implicate that PGCs are retained in the gonads of these protogynics belonging to Labridae, Pomacanthidae and Parapercidae (see below) and protandrics (Pomocanthidae), even after they have undergone the first and second sex changes.

When multiple females of protogynic *Parapercis snyderi* (Kobayashi et al., 1993a,b) and serial *Epinephelus akaara* (Tanaka et al., 1990) are housed in isolation, the expected protogynic sex change to males occurs. The same holds true for the serial *Trimma okinawae* characterized by the polygynous mating system. However, when unequal sized pairs of female and male of the serial *Cirrhitichthys aureus* are held, the larger female and male change to male and female, respectively (Table 35). This observation implies that size may not be the selecting criterion for the sex changing individual. This was also confirmed by Morrey et al. (2002). Irrespective of the sizes of the TP male and female as components of the tested social system of *Thalassoma duperrey*, the TP male inhibits sex change in his pairing partner female. Again, irrespective of its size, the IP male fails to inhibit sex change in the female. Nevertheless, the importance of size in social induction of sex change can not totally be ruled out. The socially mediated sex change seems to require the following: 1. Removal of the inhibiting factor, namely the dominant male and 2. In some species like *L. dalli*, an initiation factor, i.e., the presence of smaller individual(s) (cf. size ratio induction, Ross et al., 1983). In the absence of the inhibitor, the dominant male *L. dalli*, the larger female in the presence of a smaller female, changes to male rather quickly in 8.6 d (Table 37) than when the two females of equal size are kept together.

6.8 Sex changing chronology

Regarding the age/size at which sex change occurs, there is a need to consider sex ratio and fecundity. Sex changers maintain sex ratios from one male for every two or three females and thereby increase reproductive output. Analyzing male sex ratio among the different patterns of protogynous hermaphrodites, Pandian (2010) found reductions in male ratios to 0.31, 0.30, 0.23 and 0.22 in monoandric monochromatic, monoandric dichromatic, diandric dichromatic and haremic protogynous hermaphrodites, respectively. Quite independently, Molloy et al. (2007) estimated the male ratio as 0.29 for 14 protogynous hermaphrodites. They also estimated that the sex changing males invested just 20% of that of gonochores, while the females about 50% of that of the gonochores. Thus the protogynous sex changers seem to invest less on gonads, perhaps to neutralize the cost involved in sex change. Whereas fecundity increases with body size of fish, it does not during the period of terminal age (see Pandian, 2010). When a fish attains > 70% of its maximum body size of the total life span, it is approaching a more or less reproductively inactive menopause stage. In different populations of the guppy *Poecilia reticulata,* this period of terminal life span has been estimated to range between 12 and 15% of total life span (Reznick et al., 2006). The empirical estimation predicts that sequential hermaphrodites change sex, when they reach 80% of their body size (Allsop and West, 2003). When all the recent information is taken together, the protogynous hermaphrodites switch to males, when they reach a body weight of 80% of their maximum, corresponding to the completion of 80–85% of their life span, and when they produce negligible number of eggs (in relation to body weight and time) or cease to produce eggs.

In a limited sense, the sex change may be defined as the proliferation of gonadal tissue of secondary/tertiary sex and simultaneous degeneration of gonadal tissue of primary/secondary sex (Frisch, 2004). To accomplish these structural gonadal changes, the sex changing fishes do require a period of time. The duration required for completion of gonadal sex change at an individual level of protogynics vary widely from just 6 d in *Centropyge fisheri* (Table 36) to 54 d in *Epinephelus adscensionis.* Most of these data are reported for haremic protandrics. Unlike the simple removal of the critically important inhibitor, the dominant male at one stroke in an experimental system, the socially mediated sex change in a large number of individuals in natural populations may have to compromise with more than one environmental cue, which ultimately affects the sex change. Understandably, the duration required for 50% females in a natural population to change gonadal sex requires a longer duration that ranges from 4 mon in *Thalossoma bifasciatum* to 36 mon in *E. marginatus* (Table 36).

Table 36. Days/months required for completion of gonadal sex change in protogynous fishes (from Pandian, 2010, 2011 updated, for references see Pandian, 2010, 2011).

Species	Gonadal sex change (days/month)	Heramic
	I Experimentals at individual level (days)	
	Labridae	
Bodianus rufus	7–10	Yes
Labroids dimidiatus	17	Yes
Halichoeres melanurus	14–21	Yes
Thalassoma bifasciatum	22*	Yes
T. duperrey	49**	Yes
	Scaridae	
Sparisoma radians	12–18	Yes
	Pomacanthidae	
Genicanthus watanabei	25	Yes
Apolemichthys trimaculatus	25	Yes
Centropyge vroliki	10–16	Yes
C. fisheri	6	Yes
C. acanthrops	8	Yes
C. bicolor	<20	Yes
	Gobiidae	
Coryphopterus personatus	20	No
C. glaucofraenatum	20	No
	Serranidae	
Epinephelus rivulatus	21†	
E. adscensionis	54††	Yes
Cephalopholis boenak	49 ☺	
	II Natural at population level (months)	
	Serranidae	
Mycteroperca bonaci	24	No
Hypoplectrodes maccullochi	24	No
Epinephelus tauvina	≈18	No
E. marginatus	≈36	No
	Sparidae	
Pagrus pagrus	≈20	No
	Labridae	
Coris julis	≈8	No
Thalassoma bifasciatum	≈4	No

*Warner and Swearer (1991), **Money et al. (2002, however 60 days, Ross, 1983), †Macki (2003), ††Kline et al. (2011), ☺ Liu and Sadovy de Mitchesson (2009)

Among protandrics, the required duration also ranges from 23–63 d in *Amphiprion akallopisos* (Fricke and Fricke, 1977) to as long as 15–22 mon in *A. ocellaris* in the field (Madhu et al., 2010). Table 32 lists the relatively longer durations required for sex differentiation in *Amphiprion* spp. following the pair formation in the laboratory. In general, the protandric clownfishes require longer durations for sex differentiation, despite their testes possessing ovarian tissues. Among the serials, the duration required for the first protandric sex change is about 5 d in *Trimma okinawae* and *Lythrypnus dalli* (Table 37) but for the second one it is as much as 3 ½-times longer. Protogynic sex change also requires nearly 2-times longer duration in both *T. okinawae* and *L. dalli*. On the other hand, the reverse is true of *Paragobiodon echinocephalus*. Further research is required to comprehend why the second protandric sex change requires a longer duration than the first one, and why the durations required by *P. echinocephalus* contrastingly differ from those of *T. okinawae* and *L. dalli*.

Figure 41 shows the manifestation of the chronological sequence of behavioral, morphological and gonadal sex change in the rock hind *E. adscensionis*. Immediately after the removal of the male, behavioral expression of aggressiveness becomes apparent in the sex changing female. Behavioral changes occur at two points of time in a transitional. The first one is the establishment of dominance by aggression and the second is the courtship display (Fig. 41). Of these, the first one is reversible. When the

Table 37. Days required for completion of gonadal sex change in serially sex changing fishes. L = large, S = small, E = equal sized, R = removed.

Species, Reference	Direction of sex change	Days required
Paragobiodon echinocephalus Kuwamura et al. (1994)		
	♀ → ♂	≈25
	♂ → ♀	≈45
Trimma okinawae Grober (1998)		
	♂ → ♀	5
	♀ → ♂	9
	♀ → ♂ → ♀	27*
	♂ → ♀	18
Lythrypnus dalli Grober and Sunobe (1996)		
	♂ → ♀	4.5
	♀ → ♂	8.2
	♀ → ♂ → ♀	26.6*
	♂ → ♀	18.4
L. dalli Reavis and Grober (1999)		
	♂ R	5.5
	♂ R, L♀ + S♀	5.7
	No ♂, L♀ + S♀	8.6
	No ♂, E ♀♀	9.2

*Days required for the second sex change

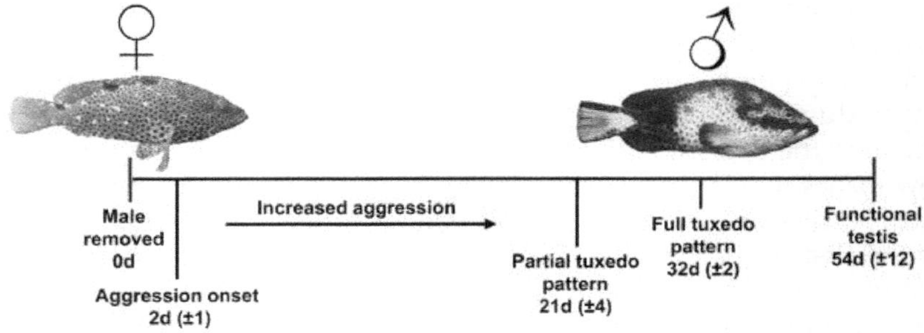

Fig. 41. Chronological sequence of sex change in rock hind *Epinephelus adscensionis* (from Kline et al., 2011).

Color image of this figure appears in the color plate section at the end of the book.

removed dominant male of *Labroides dimidiatus* is returned to the same harem after the potential female has already established dominance and even behaviorally 'spawned' with other female member, the potential sex changing female reverses to her original submissive female status and spawns with the returned dominant male (Nagashima et al., 2000). Then a morphological change to the male characteristic tuxedo color pattern requires 32 d. By the 54th day, i.e., 22 d after the manifestation of morphological changes, the gonadal sex change is completed. Apparently, these behavioral, morphological and gonadal sex changes are affected at different periods of time during the transistion period and are under the control of different systems like the hypothalamus → neurophypophysis → AVT-ir, hypothalamus → pituitary—gonadal axis and so on. Figure 42 represents a schematic illustration of temporal sequence of events that are likely to take place in sex changing transitional.

6.9 Gonadectomy

The credit for compending the widely scattered information on the effects of gonadectomy in fishes goes to Oliveira and Goncalves (2008). However, most of these surgical experiments have been performed to study the restoration of courting and nesting behavior in the revived fishes. It is more important to know whether or not the gonadectomized fishes are able to regenerate their gonad (Pandian, 2011). Reports available on about half a dozen species on successful gonadectomy and regeneration of gonad are summarized in Table 38.

Following castration or ovariectomy, the testis or ovary is regenerated in primary gonochores like *Oncorhynchus nerka*, *Oreochromis niloticus* and *Ctenopharyngodon idella*. Apparently, one or more of their organ systems other than gonads have retained PGCs and are able to regenerate their

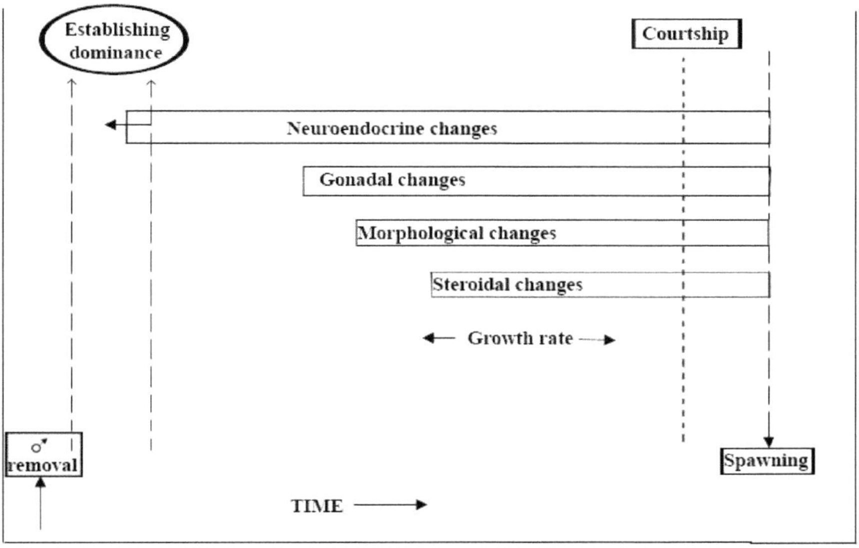

Fig. 42. A tentative generalized temporal sequence of events that take place during the transition period in a sex changing hermaphrodite. The reversible dominance establishment phase is circled and marked by reversed arrow (compiled from Reavis and Grober, 1999; Black et al., 2005a).

Table 38. Gonodectomy and regeneration in fishes (from Pandian, 2012, modified, for references also see Pandian, 2012).

Species/Reference	Reported observations
Gambusia affinis Turner (1947)	Castrated male apparently regenerates testis and complete gonopodium in the presence of small fraction 4% testis
Oryzias latipes Okada and Yamashita (1994)	Treated with E_2, ovariectomized females fails to regenerate ovary
Oncorhynchus mykiss Kersten et al. (2001)	Castrated males regenerate testes, irrespective of receiving E_2-treatment. Gonadectomized neomales (XX) regenerate testes
O. nerka Robertson (1961)	Castrated males regenerate functional testes at 'wrong' sites, i.e., adjacent to spermatic duct. Ovariectomized females develope ovaries but with reduced fecundity
Oreochromis niloticus Akhtar (1984)	67% of ovariectomized females regenerate ovaries within 119 d
Ctenopheryngodon idella Underwood (1986)	On removal of the gonads including surrounding mesentery, the testes and ovaries are completely regenerated in 91% males and 38% females, respectively
Acanthopagrus schlegeli (Wu et al., 2008a)	Following surgical removal of testicular zone alone from the ovotestis, a functional ovary is completely regenerated
Thalassoma bifasciatum Semsar and Godwin (2003)	Ovariectomy fails to regenerate ovary/gonad. Castration also fails to regenerate testis

original sex; incidentally, the germ cells supporting somatic cells (GCSSCs), which have lost bisexual potency in primary gonochores, facilitate the restoration of the original sex. However, the other gonochore *Orizias latipes*, on ovariectomy, is unable to restore its ovary even with E_2-treatment; in the absence or adequate number of PGCs, the ovariectomized medaka is unable to regenerate an ovary (see p 177). The hormonal masculinization of sex reversal appears to have also induced the loss of feminization potency of the GCSSCs. Hence both castrated, ovariectomized or endocrine sex reversed neomale (XX) *O. mykiss* regenerates only the testis. It may be recalled that estrogenic protogynic sex reversal is possible but not androgenic protandric sex reversal even in a secondary gonochoric zebrafish (see Larsen et al., 2009; Larsen and Baatrup, 2010; see also p 139–140).

Interestingly, following the removal of the testicular zone from the ovotestis, the black porgy *Acanthopagrus schlegeli* is able to develop its ovary (Wu et al., 2008a). It is known that PGCs are present at the junction of the ovary and testis in many sparids (see Fig. 35, Table 53). With the presence of adequate number of PGCs, the black porgy is able to develop a functional ovary. Incidentally, it will be informative to know whether a testis is developed following ovariectomy in sparids possessing ovotestis. Contrastingly, *Thalassoma bifasciatum*, a protogynic diandric hermaphrodite, whose brain is not sexualized (see Pandian, 2012) and presumably do not retain PGCs in any other organs other than gonad [in the absence of PGCs along with the ovariectomized ovary (Fig. 35)], is unable to regenerate its gonad. It may be recalled that the gonad of secondary sex is simply derived from the existing walls of the primary gonad in almost all the diandrics (see p. 143, Pandian, 2012). Clearly, the presence of PGCs in the gonad (e.g., ovotestis) or in other organs is obligately required to regenerate the gonad and in the primary gonochores, the original sex alone is revived, as facilitated by the GCSSCs, which have lost the bisexual potency.

Sixteen species of fishes have been successfully bilaterally gonadectomized (Table 12, 24). Barring the species listed Table 38, no observation has been reported on regeneration of the testis or ovary. Unfortunately, most of the reported observations are limited to courting, nesting and/or aggressive behavior alone. However the fact that many fishes are able withstand castration or ovariectomy is important and it may facilitate incoming researchers to report on gonadal regeneration also.

7

Sex Reversal

Introduction

The estrogen 17β-estradiol (E_2) was synthesized and became available for research in 1936. This event accelerated research activity on endocrine mechanism of sex differentiation by administration of one or another steroid initially and other non-steroid chemicals subsequently. For example, Berkowitz (1937); Castelnuova (1937) and Padoa (1937) attempted to manipulate sex differentiation in *Poecilia reticulata, Cyprinus carpio* and *Salmo irdeus,* respectively. However, Yamamoto was the first to achieve complete sex reversal in the medaka *Oryzias latipes* from genotypic male to phenotypic female (Yamamoto, 1953) and genotypic female to phenotypic male (Yamamoto, 1958). In 1969, Yamamoto postulated that sex steroids are in fact the natural male and female sex inducers and established a framework for research that has continued till today. Since 1969, there have been numerous publications, which have been reviewed from time to time (Hunter and Donaldson, 1983; Yamazaki, 1983; Dunham, 1990; Pandian and Sheela, 1995; Piferrer, 2001). The high citation index of Pandian and Sheela (>268) and Piferrer (>236) is a key evidence to the hectic research activity in this academically fascinating and commercially important field of research. Table 39 lists the landmark events in the history of chemical induction of sex reversal in fishes.

7.1 Gonadal differentiation

In fishes, gonochorism, hermaphroditism and unisexualism are the recognized patterns of sexuality. In many unisexuals, there are only females

Table 39. Landmark events in the history of chemical induction of sex reversal in fishes.

Authors	Events
1936	Estradiol synthesized and becomes available for research
Yamamoto (1953)	Steroid induction of complete sex reversal of medaka from ♂ → ♀
Yamamoto (1958)	Production of medaka YY ♂
Hishida (1965)	Intraperitoneal injection of steroid reduces hormone requirement to a tenth of that through dietary administration
Muller (1969)	Paradoxical endocrine sex reversal in tilapia
Fagerlund and McBride (1978)	Methylation slows elimination of testosterone
Johnstone et al. (1983)	Dietary administration leads to 99% excretion of administered hormone through diet
Piferrer and Donaldson (1991)	Aromatization of exogenous androgen as the cause for paradoxical sex reversal
Varadaraj and Pandian (1991)	DMSO ensures ten times better solubility of androgen in diet preparation
Devlin et al. (1991)	Use of DNA Y probe to distinguish genotypic sex after hormonal masculinization
George and Pandian (1995)	Production of (ZZ) ♀ molly
Muller-Belecke and Horstgen-Schwark (2007)	Production of 100% F_1 males by crossing Nile tilapia YY♂ x YY ♀
Cek et al. (2007a,b)	The use of herb *Tribulus terrestris* for masculinization of a cichlid and guppy

(e.g., *Poecilia formosa*). In the others, in which males are present, they are sterile. However, there are functional males in the unisexuals *Squalius alburnoides* and *Cobitis granoei taenia* (see Chapter 4). With reference to endocrine sex reversal in unisexuals, there is a single publication by Schartl et al. (1991) who reported restoration of male characteristics limited to morphological features in masculinized *P. formosa*. Virtually, endocrine sex reversal among unisexuals remains a virgin field. Pandian (2011) suggested the possible gamety in a few unisexuals. However, the results reported from experimental crosses in *S. alburnoides* are very confusing and it is difficult to assign it to any known gamety. Hence induction of endocrine sex reversal in unisexuals as well as sperm donor species may provide clues to assign one or another gamety to unisexual fishes. Chapter 6 has covered available information on endocrine sex reversal in hermaphrodites. Hence the coverage in this section on endocrine sex reversal is limited to gonochores.

"Sex differentiation relates to the events that occur during development and allows the expression of the genetic sex into appropriate phenotypic sex" (Piferrer, 2001). It is a conserved process across vertebrates with identical/similar factors and signaling pathways in the different groups.

Sex steroids particularly androgen and estrogens play a prominent role in the regulation of the sex differentiation process (Pifferrer and Guiguen, 2008). The differentiation process commences with the formation of gonadal primordium on arrival of PGCs and their colonization of the genital ridges, it extends until the first wave of gametogenesis and release of gametes in non-sex changing gonochores but continues until the sex changing processes are completed in hermaphrodites. The endocrine events and their genetic control during early development in primary and secondary gonochores suggest a major role for *foxl2, aromatase* and *amh,* as prime inducers of ovarian and testicular differentiation, respectively.

In the gonochore *Oryzias latipes* and *O. curvinotus,* sex is determined by *Dmy/Dmrt1bY* (Matsuda et al., 2002; Nanda et al., 2002). But it is not the sex determining gene in other species including those belonging to *Oryzias* and others (Konda et al., 2003). However, Suzuki et al. (2004); Hirai et al. (2006) and others provided detailed information on the role played by estrogen in early sex differentiation in *O. latipes* and the same is briefly summarized in Table 40. Experimental fadrozole exposure of embryos/hatchlings has revealed that 1. Estrogen has no role in proliferation of PGCs and in sex differentiation prior to hatching. Aromatase is detectable in testis of XY sex reversing *O. latipes* exposed to 10 ng EE_2/l (Scholz and Gutzeit, 2000). In *Oreochromis niloticus,* hormone-dependent sex reversal is accompanied by sex specific *tDmrt1* expression and proliferation of PGCs (Kobayashi et al., 2008b). The ovarian *aromatase* gene expression, a distinctive marker of sexual dimorphism, commences from the 4 dph, compared to the first appearance of steroid producing cells (SPCs) on the 20–30 dph. In *O. niloticus* too, the aromatase begins to express from the 11–27 dpf (days post-fertilization) but the presence of SPCs becomes apparent by 27–30 dpf (Kwon et al., 2001a,b). 2. However, endogenous estrogen is specifically required for formation of the ovarian cavity in *O. latipes*. 3. Yet, exogenous estrogen is capable of initiation and completion of folliculogenesis and vitellogenesis.

In genetic male hatchlings of medaka, the proliferation of PGCs commences 10 d later than in the female. Consequently, the androgen producing Leydig cells in testis appear after the 10th dph. Immersion of the newly hatched medaka to appropriate E_2 dose and duration induces the appearance of testis + ova in not only genetic male alevins but also in adults (Egami, 1955). Histological examination has shown that the estrogen-induced testis + ova originate from type B spermatogonia, suggesting that some type A or type A^{und} spermatogonia (see p. 71) with bisexual potency are retained in adult testis (Shibata and Hamaguchi, 1988). However, the adult female is not amenable to androgen induction to produce ovo + testis indicating that the female has lost the bisexual potential prior to puberty. Incidentally, this may be the reason why an ovariectomized medaka is unable to regenerate ovary even when E_2 is administered (Okada

Table 40. Hormonal induction of early sex differentiation in the medaka *Oryzias latipes* (compiled form several sources cited in Suzuki et al., 2004; Hirai, et al., 2006).

Day post-hatch (dph)	Reported observations
Ovarian differentiation	
0	Proliferation of PGCs increases. Exposure to fadrozole prior to hatching does not affect proliferation of PGCs; hence estrogen may not have a role in proliferation of PGCs Ovarian differentiation prior to hatching is not dependent on estrogen
4	Aromatase is first detected on the ventral side of the ovary from 4 to 10 dph (day post hatching), when pachytene-diplotene oocytes are also detected. Aromatase positive cells increase in number upto 60 dph and most of them are located around the developing oocytes
6	Fadrozole-exposure prior to hatching has no effect of gonadal sex
20–30	Steroid producing cells (SPCs), identified on the basis of subcellular organalles, appear among interstitial cells from 20 dph but are subsequently detected around oocytes from 30 dph
60–90	Ovarian cavity is formed
90	0.5 mg fadrozole/g is adequate to inhibit estrogen synthesis. Hence endogenus estrogen is obligately required to induce the formation of ovarian cavity folliculogenesis and oogenesis
Testicular differentiation	
10	The androgen producing Leydig cells in testis have not yet appeared
10–15	Proliferation of PGCs In genetic male, vitellogenin and testis-ova can be induced on exposure to 8 ng and 29 ng E_2/l, respectively Induction of testis-ova occurs on exposure to higher dose of 4 µg E_2/l for 48 hrs or 36 ng E_2/l for 2 weeks Exposure to 140 ng E_2/ml reverses the sex to female

and Yamashita, 1994, see also, Table 38). This could also be the cause for greater stress and higher mortality of the estrogen-induced sex reversed XY neofemales as reported (Pandian et al., 1994).

Characterizing 2,474 genes that are either up-regulated or down-regulated in the androgen-induced female (AIF) *Oncorhynchus mykiss*, Baron et al. (2007) selected 1,133 unique annotated genes that are associated with at least one gene ontology, and grouped these genes into 13 k-means clusters of temporally correlated expression profiles. Among them, cluster 6 contains genes with increasing expression profiles and commences from day 16 in AIF but from day 27 in normal female (nf) and thereby accounts for the observed precocious testicular development in AIF (cf. p. 158–160). Clusters 7 and 8 are of particular interest, as they are up-regulated only in AIF. Cluster 11 contains genes that are down-regulated in AIF from day 27 but in females from day 12. The genes in cluster 13 are all down-regulated in AIF from day 27. From this first comprehensive study of gene expression during androgen-induced masculinization, Baron et al. (2007) showed that the treatment in *O. mykiss* results in dysregulation of gene expression levels. The androgen treatment induces the complete down-regulation of female-specific genes but not of the complete restoration of the male-specific expression patterns.

On administration of 10 mg hydroxyandrostenedione/kg feed for a period of 3-mon to 55-dpf fry of *O. mykiss*, masculinization acts mainly through an early inhibition of ovarian differentiation rather than through a direct testicular differentiation. It commences with strong repression of granulosa-related genes *aromatase (cyp19a1a), follistatin (fst)* and *foxl2a,* resulting in abortion of early oogenesis. Incidentally, Wang et al. (2007) showed that *foxl2* has a decisive role in early ovarian differentiation by activating the *cyp19a1* transcription either directly or in conjunction with Ad4BP/SF-1 to enhance estrogen production. Hence repression of *foxl2* seems to be more important. In *O. mykiss,* the repression of another cluster of genes inhibits meiosis and previtellogenesis (Fig. 43). Folliculogenesis fails with the repression of *fst* and *gdf9* and steroidogenesis with *sox23, sox24* and *nup62.* With the expression of *tgfb* and *dax1,* Sertoli cells and subsequently steroidogenic Leydig cells appear (Baron et al., 2008).

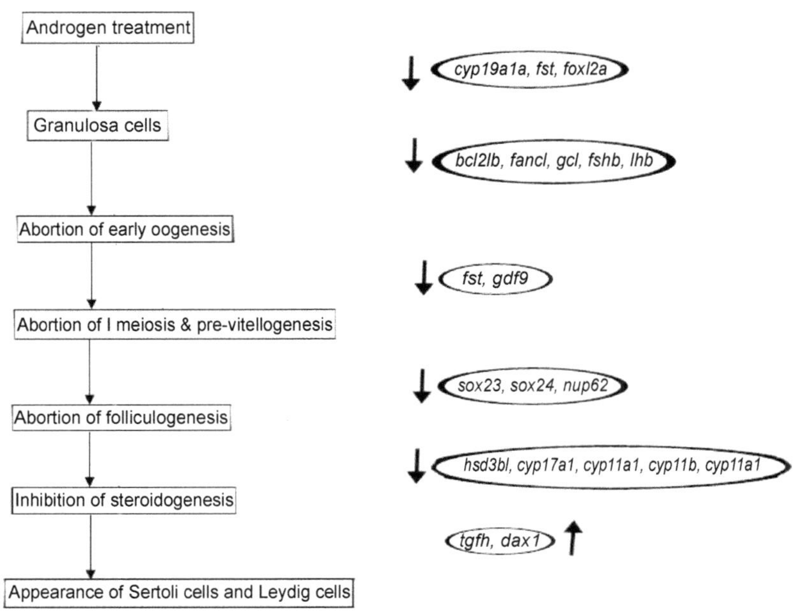

Fig. 43. Suggested sequence (→) of repressed (↓) ovarian gene clusters during androgen-induced sex reversal from female to male in a fish. (↑) indicates upward regulation (compiled from Baron et al., 2008).

Regarding the secondary gonochores, the zebra *Danio rerio* serves as a good model. Gonadal differentiation in the male involves a transformation from a 'juvenile-ovary' to testis and testicular differentiation in zebrafish is similar to that in a sex changing protogynic hermaphroditic fishes (Wang XG et al., 2007). Using a *vasa*: : *egfp* transgenic to identify sex transforming juvenile zebrafish, Wang and Orban (2007) showed that the expression of *cyp19a1a* commences in the non-transforming juveniles; it is up-regulated around the oocytes in the ovary until 3 wpf. P450 11β-hydroxylase (p450 11β), also known by its gene *cyp11b,* is a key enzyme in the synthesis of KT. In the presumptive male, *amh* positive cells are identifiable in its undifferentiated gonad (Rodriquez-Mari et al., 2005). In the 3 (week post fertilization)-wpf-presumptive male fry, transcription of *amh* and *cyp11*β shows a sudden increase, when it enters into the ovary-to-testis transformation process. The expression of *cyp19a1a* is, however, down-regulated from 3 wpf onwards but the expression level of *amh* is increased 300–600 times by up-regulation in the region, where *cyp19a1a* has been expressed earlier. *amh* is expressed in Sertoli cells (Rodriquez-Mari et al., 2005) but *cyp11b* mRNA are localized in Leydig cells of the adult testis and their level of expression is four times higher in the testis than in the ovary (Wang and Orban, 2007). Notably, *amh* expression is up-regulated earlier and to a higher level than that of *cyp11b,* suggesting that KT, the product of *cyp11b,* is not the earliest factor during the testicular differentiation. Hence the presence of KT is not considered as the prime inducer of male differentiation in the secondary gonochore *Danio rerio* (Wang and Orban, 2007).

7.2 Chemical inducers

In fishes, a wide range of structurally diverse chemicals have been used to successfully induce sex reversal. The masculinizing chemicals may be classified into (1) aromatizable (i) natural and (ii) synthetic steroids, (2) non-aromatizable steroids, (3) non-steroids, and (4) unidentified active principle from the herb *Tribulus terrestris* (Table 41). As many as 31 natural and synthetic steroids have been used to induce sex reversal in fishes (Pandian and Sheela, 1995). KT is a key androgen involved in testicular differentiation in most fishes (Pandian, 2012). It is synthesized from the substrate T. For instance, 11β-hydroxyandrostenedione (OHA), a precursor of KT, is used as substrate for the synthesis of KT in testis of many fishes but in liver of *Clarias gariepinus* (Cavaco et al., 1997) and in blood cells of *Gasterosteus aculeatus* (Mayer et al., 1990b) catalyzed by 11β-hydroxysteroid dehydrogenase (11β-HSD) and 17β-hydroxysteroid dehydrogenase (17β-HSD). For example, female gonads of *Cyprinus carpio* synthesize estrogens from 60 dph onwards; in contrast male XY gonads synthesize 11-oxygenated androgens alone. However, the sex reversing female gonads initially

Table 41. Chemical inducers used for sex reversal in fishes.

Classification of inducer chemicals
I Androgens
1 Aromatizable (i) Naturals (a) Testosterone (T) (b) 11-Ketotestosterone (KT) (c) Androstenedione (AST)
(ii) Synthetics (a) 17α-methyltestosterone (MT) (b) Mibolerone (c) Fluxymestrone (d) 17α ethyltestosterone (ET)
2 Non-aromatizable (i) Steroids (a) 1,4,6-androstatrien 3, 17-dione (ATD) (b) 17α methyldihydrotestosterone (MDHT) (c) 4 hydroxy-4-androstene- 3, 17 dione (4OH) (d) 4 androsten-4-ol-3, 17 dione (formestone) 3 Non-steroids (a) Fadrozole (Fz) (*aromatase* inhibitor) (b) Letrozole (c) Tamoxifen (Tx) (estrogen antagonist) (d) Flutamide
4 Medicinal herb (a) *Tribulus terrestris*
II Estrogens
1 Naturals (a) Estrone (E_1) (b) 17β-Estradiol (E_2) (c) Estriol (E_3)
2 Synthetics (a) Diethylstilbestrol (DES) (b) 17α-Ethynylestradiol (EE_2) (c) Estradiol benzoate (EB)
3 Estrogen receptor agonist (a) 4-Nonyphenol (NP) (b) Spironolactone aldosterone antagonist

synthesize both estrogens and 11-oxygenated androgens but its capacity to synthesize estrogens subsequently decreases (Komen et al., 1995). In *Betta splendens,* KT synthesis from 11-oxygenated precursors increases in the following order: 11β-hydroxytestosterone < 11β-hydroxy androstenedione < androstenetrione (Leitz and Reinboth, 1987).

Synthetic steroids are more potent than the natural ones. For example, MT is a methylated form of T and methylation slows the degradation of the molecule and reduces its clearance from the fish (Fagerlund and McBride, 1978). MT is estimated to have < 10,000–100,000 times less affinity to the *er* than E_2 (Hornung et al., 2004). The methylated form of 5α-dihydrotestosterone (DHT) is the 17α-methyldihydrotestosterone (MDHT). The pathways, through which the androgens and AI masculinize the treated fishes, are different. For example, the treatment with MDHT decreases not only *cyp19a* but also *er* expression in *Dicentrarchus labrax,* whereas Fz does not affect *er* expression. Apparently, the androgens and AI use different pathways to induce masculinization (Navarro-Martin et al., 2009). The effective dose for 100% masculinization in *Betta splendens* is 8, 15, 60 and 90 mg/kg feed for 19-nor-ethyltestosterone (19-NET), 17α-methyltestosterone, 11-ketotestosterone and androsternedione, respectively. From the point of potency, the androgens may be arranged in the following order: mibolerone > 19-nor- ethyltestosterone > 17α-methyltestosterone > testosterone. However, MT is the most widely used androgen to induce sex reversal among gonochores (Pandian and Sheela, 1995) (Table 42) and sequential hermaphrodites (see Table 31).

Recently, it has been brought to light that an extract of the herb *Tribulus terrestris* can successfully induce masculinization in both oviparous cichlid *Cichalasoma nigrofasciatum* (Cek et al., 2007a) and viviparous poeciliid *Poecilia reticulata* (Cek et al., 2007b) and *P. latipinna* (Kavitha and Subramanian, 2011a,b).

The estrogen 17β-estradiol (E_2) is synthesized from the substrates T, KT and/or MT and its synthesis is catalyzed by the enzyme aromatase. Hence these precursor-androgens are classified together as aromatizable steroids. The other inducers are either steroids like MDHT or non-steoids like fadrozole (Fz), tamoxifen (Tx) and so on. Among the non-aromatizable steroids, Fz competes with other enzymes for the same sites in the substrate. Being a non-steroid type 1 antiestrogen, Tx binds competitively to the *er.* However, the receptor may still remain partially active, when the Tx-*er* complex is formed. Hence Tx is capable of acting both as estrogenic and anti-estrogenic inducer (MacGregor and Jordan, 1998). For example, Tx acts as antiestrogenic at female-producing temperatures, but as estrogenic at female-producing temperatures in alligator with temperature-sensitive sex differentiation (Lance and Bogart, 1991). In fishes too, Tx masculinizes sexually undifferentiated gonad by suppressing *cyp19a* expression in *Paralichthys olivaceus* (Kitano et al., 2007) and *D. labrax* (Navarro-Martin et al., 2009). But it fails to masculinize the all-female progenies of *Oncorhynchus mykiss* and *Oreochromis niloticus* (Guiguen et al., 1999).

Feminizing inducers may be classified into (1) natural- (2) synthetic estrogens and (3) estrogen receptor agonists like the 4-nanophenyl (NP). As indicated, the synthetic androgens are more potent than the naturals. For example, 17α-ethynylstilbestrol (EE_2) and diethylstilbestrol (DES) are 2

Table 42. Protocols for chemical induction of sex reversal in selected gonochoric fishes TT = *Tribulus terrestris*.

Species, Reference	Reported observations
Oreochromis niloticus Phelps and Okoko (2011)	14 mg MT/kg feed for 28 d induces 95% ♂♂
Paralichthys olivaceus Kitano et al. (2007)	10 mg MT/kg + 100 mg Tx/kg from 30 to 100 dph induces 100% ♂♂
O. niloticus Orlando et al. (2001)	75 mg Fz/kg from 9 to 39 dph induces 100% ♂♂
Siganus guttatus Komatsu et al. (2006)	500 g Fz/kg from 21 to 111 dph induces 100% ♂♂
Hippoglossus hippoglossus Hendry et al. (2003)	5 mg MDHT/kg fed to post-larvae for 45 d induces 97–100% ♂♂
Betta splendens Kirankumar and Pandian (2002)	9 µg MT/l for 3 hr/d on 2, 3, 4, 5 and 8 dph discrete immersion induces 98% ♂♂
Pseudobagrus fulvidraco Park et al. (2004)	20 mg Tx/l immersion from 3 to 9 dph induces 90% ♂♂
Cichlasoma nigrofasciatum Cek et al. (2007a)	0.3 g TT/l immersion for 2 months induces 87% ♂♂ (8 times water changed)
Poecilia reticulata Cek et al. (2007b)	0.15 g TT/l immersion for 8 wk induces 80% ♂♂
P. latipinna Kavitha and Subramanian (2011b)	0.50 g TT/l immersion for 60 d induces 97% ♂♂
Salvelinus fontinalis Haffray et al. (2009)	400 µg MT/20 l, 5 immersions of eggs 7 d before hatching and feeding larvae with 3 mg MT/kg for 800° days induces 100% ♂♂
H. hippoglossus Hendry et al. (2003)	10 mg E_2/kg fed to post-larvae for 45 d induces 72% ♀♀
Lepomis macrochirus Wang et al. (2008)	150 mg E_2/kg for 60 d induces 100 ♀♀, of which 5% are intersex
P. fulvidraco Park et al. (2004)	40 mg E_2/l immersion from 3 to 9 dph induces 100% ♀♀
Silurus asotus Kim et al. (2001)	10 µl E_2/l immersion of 1-day old fry induces 92% ♀♀
Puntius conchonius Kirankumar et al. (2003)	600 µg E_2/l discrete immersion on 2, 4 and 6 dph induces 98% ♀♀
Gobiocypris rarus Liao et al. (2009)	100 ng E_2/l immersion of larva and juvenile for 21 to 42 dph induces 100% ♀♀ but testis + ova is detected in 9% ♀♀
Oryzias latipes Scholz and Gutzeit (2000)	100 µg EE_2/l immersion for 2 mon induces all ♀♀
Morone chrysops ♀ x *M. saxatalis* ♂ *hybrid* Davis and Ludwig (2004)	30 mg E_2/kg feed from 60 to 103 dph induces 100% ♀♀ but 30 mg MT/kg from 60 to 103 dph produces intersex only

½ and 6 times more potent than 17β-estradiol (E_2), respectively in inducing 100% feminization of *Betta splendens* (Fig. 44). Comparative potencies of tested estrogens are arranged in the following order: EE_2 > DES > E_1 > E_3, i.e., EE_2 is about 3 and 7 times more potent than E_2 and E_3, respectively (Piferrer, 2001). Yet, E_2 is the most widely used feminizing inducer both among gonochores (Pandian and Sheela, 1995) (Table 42) and sequential hermaphrodites (see Table 35). Until 2000, one or other of a dozen estrogens (3 naturals and 9 synthetics) were administered to control sex differentiation in not less than 56 species belonging to 24 families (Piferrer, 2001).

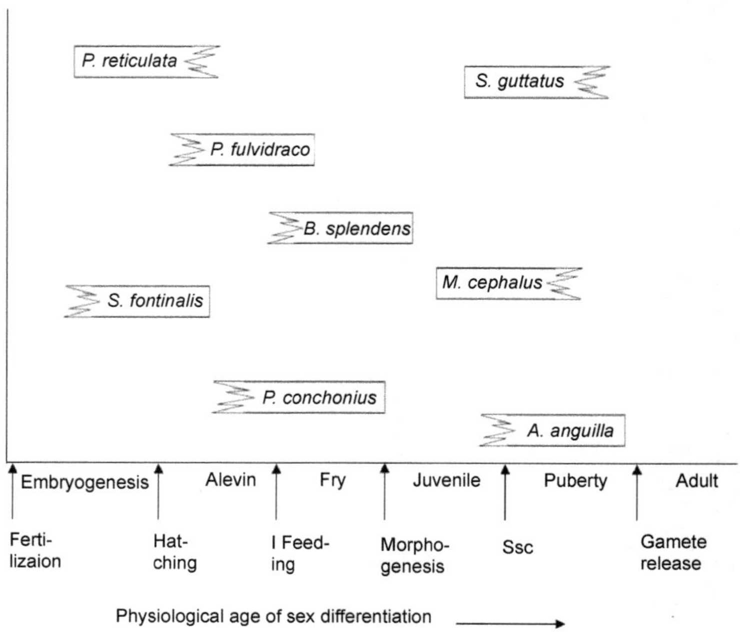

Fig. 44. Temporal locations of labile period during physiological age of differentiation in selected fishes.

7.3 Labile period

In teleost fishes, the process of differentiation is diverse and labile, rendering chemical induction of sex reversal possible (Francis, 1992). However, the lability is restricted to specific ontogenetic stage(s). The labile period is defined as the period of time covering one or more life stage(s), when the sexually undifferentiated or differentiating gonad is responsive and amenable to the desired high percentage of sex reversal on exposure to an optimal dose of one or another sex inducing chemical. In most teleosts, ovarian differentiation preceeds testicular differentiation, implying that the labile period for feminization is located earlier in development than that for masculinization (Piferrer, 2001). The time point through which a fish embryo passes through alevin, fry, juvenile, puberty and adult stages is marked by a specific event. Accordingly, embryogenesis is terminated by hatching in oviparous and parturition in viviparous fishes. The alevins depend on yolk nutrition and do not feed. The opening of the mouth marks the entry of an alevin into the fry stage. The completion of morphogenic features like the fin(s), eye, lateral line and so on indicates the entry into the juvenile stage. The appearance of secondary sexual characters (sscs), such as, gonopodium (Ogino et al., 2004) marks the entry into puberty. Maturity and the adult stage commence with the first wave of gametogenesis and release of gametes. Each of these described life stage may run for a short or longer duration, depending upon the life history and life span characteristics (Winemiller and Rose, 1992).

Formation of early gonadal primordium includes the development of the genital ridge (GR), and arrival and colonization of GR by PGCs. In *Oryzias latipes*, sex is determined, and the dimorphic sex differentiation is commenced prior to hatching, but after hatching in majority of the fishes. The precocious initiation of testicular differentiation, as induced by androgen, in embryonic *Poecilia reticulata* has been indicated (Takahashi, 1975; Kavumpurath and Pandian, 1993c) (Fig. 44). In others like *Cyclopterus lumpus* (Martin-Robichaud et al., 1994) and *Oreochromis niloticus* (Kobayashi et al., 2000; Ijiri et al., 2008), gonadal differentiation is initiated within a few days after hatching, i.e., at the alevin stage. More than age, body size seems to limit the labile period; for example, within the fry stage, smaller *Pomoxis nigromaculatus* fry (2 cm) are more amenable to masculinization than the larger (2.6 cm) ones (Arslan and Phelps, 2004). However, the gonad of *Siganus guttatus* remains undifferentiated even at the juvenile stage (Komatsu et al., 2006). The entire process of gonadal differentiation is completed during the fry stage in *Betta splendens* (Kavumpurath and Pandian, 1993a, 1994), during the juvenile stage in *Mugil cephalus* (Chang et al., 1995c) but at puberty in *Anguilla anguilla* (Colombo and Grandi, 1990) and *S. guttatus* (Komatsu et al., 2006).

To delineate the labile period, the test individuals are treated for different durations but with the same chemical inducer and dose; the period during which a high percentage of the treated individuals have undergone the desired sex reversal is taken as the labile period. The long tables prepared by Hunter and Donadson (1983), Devlin and Nagahama (2002) and Piferrer (2001) provide detailed information on the labile period for many fishes. Table 42 lists the periods for other species from recent publications. The available information on the labile period hints at the possibility of some generalizations. Briefly, three major groups of fishes may be recognized: the first group includes relatively smaller species, in which gonadal differentiation occurs within just 3–40 d following hatching. For example, the labile period falls between 10 and 30 dph in cichlids and cyprinodontids, and between 3 and 40 dph in anabantids (21 d in *Trichogaster pecteralis*, Pongthana et al., 1996). The second group comprises of species, in which gonadal differentiation occurs during the late juvenile stage and lasts for a period of 150 d (e.g., *Oncorhynchus mykiss*, Yamazaki, 1976). The third group includes species like *Ctenopharyngodon idella*, which is not amenable to sex reversal even after an induction period of 500 d (Shelton and Jensen, 1979).

Apart from these groupings, there are others like the golden rabbitfish *S. guttatus*, in which there is a considerable time lag between sexual differentiation of somatic element (i.e., the ovarian cavity formed on 45 dph) and onset of oogenesis by 100 dph. Consequently, the labile period is a prolonged one. The presence of a completely formed ovarian cavity alone, as revealed by an histological study, is a definite indicator of the gonad developing into an ovary. Komatsu et al. (2006) masculinized the rabbitfish using Fz supplemented diet. Administration of 500 mg Fz/kg feed for 30 d induced the development of testis in 87% treated rabbitfish, against 100% in rabbitfish treated for 90 d.

Within the determined labile period, the chosen chemical, its dose and route of administration may alter the effective duration of the labile period. Acute immersion sharply reduces the effective duration. For example, *O. kisutch* requires just two immersions in 400 µg E_2/l for 2 hr each on 4 and 6 dph to induce 100% females (Piferrer and Donaldson, 1992). However, chronic immersion for 18 d is required at the dose of 5.0 µg E_2/l from 5 dph to achieve 100% females (Nakamura, 1981a). Dietary administration of MT for 40 d during larval period produces 100% males in *B. splendens* (Kavumpurath and Pandian, 1994). But Kirankumar and Pandian (2002) achieved equal success through immersions of 900 µg MT/l on 2, 5 and 8 dph, each immersion lasting for 3 hr/d. Clearly, the effective labile period is reduced to just 3 d, when the immersion method is chosen. Notably, the immersion advances the labile period to the initial duration.

Attempts have been made to induce sex reversal during stages other than the labile period. Unsuccessful attempts have been made to induce sex reversal in mature adults. For example, Liao et al. (2009) successfully induced feminization of fry and juveniles of *Gobiocypris rarus* immersed in E_2 dissolved water for 21 d; by extending their experiments to adult male, they induced only an testis + ova with signs of vitellogenin. Swapna and Senthilkumaran (2009) also showed that in the adult *Clarias gariepinus* MT and EE_2 affect gonadal recrudescence only; MT treatment led to precocious ovarian recurdscence, perhaps with availability more substrate for conversion into estrogen; in the EE_2-treated males, the male reproductive axis is more disturbed than that in MT-treated females. However, no sex reversal has been achieved. Ogawa et al. (2008) sensationally titled their paper "Post-sex differentiational sex reversal..." but ended stating that "sexual bipotentiality of the ovary is maintained in the common carp, as in tilapia, at least before puberty". Hence adult gonochores are not amenable for sex reversal and the labile period is terminated before/during puberty.

7.4 Methods of administration

The routes, through which the chemical is administered into the fish, are: 1. Continuous (chronic) or discrete (acute) immersion in static or running aquarium water, 2. Feed/feed organism, 3. Intramuscular (Schreck, 1973) or intraperitoneal (Oliveira and Oliveira, 1979) injection and 4. Implantation of silastic capsule or cholesterol pellets (Table 43). Advantages and limitations of these methods are briefly summarized in Table 44. The dietary route is the more appropriate method for species, in which the labile period coincides with the external feeding fry and juvenile stages. The method involves the homogenous mixing of the chemical in the diet. The most widely used solvent for steroid application is the alcohol evaporation method (Guerrero, 1975). Comparison of the solubility of ET in (i) dimethyl sulfide (DMSO) (200 mg ET dissolved in 20 ml DMSO and mixed with 180 ml of 95% ethanol) + ethanol, (ii) ethanol (200 mg ET dissolved in 200 ml of 95% ethanol) and (iii) acetone + ethanol (200 mg ET dissolved in 20 ml acetone and mixed with 180 ml of 95% ethanol) has revealed that the solubility of ET in DMSO

Table 43. Routes of administration of chemicals to induce sex reversal in gonochoric fishes.

Routes of administration	Examples
I Dietary supplementation	
1. Steroid	MT, *Cichlasoma nigrofasciatum* (George and Pandian, 1996)
2. Non-steroid	Fz, *Siganus guttatus* (Komatsu et al., 2006)
II Immersion	
1. Discrete (discontinuous)	MT, *Betta splendens* (Kirankumar and Pandian, 2002)
2. Continuous	
a) Static	Tx, *Pseudopagrus fulvidraco* (Park et al., 2004)
b) Running water	*Oncorhynchus tshawytscha* (Hunter et al., 1980)
III Dietary cum immersion	MT, *Salvelinus fontinalis* (Haffray et al., 2009)
IV Injection	E_2, *Cyprinus carpio* (Castelnuovo, 1937)
V Implantation	
1. Silastic capsules	MT, *Ctenopharyngodon idella* (Shelton, 1982)
2. Cholestrol pellets	*O. kisutch* (Higgs and Donaldson, 1975)
3. Osmotic pumps	*O. kisutch* (Down et al., 1988)

Table 44. Advantage and limitations of methods to administer chemicals used for induction sex reversal in fishes.

Advantages	Limitations
Dietary supplementation	
Most commonly used and cheapest method. Requires almost no skill	Steroids suffer in digestive tract[1]. Hence it may lead to pollution[2]. Its purity varies[3]. Its solubility also varies with the solvent used[4]. Uniformity of steroid distribution in feed may vary. Size hierarchy may lead to differential feed uptake and hence steroid uptake. Intensive treatment with aromatizable androgens leads to paradoxical sex reversal[5]
Immersion	
Ensures synergic induction within shorter duration of treatment. Immersion can also be used for alevins, which do not feed. It is cheaper than dietary supplementation. Also requires almost no skill	Its usability is limited to embryos and post-hatchlings. Almost not useful in farm situation. Freshwater fishes do not drink water; hence may require longer durations than marine fishes, which drink water for osmoregulation
Injection	
Required quantity of chemical for induction may be the lowest. Ensures quicker results	Most laborious, expensive and skilled technique. Might lead to injury and infection. Not useful for induction of embryos and post-hatchlings
Implantation	
Improved implants deliver release of uniform dose of hormones. Ensures sex reversal in silver carp[6], which otherwise resists it	Costlier technique requiring skill for implantation. Can be used only in larger fish with a late onset of sexual differentiation

1. Hishida (1965), 2. Johnstone et al. (1983), 3. Pandian and Varadaraj (1991), 4. Varadaraj and Pandian (1991), 5. Muller (1969), 6. Hunter and Donaldson (1983)

+ ethanol is the highest, resulting in reduction of hormone requirement from 50 mg ET dissolved in DMSO + ethanol/kg feed (to 5 mg ET dissolved in ethanol or acetone + ethanol/kg to ensure 100% masculinization of 9 d old *Oreochromis mossambicus* larvae from 11 to 20 dph). Hence the cost of hormone-treated feed preparation with DMSO + ethanol as solvent can the reduced by 10% (Varadaraj and Pandian 1991). Testing the effect of different storage conditions on the potency of MT-treated feed, Varadaraj et al. (1994) recommended that MT-treated feed remains stable and active, when stored in a darkened desiccator at room temperature; apparently the desiccator eliminates bacterial degradation of the MT in feed.

The scope for using live feed organism *Artemia* as a vehicle for steroid has been studied in *Micropterus salmoides* (Garrett, 1989) and *Cyclopterus lumpus* (Martin-Robichaud et al., 1994). Steroid loadings vary considerably even at the same dosage and exposure duration. Further, the steroid-loaded live feed, when used in large scale, as in fish farms, may lead to pollution. Hishida (1965) is the first to show that for *O. latipes,* the required estrone dose, when administered by an injection, can be reduced to a tenth of that required through dietary supplementation. Over 99% of the hormone administered to *O. niloticus* and *Oncorhynchus mykiss* is released into water in less than 24 hr. Hence large scale farm application of steroids may lead to a synergic effect on sex reversal in fish farms and natural aquatic systems (Johnstone et al., 1983). In fact, Gomelsky et al. (1994) demonstrated that MT metabolites present in culture water are more potent than dietary administration of MT to masculinize *Cyprinus carpio* and the culture water containing MT metabolites reverse sex in a higher percentage of carps than that in MT-treated carps (Table 44). Nevertheless, a glance at Tables 1 and 3 of Piferrer (2001) listing the methods used for feminization shows that over 80–90% of authors chose dietary administration to induce sex reversal in gonochoric fishes.

Immersion is more suitable for species, in which the labile period coincides with embryogenesis or occurs during the early non-feeding alevin stage. An approximate calculation of MT administered through diet and immersion to ensure 100% males in *O. mossambicus* (Nakamura, 1975) and *O. mykiss* (Simpson, 1976) indicates that the MT required is 200–1,000 times less, when immersion is chosen. Clearly, the immersion method is cheaper, consumes less time and produces far less steroid pollutants into the culture water (see Pandian and Kirankumar, 2003). Yet the treatment involving periodic immersion alone has not ensured 100% sex reversal in many salmonids (e.g., Johnstone et al., 1978). Juvenile salmonids, which have undergone discrete immersion in aqueous steroid solution during the pre-feeding stages respond more favorably to the dietary administration than those fed on endocrine supplemented diet alone. For example, Goetz et al. (1979) have achieved 60 and 100% sex reversal in *O. kisutch,* when the salmons were fed with steroid diet alone and a steroid-supplemented diet preceeded by periodic immersion. An attempt to induce sex reversal in eggs and embryos of *O. mossambicus* by estrogen immersion proved ineffective (Rosenstein and Hulata, 1994). In contrast, the method ensures feminization of *O. niloticus* (Gilling et al., 1996). A combination of discrete immersion and dietary administration of hormone induces the desired sex reversal in not readily amenable fishes (e.g., *Salvelinus fontinalis*). It also induces complete sex reversal including the formation of gonadal duct in *O. mykiss,* in which the sex reversal by dietary administration is limited to gonads and is not extended to gonadal ducts (Haffray et al., 2009).

7.5 Dosage

In general effective steroid dosages to induce sex reversal in fishes range within a few micrograms for the immersion technique but in milligram levels for dietary supplementation. With increasing dosage for both immersion and dietary administration, the process of sex reversal commences with an initial acceleration phase, followed by a log phase, when the treated fish is more responsive to the dose(s), and is levelled off at 90–100% (Fig 45). For example, the responsive doses are 300 and 400 mg/kg feed for E_2 and 3-benzoate estradiol (EB) but is just 200 and 250 mg/kg for the more potent synthetic EE_2 and DES for ferminization of *Poecilia reticulata*. They are 100 mg/kg for NET but 400 mg ET/kg for masculinization (Table 45). Figure 45 shows the trends for the dose vs sex reversal induced by selected estrogens and androgens of different potencies, respectively in *Betta splendens* and *P. reticulata*. With decreasing potency of the selected steroid, the trend shifts towards the right, indicating the need for increasingly higher doses required to induce 100% sex reversed individuals. With reference to the dietary treatment alone, the required dose for sex reversal in 100% treated individuals increases in the following order: Cichlidae > Cyprinodontidae, Anabantidae, Salmonidae, Poeciliidae and Cyprinidae. Notably, the gestating females require a higher dose than post-hatchlings; the effective dose for the gestating female *P. reticulata* to feminize her embryos is 400 mg E_2/kg (Kavumpurath and Pandian, 1992) but 200 mg E_2/kg is adequate for the post-hatchlings of *P. sphenops* (George and Pandian, 1995).

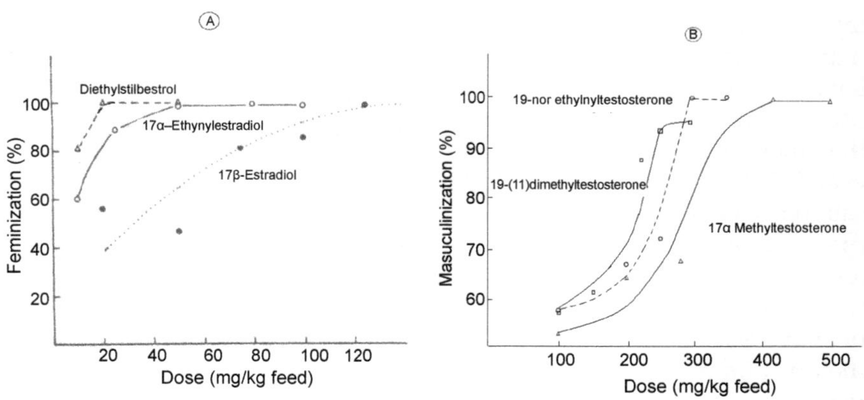

Fig. 45. Effects of steroid dose on sex reversal in fish. A Left panel: feminization by increasing doses of selected estrogens in *Betta splendens* (from Kavumpurath and Pandian, 1993a). B Right panel: masculinization by increasing doses of selected androgens in *Poecilia reticulata* (drawn from data of Kavumpurath, 1992).

Table 45. Effects of optimal steroid doses for masculinization on the number born and their survival of *Poecilia* spp. – = data not reported, * = dose in g/l, † = 88% masculinization, †† = 97% masculinization.

Steroid	Optimal dose (mg/kg)	Progenies born (no)	Survival of progenies (%)	Responsive dose (mg/kg)
Poecilia reticulata (Kavumpurath and Pandian, 1992)				
E_2	400	38	84	300
DES	300	30	83	250
EE_2	300	31	84	200
EB	600	15	93	400
P. sphenops (George and Pandian, 1995)				
E_2	200	28	70	100
DES	200	22	50	25
P. reticulata (Kavumpurath and Pandian, 1993b)				
NET	300	13	85	100
AST	300	10	60	200
ET	500	11	73	400
DMT	300	10	73	150
P. reticulata (Cek et al., 2007a)				
TT*	0.15†		86	0.15
P. latipinna (Kavitha and Subramanian, 2011b)				
TT*	50††	–	–	0.15

There are considerable variations in the sex reversing potency of commercial grade MT, when it is tested in cichlids like *Oreochromis mossambicus* and *O. niloticus*. The feed supplemented with 10–52 mg MT/kg is reported to reverse the sex from female to male (Clemens and Inslee, 1968) (Fig. 46). At the optimum MT dose, the proportion of treated individuals undergoing sex reversal varies between 75 and 85% (Nakamura, 1975; Das et al., 1987), indicating the interference from impurities of MT. Pandian and Varadaraj (1991) achieved 100% males with only 5 mg MT/kg. Differences in quality and purity of MT by different manufactures may explain some of these variations.

At an ineffective dose or overdose of a chemical inducer, intersexes are produced in many species (e.g., *O. niloticus*, Phelps and Okoko, 2011). For details on the hormone inducing intersex in selected fishes, Pandian and Sheela (1995) may be consulted. Reports on induction of intersex at sub-optimal or super-optimal doses are continuously published; for example *Misgurnus mizolepis* (Kim et al., 1998) the red tilapia, the Thai hybrid between *O. niloticus* and *O. mossambicus* (Manosoroi et al., 2003), *Betta splendens* (Kirankumar and Pandian, 2002) and *Tinca tinca* (Hulak et al., 2010). Low

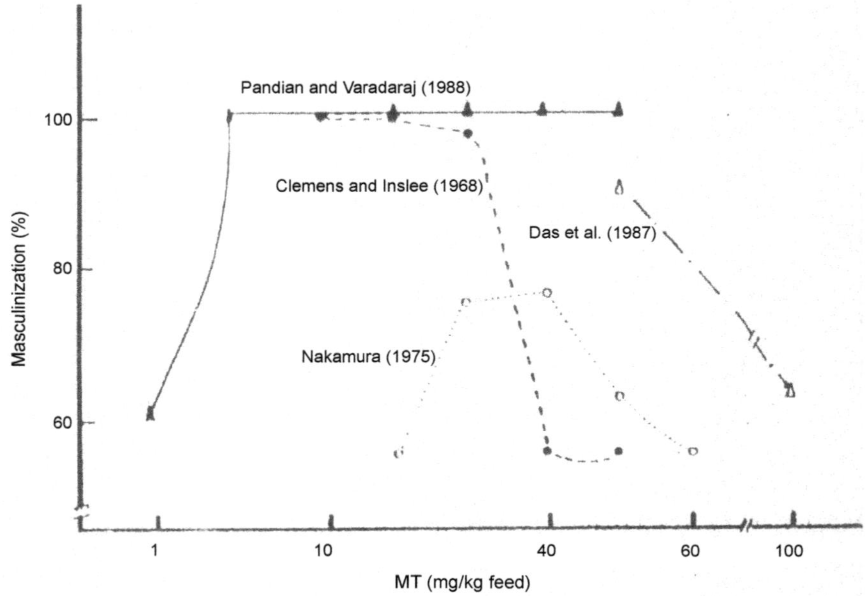

Fig. 46. Differences in the levels of masculinization and optimal doses to produce all male progenies in *Oreochromis mossambicus* and *O. niloticus* by different authors (from Pandian, 1995).

percentages of intersex are produced at the sub-optimal doses and they have a higher proportion of their gonad occupied by testicular tissues. In contrast, the gonads of those produced at the super-optimal doses are comprised more of ovarian tissues. Briefly, intersexes produced at sub-optimal doses have testis+ova but those at higer doses have ova+testis. Like mutants, intersexes may be important to have a greater insight into the mechanism of endocrine sex differentiation.

Hormonal sterilization is one of the potential methods for large scale sterilization of salmonids and cyprinids. It can be induced by administration of androgen at high doses and for longer durations than those required for masculinization. Administration of high doses of MT for longer durations induces 100% sterility in *Cyprinus carpio* (Basavaraj et al., 1997) and *O. kisutch* (Shelbourn et al., 1992). Dietary administration of E_2 (200 mg E_2/kg) for 131 d induces high sterility in *C. carpio* (Rao and Rao, 1983). For more details on the hormonal induction of sterility in selected fishes, Pandian and Sheela (1995) may be consulted. Piferrer et al. (1994b) have shown that the induction of sterility increases with increasing MT dose in the following order of hormone administration route: dietary administration > continuous immersion > discrete immersion with MT > discrete immersion with MDHT. Since the fully matured ovarian mass constitutes 20–25% of the body mass, it has been considered that in the absence of ovary, the sterile fish may have more muscles. However, the available experimental evidence does not support this hypothesis (Table 48). Incidentally, Hsu et al. (2010) developed a technique to induce male specific sterility in transgenic *eGFP-Ntr* zebrafish. The gene nitroreductase (*Ntr*) is expressed only in the presence of spermatocyte specific Asp, old and sam promoters. In the transgenic line, females are therefore fertile but the males have smaller testes that are devoid of sperm.

Muller (1969) is the first to report evidence for paradoxical sex reversal in tilapia. With increasing doses of MT (1 to 100 mg/l) treatment of *Clarias gariepinus* fry from 14 to 42 dph results in paradoxical sex reversal of upto 90% females (van den Hurk et al., 1989). A similar paradoxical feminization has been reported for cichlids (see Nakamura, 1975), *Ictalurus punctatus* (Goudie et al., 1983), *Oncorhynchus mykiss* (Solar et al., 1984) and *Heteropneustes fossilis* (Tiwary et al., 1998) and *Esox masquinongy* (Rinchard et al., 1999). Piferrer and Donaldson (1991) are the first to trace aromatization of exogenous androgens as the cause for the paradoxical sex reversal. However, aromatization alone may not explain paradoxical feminization. *I. punctatus* has been reported to be susceptible to paradoxical feminization (upto 78%), when its fry are fed from 1 to 21 dph the non-aromatizable 17α-methyldihydrotestoserone (MDHT) at the dose of 60 mg/kg (Davis et al., 1990) or when immersed in aqueous solution containing the non-aromatizable dihydrotestosterone (DHT) at the dose of 200 mg/l (Davis et al., 1992). Iwamatsu et al. (2006b) immersed *Oryzias latipes* fry at increasing doses from 1 ng MDHT/l to 100 ng/l and found 96% males in fry treated at 1 ng/l but paradoxically 36% males in fry treated at 100 ng/l. Hence it remains to be explained how these non-aromatizable chemicals induce paradoxical feminization.

The responsiveness to chemical induction to effect sex reversal differs from species to species (Table 45) suggesting that the pathways through which the introduced chemical inducers of sex reversal differ and the door to one or other pathway may remain closed or it may not exist, resulting in ineffectiveness of the chemical inducer(s). For example, gestating mothers of male heterogametic guppy *P. reticulata* are amenable to reverse the sex of their embryos from males to females (Kavumpurath and Pandian, 1992) or females to males (Kavumpurath and Pandian, 1993b,c). But the treated gestating females of female heterogametic molly *P. sphenops* abort their progenies. On administration of E_2 or DES, the newly hatched fry are feminized (George and Pandian, 1995). However, the fry are not amenable to masculinization by either MT or AST (George and Pandian, 1998). Secondly, the cichlids belonging to the same genus are almost readily amenable for 100% masculinization by either steroid (e.g., *O. mossambicus*, Pandian and

Varadaraj, 1990) or chemicals like fadrozole (e.g., *O. niloticus*, Alfonso et al., 2001), or for 100% feminization (e.g., *O. mossambicus*, Varadaraj and Pandian, 1989). However, the zebra cichlid *Cichalasoma nigrofasciatum* is not amenable to the induction of 100% males by either MT (e.g., George and Pandian, 1996) or TT (Cek et al., 2007a), although the cichlid is amenable for 100% feminization by E_2. At the super-optimal doses or extended durations of treatment by MT, the treated fry suffer 100% mortality rather than increasing the percentage of males above 82% (George and Pandian, 1996).

The response of the silurids to inducers differs from others. In *Ictalurus punctatus*, the treatment with either aromatizable or non aromatizable androgen fails to induce sex reversal from female to male (Davis et al., 1990). Similarly, MT treatment from 5 to 25 dph with 50 mg/kg fails to induce more males in the southern catfish, *Silurus meridionalis* (Liu Z et al., 2007, 2008). The induction by 100 mg Fz/kg or 25 mg Tx/kg is able to induce 56 or 70% males, although a combination of Fz and Tx produces 80% males. Apparently, the door to the pathway, through which MT induces sex reversal from female to male, is closed in these silurids. At gene expression level, *Dmrt1a* and *Dmrt1b* are up-regulated in the gonad and *foxl2* and *cyp19a* are down-regulated, but the expression level of *cyp19b* remains unchanged in Fz-, Tx- and Fz+Tx-treated groups of *S. meridionalis*. The down-regulation of *foxl2* and *cyp19a* is highly correlated with the female to male sex reversal resulting in incomplete formation of ovarian cavity, degeneration and reduction of follicles and appearance of testicular structures. Incidentally, E_2 is also unable to up-regulate *cyp19a* expression.

7.6 Sex reversal and survival

The majority of fishes that have been experimentally induced to reverse sex, has undergone the desired sex reversal. However, a few are not amenable to chemical induction of sex reversal. For example, all the six protocols employed by different authors including the one releasing a maximum of 16.6 µg MT/day for a period of more than 300 d have failed to induce the desired sex reversal in the grasscarp *Ctenopharyngodon idella* (see Hunter and Donaldson, 1983). The tench *Tinca tinca* too is not readily amenable for chemical induction of sex reversal. Dietary treatment of (50-dpf) gynogenic juveniles from 55 to 105 dpf with 150 mg MT/kg or 150 mg Tx/kg or 150 mg MT/kg + 200 mg Tx/kg produces 26, 25 or 13% males, respectively (Hulak et al., 2010). In others like the cichlid *Cichlasoma nigrofasciatum*, in which either MT or TT is unable to induce more than 82 or 87% males (Table 46), attempt to increase the induction of males above 88% in *Poecilia reticulata* by increasing the dose from 300 to 400 mg dimethyltestosterone (DMT)/kg feed results in the death of the treated gravid mothers (Table 46, Kavumpurath and Pandian, 1993c). This is also true of *Betta splendens*, in which the increase

Table 46. Effect of increasing doses of sex reversing inducers on survival after treatment and at sexual maturity of *Cichlasoma nigrofasciatum, Poecilia reticulata* and *Betta splendens*. 0 = dead, – = data not reported (compiled from George and Pandian, 1996*; Cek et al., 2007a**, b*****; Kirankumar and Pandian, 2002***; Kavumpurath and Pandian, 1993c****).

Steroid and Dose	Sex reversal (%)	Survival after treatment (%)	Survival at sexual maturity (%)
	Cichlasoma nigrofasciatum (mg MT/kg)*		
0	49	95	92
5	51	90	82
10	56	87	80
25	60	82	75
50	65	82	72
75	74	77	67
100	77	75	67
200	82	67	55
	C. nigrofasciatum (g TT/l)**		
0.1	80	89	–
0.2	85	90	–
0.3	87	90	–
	C. nigrofasciatum (mg E_2/kg)*		
0	53	97	95
25	70	90	65
50	64	90	55
100	80	82	57
200	100	75	50
300	100	60	50
	Betta splendens (µg MT/l)***		
0	48	99	97
100	55	93	87
200	72	92	83
500	75	90	83
700	82	85	79
900	98	76	71
1000	92	46	33
	Poecilia reticulata (mg DMT/kg)****		
0	54	95	–
100	48	87	–
150	60	83	–
200	65	57	–
250	86	72	–
300	88	73	–
400	0	0	–
	Poecilia reticulata (g TT/l)*****		
0.0	41	82	–
0.05	58	86	–
0.10	60	82	–
0.15	80	86	–

in immersion dose from 900 µg/l to 1000 µg/l reduces the sex reversal from 98 to 92% (Table 46, Fig. 47). The reason for the occurrence of these non-amenable individuals is now traced to mutations. For example, Wang DS et al. (2007) traced the non-amenability to mutations in *O. niloticus*: (i) the disruption of endogenous *foxl2* in inducing varying levels of testicular degeneration in XX females, which may not undergo sex reversal and (ii) the overexpression M_3, a mutation, in XX females leading to low levels of both E_2 and KT. In them, the E_2 level is seven times lower than that of the XX control female but two times higher than that of the XY control male. And the KT level is much lower than that of both XX and XY controls. As a consequence of this unusual opposing sexual tendencies, these mutant individuals are perhaps unable to decisively move into the direction of the male and may become a non-functional intersex.

In many treated fishes, the percentage of surviving individuals after completion of the treatment progressively decreases with increasing dosage. For example, survival of the treated individuals of *C. nigrofasciatum* progressively decreases from 95% in control to 67% in individuals treated at 200 mg MT/kg, a treatment level, at which the highest 82% males are produced (Table 46). The same trend holds for dietary E_2 treatment in oviparous *C. nigrofasciatum* and DMT-treated viviparous *P. reticulata*. It holds also true of MT immersion treated *Betta splendens*. Interestingly, survival of TT-treated oviparous *C. nigrofasciatum* and viviparous *P. reticulata* remains

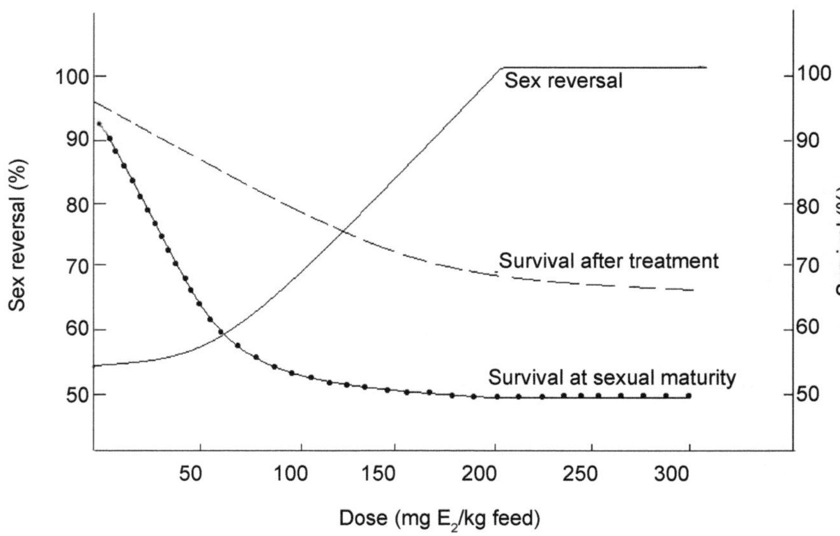

Fig. 47. Free hand drawing to illustrate the effect of estrogen dose on sex reversal (—), survival at after treatment (— —) and at sexual maturity (• • •) of *Cichlasoma nigrofasciatum*. For data see Table 46.

as high as 86–90%, against 67–73% survival of MT-and DMT-treated individuals of these two species. The credit of using extract of *Tribulus terrestris* for the first time to masculinize fishes goes to Dr S Cek of Turkey. The use of the extract of such medicinal herbs to induce sex reversal in fishes may prove effective, cheaper, ensure a higher survival and may leave no residual. Hence this field of research may be rewarding, especially for tropical aquaculturists.

Very few authors, for example the author and his group, have provided information on survival upto sexual maturity of the treated fishes. Survival following treatment and at sexual maturity is 75 and 50% for feminization by E_2 in *C. nigrofasciatum* and 76 and 71% for masculinization by MT in *B. splendens,* respectively (Table 46). The net survival values for these sex reversed fishes have also been calculated considering (i) percentage of sex reversal, survival after (ii) treatment and (iii) at sexual maturity. These values are 38–39% for feminization in *C. nigrofasciatum* and 53% for masculinization in *B. splendens.* At first glance, the calculated values suggest the death of almost all the presumptive males in the feminization process and presumptive females in masculinization process. But both the treated females and males suffer mortality, following the treatment with one or other chemical to induce sex reversal. Progeny testing alone facilitates the estimation of mortality of treated males and females. Incidentally, progeny testing involves one more generation to produce all-male or all-female progenies. Hence it is a time-consuming indirect method. A direct method involving sex specific DNA probe that identifies the genetic sex of the phenotypic XX neomales and XY neofemales has been developed for 14 fish species (see Pandian 2011, 2012). This highly skilled method is in use on a commercial scale in the culture of long-living rainbow trout in UK and Canada (see Piferrer, 2001). But the non-skilled indirect method of progeny testing is more suitable for tropical fishes like the cichlids. The results reported on progeny testing by the author and his group clearly show that the presumptive females in masculinization and presumptive males in feminization processes have withstood and survived the treatment. Table 47 shows the ratios of XX♀ : XY♀ and XX♂ : XY♂ in the estrogen or androgen treated male heterogametic fishes, respectively and ZW♀ : ZZ♀ and ZW♂ : ZZ♂ in the estrogen or androgen treated female heterogametic fishes, in which progeny testing has been undertaken. The tested androgens AST, MT, NET and DMT reduce neomale (XX♂) ratio to 0.42, 0.33–0.44, 0.17 and 0.10 in treated individuals of these species, respectively. DMT inflicts heavy mortality in treated females during sex reversal to neomales (Pandian et al., 1994). E_2 reduces the neofemale (XY♀) ratio, which ranges from 0.31 in *P. reticulata* to 0.44 in *B. splendens*. Strikingly, the homogametic ZZ neofemales of *P. sphenops* suffer heavy mortality and its male ratio is 0.08 only. Briefly, (i) both males and females suffer mortality but unequally,

Table 47. Altered sex ratio of F_1 progenies due to differential sex-dependent mortality of fishes treated to induce sex reversal. Progeny testing identifies the genotype of sex reversed XX neomales and XY neofemales.

Species and reference	Sex ratio	Steroid
Oreochromis mossambicus Varadaraj and Pandian (1989)	0.60 XX♀ : 0.40 XY♀	DES
Cichlasoma nigrofasciatum George and Pandian (1996)	0.50 XX♀ : 0.50 XY♀	E_2
Betta splendens Kavumpurath and Pandian (1993a)	0.56 XX♀ : 0.44 XY♀ 0.60 XX♀ : 0.40 XY♀ 0.57 XX♀ : 0.43 XY♀	E_2 EE_2 DES
Puntius conchonius Kirankumar et al. (2003)	0.54 XX♀ : 0.46 XY♀	E_2
Poecilia reticulata Kavumpurath and Pandian (1992, 1993b)	0.39 XX♀ : 0.61 XY♀ 0.56 XX♀ : 0.44 XY♀ 0.69 XX♀ : 0.31 XY♀	EE_2 DES E_2
O. mossambicus Pandian and Varadaraj (1990)	0.44 XX♂ : 0.56 XY♂	MT
B. splendens Kirankumar and Pandian (2002)	0.33 XX♂ : 0.67 XY♂	MT
B. splendens Kavumpurath and Pandian (1994)	0.62 XX♂ : 0.38 XY♂ 0.70 XX♂ : 0.30 XY♂ 0.20 XX♂ : 0.80 XY♂	AST ET NET
C. nigrofasciatum George and Pandian (1996)	0.33 XX♂ : 0.67 XY♂	MT
Oncorhynchus mykiss Johnstone et al. (1979)	0.15 XX♂ : 0.85 XY♂	MT
P. reticulata Kavumpurath and Pandian (2003b)	0.10 XX♂ : 0.90 XY♂ 0.17 XX♂ : 0.83 XY♂ 0.42 XX♂ : 0.58 XY♂	DMT NET AST
P. sphenops George and Pandian (1995)	0.92 ZW♀ : 0.08 ZZ♀	E_2
O. aureus Melard (1995)	0.27 ZW♀ : 0.63 ZZ♀ 0.46 ZW♂ : 0.54 ZZ♂	E_2 MT

(ii) androgens inflict a higher mortality among the treated females than estrogens among the treated males and (iii) E_2 seems to inflict the highest mortality of homogametic ZZ neomales.

7.7 Growth and reproduction

Growth: As many authors have not undertaken long-term studies on growth of the sex reversed fishes, information on this aspect is scanty and inconsistent. Table 48 lists the available information on accelerated or retarded growth of sex reversed fishes. Induction of sex reversal either by an androgen (MT) or estrogen (E_2) retards growth in *Cichalasoma nigrofasciatum*

Table 48. Growth responses (+ = positive, – = negative) observed for hormonally sex reversed fishes. For references consult Pandian and Sheela (1995). For additional species, references are cited below.

Species	Growth
Salmonidae	
Oncorhynchus gorbuscha	–
O. mykiss	+
O. mykiss	–
O. kisutch	+
O. tshawytscha	+
Cyprinidae	
Cyprinus carpio	+
Cichlidae	
Oreochromis mossambicus	+
O. niloticus	+
O. niloticus	–
O. aurea	+
O. spirulurus	–
Cichlasoma nigrofasciatus[1]	–
Sparidae	
Dicentrarchus labrax[2]	–
Cyprinodontidae	
Oryzias latipes[3]	–
Siluridae	
Silurus asotus[4]	+
Anabantidae	
Betta splendens[5]	–
Centrarchidae	
Lepomis macrochirus[6]	–
Moronidae	
Morone chrysops x *M. saxatalis*	–

1 George and Pandian (1996), 2 Blazguez et al. (1999), 3 Scholz et al. (2000), 4 Kim et al. (2001), 5 Kirankumar and Pandian (2002), 6 Wang et al. (2008)

(George and Pandian, 1996). However, the MT induction in *Oreochromis mossambicus* accelerates growth of the treated males 1 ½–2 times (Varadaraj and Pandian, 1987). But the initially accelerated growth is not sustained subsequently (e.g., *Oncorhynchus tshawystcha,* Johnstone et al., 1978). In some fishes, even the condition factor is negatively affected (e.g., *Misgurnus mizolepis,* Nam et al., 1998, *Stizostedion luciperca*, Zakes and Zakes, 1999). The inconsistent results reported by different authors can be resolved only by long-term studies.

The observations from long-term growth studies by George and Pandian (1998) on *Poecilia sphenops* treated at different MT doses are relevant. The authors have estimated the weight gain as a measure of growth in treated and control males and females from birth to 18 mon. The MT induction accelerates growth during the first 3 mon with increasing doses up to the optimum level at which a highest percentage of sex reversal occurs in both males and females, beyond which the growth acceleration begins to diminish (Fig. 48). Notably, growth is consistently retarded in 18 month-old males and females. Clearly, steroid treatment may accelerate growth in juveniles and the accelerated growth may afford an advantage to escape from predators but with the penalty of retarded growth of adults. From available information on growth of equally aged or sized treated fishes, generalized trends observed for hormonally sex reversed fishes belonging to salmonids (e.g., *O. kisutch,* Goetz et al., 1979), cyprinids (e.g., *Cyprinus carpio,* Rao and Rao, 1983) and cichlids (e.g., McIntosch et al.,1985, observations in the author's laboratory) are drawn; accordingly, the cyprinids show higher growth response of two-three times that of the control, while that for cichlids is one to two times faster. The accelerated growth of hormonally sex reversed cyprinids and cichlids may prove an advantage for tropical aquaculture (Pandian and Sheela, 1995).

Reproduction: In general ovaries are larger as they accommodate large oocytes/ova than the testes with numerous but microscopic spermatocytes/sperm. For some reason, most authors have not described the morphological changes undergone by the gonad of the endocrine sex reversed fishes. When treated with an androgen MT, the roundish ovary of the hybrid bass (*Morone saxatilis* x [*M. chrysops* x *M. saxatilis*] F_1) is transformed to an oval shaped testis, a shape between a roundish ovary and linearly oval testis (Fig. 49). With completion of feminization by E_2 induction, the ovary of *Silurus asotus* also assumes a shape that is intermediate between the ovary and testis. In fact, Kim et al. (2001) provided colored pictures of the different shapes including one with a smaller testicular portion. The paired ovaries of *Perca flavescens* become a single unit, during ontogenetic development, while the testes remain paired. Interestingly, Malison et al. (1986) identified the genetic XX neomales (44%) by the presence of a single testis among the sex reversed males (induced by 60 mg MT/kg) but the XY males (50%) by the paired testes. Similarly, the genetic XY neofemales (45%) were identified by the presence of paired ovaries among the sex reversed females (including by 120 mg E_2/kg) but the XX female (55%) by a single ovary. Following these morphological changes in the gonads of the sex reversed individuals, the GSI values are: 0.25 for the treated XX♀ but 0.05 for the sex reversed XY♀ and 0.02 for the treated XY♂ but 0.07 for the sex reversed XX♂ (Malison et al., 1986). MT induced *Betta splendens* males have GSI of 0.18, against 0.23 for

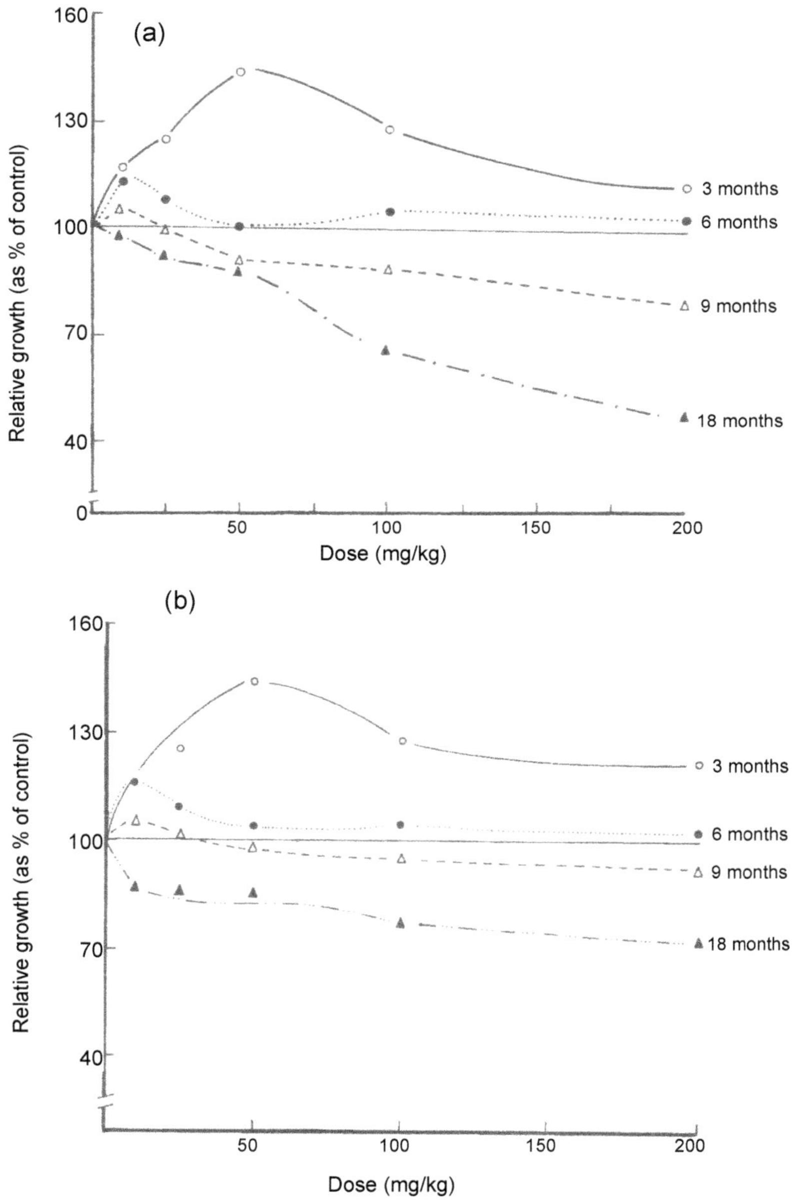

Fig. 48. Growth (as percentage of control) trends of *Poecilia sphenops* expressed as functions of increasing dose and age of (a) males and (b) females treated with 17α-methyltestosterone for 30 d from birth. Data obtained for the different dose groups at the ages of 12 and 15 mon fall between those observed at the ages of 9 and 18 mon (from George, 1995).

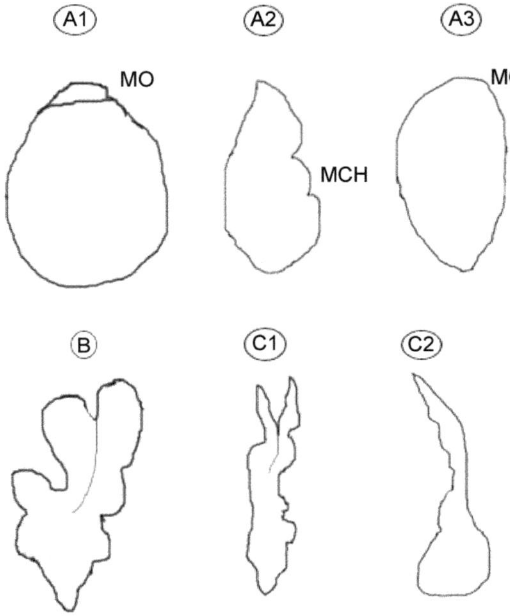

Fig. 49. Free-hand drawings to illustrate the morphological changes undergone by gonads of sex reversed fishes. A1 ovary, A2 testis and A3 ovary/ovotestis of the hybrid bass *Morone saxatalis* x (*M. chrysops* x *M. saxatalis*) (drawn from Gomelsky et al., 1999). MO = mesoovarium, MCH = mesorchium. B ovary/ovotestis of *Silurus asotus* (drawn from Kim et al., 2001). C1 paired and C2 single ovaries of *Perca flavescens* (drawn from Malison et al., 1986).

normal males (Kirankumar and Pandian, 2002). But the sex reversed female of *Oreochromis niloticus* has a GSI of 4.4, against 3.7 for the normal female; as the former spawns more frequently than the latter, the higher GSI of the latter may represent more a compensation (Desprez et al., 1995).

Unfortunately all reviewers, except Pandian and Sheela (1995), are silent on reproductive performance of the sex reversed fishes, perhaps for want of information. Even today, the publications available on this aspect are only a few. But thanks to the groups led by Dr C. Melard and the author, rich data are available on the reproductive performance of at least a few sex reversed fish species from both laboratory and farm. Firstly, natural and synthetic estrogens seem to delay the age at sexual maturity in male heterogametic fishes, by about 20–35 d in *Betta splendens* (see Pandian and Sheela, 1995; Kirankumar and Pandian, 2002) and by 15 d in sex reversed XX, XY and YY females and by seven days in XX males of *Danio rerio* (Fineman et al., 1975). But it advances in the female heterogametics like *Poecilia sphenops* and *O. aureus*.

Many authors have described one or more elements of the courtship behavior of castrated fishes following steroid treatment (e.g., *Platypoecilus maculatus, Cymatogaster aggregate, Hemigchromis bimaculatus,* see Pandian and Sheela, 1995, others listed in Table 22). Briefly, the courtship behavior of the castrated males is not comparable to intact males. The publication by Kirankumar and Pandian (2002) represents among the available literature which describes courting and breeding in a steroid-treated sex reversed fish. *B. splendens* commences courting behavior by building a bubble nest (stage 1) and begins to attract a responsive female by erecting his fins (stage 2). On being accepted, she follows him to the nest (stage 3). The critical stage 4 includes the male embracing the female's body to bring the sexual openings to juxtaposition, and simultaneous spawning and spermiation. The fertilized eggs are then carried to the nest (stage 5). Stages 4 and 5 are repeated as many times as possible. Table 49 lists selected morphological, physiological and behavioral features of control male and MT-treated male. The reductions in each of these features ultimately result in the production and storage of 78 fertilized eggs by a treated male, against 130 eggs by control male. This observation is also supported by the report on the MT-treated *C. nigrofasciatum*; for example, the treated XX and XY males are able to fertilize only 123 and 119 eggs/brood, against 170 eggs fertilized by untreated control XY male (George and Pandian, 1996).

Further, the MT-treated XX and XY *P. reticulata* males are not comparable to the control male, as 86% of the treated males do not participate in mating (Clemens et al., 1966). In *O. aureus*, 40% of the sex reversed ZZ females do not spawn. The inter-spawning intervals in the treated females are prolonged;

Table 49. Reported observations on courtship behavior and reproductive performance of normal and sex reversed males of *Betta splendens* (compiled from Kirankumar and Pandian, 2002).

Parameter	Courtship and Reproduction	
	Control male	Treated male
Age at maturity (day)	110	145
Milting cycle (day/cycle)	4	13
Body length (cm	6	3.5
Dorsal fin (cm)	4	2.8
Caudal fin (cm)	3.5	2.3
Anal fin (cm)	4.2	2.9
Pairing duration stage 4 (sec)	6	3
Completion stage 5 (sec)	19	8
Sperm count (x 10^5/ml)	7.7	5.9
Motility duration (sec)	100	78
Male induced spawning pulse (no)	19	16
Spawned eggs (no)	130	78

for example, from 15 to 20 d in *Puntius conchonius* (Kirankumar et al., 2003) and from 15 to 18 and 22 d in XX and XY E_2-treated *C. nigrofasciatum* females, respectively (George and Pandian, 1996). In *Oncorhynchus tshawystsha*, MT-treated male spermiates 0.25 ml semen, in comparison to 0.38 ml in the normal male (Baker et al., 1998). Sperm count is decreased from 5.9×10^5/ml in the control to 7.7×10^3/ml in the treated *B. splendens* (Kirankumar and Pandian, 2002). Contrastingly, the count in *P. conchonius* increases from 7.2×10^6/ml in the control X^1Y^2 male to 9.2×10^6/ml and 7.5×10^6/ml in the Y^2Y^2 supermale produced by ploidy manipulation and Y^1Y^2 supermale of hormonal manipulation, respectively. Consequently, the Y^1Y^2 supermale fertilizes 90% of the eggs, in comparison to 83 and 93% of eggs by control X^1Y^2 male and Y^2Y^2 supermale (Kirankumar et al., 2003).

Reproductive performance of sex reversed females, as measured by the number of eggs/alevin produced, is drastically affected (Table 50). The reduction in life time egg production ranges from 19–20% in *P. reticulata* and *B. splendens* to 43% in *Oryzias latipes*. In others, the mean reduction for successive broods is 31% for *C. nigrofasciatum* and 33% for a single

Table 50. Reproductive performance of sex reversed female fishes.

Species, Reference	Geno-type	Steroid	Egg/alevin (no)	Remarks
Oryzias latipes[1]				(mm)
Fineman et al. (1975)	XX♀		260	1.21
	XX♀	E_1	155	1.20
	XY♀	E_1	165	1.24
	YY♀	E_1+	150	1.17
Betta splendens[1]				
Kavumpurath and Pandian (1994)	XX♀		518	
	XY♀	E_2	415	
Poecilia sphenops[1]				
George and Pandian(1996)	ZW♀		231	
	ZZ♀	E_2	188	
Cichlasoma nigrofasiatum[2]				
George and Pandian (1996)	XX♀		170	
	F_1XX♀	E_2	143	
	F_1XY♀	E_2	118	
	F_2XX♀		65	
Puntius conchonius[3]				
Kirankumar et al. (2003)	XX♀		90	
	XY♀	E_2	70	
Oreochromis aureus[4]				
Desprez et al. (1995)	ZW♀		302	
	ZZ♀	E_2	155	

1 Life time egg production, 2 Mean of first 3 broods, 3 First brood only, 4 Alevin produced during 3–4 months in a fish farm

brood of *B. splendens*. In fish farms, the reduction in alevin production by ZZ female is 49% E_1 and EE_2 are more potent than E_2 in reducing the egg production in treated females. Incidentally, the steroid induction may also induce changes in levels of vitellogenesis, as indicated by differences in egg diameter of E_1–treated *O. latipes*. When values for survival, egg production and hatching success are considered together, the cumulative loss of egg production amounts to 65% eggs/female in *P. conchonius*. Remarkably, the first brood of the untreated F_2XY female drawn from the treated F_1 XY female *C. nigrofasciatum* produces 65 eggs only. It is not known how the negative trait of E_2 treatment is transmitted to the untreated F_2XY female.

Data presented in Table 51 are more interesting for two reasons: (i) this is the only information available on the female-heterogametic *O. aureus* and (ii) the information also originates from a fish farm. The consistent reductions in the considered parameters of reproductive performance of sex reversed fishes confirm the results reported from laboratory experiments. Briefly, endocrine treatment (i) inflicts considerable mortality after treatment and until sexual maturity, (ii) delays sexual maturity, (iii) extends inter-spawning and inter-spermiation interval, (iv) alters courtship behavior, (v) reduces sperm count, and their mobility and (vi) egg/progeny production. Hence it is desirable to establish broodstocks that may continuously produce all-male or all-female progenies.

Table 51. Reproductive performance of control (ZW) and sex reversed (ZZ) females of *Oreochromis aureus* in laboratory* and farm** (compiled from Desprez et al., 1995).

Parameter	ZW♀	ZZ♀
Spawning frequency (times)*	0–9	0–4
Females not spawning (%)*	20	40
Delay in first spawning (day)*	6	13
Inter-spawning interval (day)*	20	40
Relative fecundity (no/g fish)*	309	548
Relative fecundity (no/kg/spawning)*	1643	1668
GSI*	3.7	4.4
Stocking sex ratio**	2.1–3.1♀ : 1♂	3♀ : 1♂
Breeder (no/m^2)**	2.28	1.82
Mean spawning (times/4 months)**	3	1
Total alevin (no)**	92,430	42,750
Alevin/♀**	302	155
Alevin/kg♀**	1208	442
Alevins/m^2/day**	15.4	7.1
Male ratio of alevins**	54	91

7.8 Monosex culture

As they grow, many fishes undergo remarkable sexual dimorphism, which is a feature of economic importance (Dunham, 1990). Gender-dependent growth, faster growth to marketable size, meat taste and texture, and body shape and color pattern, especially in ornamental fishes, are all important for value addition and marketing of fishes. Monosex culture may facilitate (i) reduction in the number of broodstock required to produce a targeted number of progenies, (ii) the sex that grows faster and attains marketable size and (iii) to prevent undesired frequency of spawning by elimination of one sex. Table 52 lists some of these features that are relevant in monosex aquaculture of commercially important fishes. Five different techniques are employed to achieve monosex progenies: 1. Manual sorting, 2. Hybridization, 3. Ploidy manipulation, 4. Hormonal manipulation and 5. A combination of ploidy and hormonal manipulations. The last three techniques have received many inputs from research and the first one involves almost no technical input. Manual sorting of sexes is laborious and prone to mistakes. It is costlier, time-consuming and wasteful, as the fish must grow to a size large enough to distinguish sexes before

Table 52. Desired monosex culture in selected food and ornamental fishes (compiled from many sources including Gnani and Levavi-Sivan, 2009).

Fishes	Remarks
All—female progenies	
Salmonids	Males mature earlier and have lower flesh quality than females
Common carp	At equal age females are 15% heavier than males
European sea bass	Females grow 20–50% faster than males
Atlantic halibut	Females reach marketable size earlier than males
Striped bass	At age 2+, females are heavier than males
Thai silver bass	Females grow faster and reach marketable size earlier than males
Czech tench	Females grow significantly faster than male and attain the ultimate larger size
Acipenserids	Females alone produce expensive caviar
All—male progenies	
Tilapias	Males grow faster and monosex culture eliminates unwanted frequent spawning
Channel catfish	Male weighs 25% more than female at marketable size of 500 g
Black crappie	Males grow faster (?) and monosex culture is to eliminate unwanted prolific breeding and stunted growth
Fighter fish	Males are more attractive with more beautiful hood-like tails
All sterile progenies	
Grass carp	Introduction of exotics to control aquatic vegetation
Black carp	Introduction of exotics to control snails, intermediary hosts human parasites

manual separation and culling of the undesired sex. The process involves considerable stress and sometimes may result in death of the fish. Regarding hybridization and ploidy manipulation to achieve monosex progenies and the latter involving induction of gynogenesis or androgenesis, Pandian (2011) may be consulted.

Regarding hormonal or hormonal cum ploidy manipulations to produce monosex progenies, the following must be noted: (a) About 67% of fishes are male heterogametic and the remaining are female heterogametic (Pandian, 2011). (b) Viviparous fishes are not amenable to ploidy manipulation. These features of fishes demand changes in protocols of hormonal manipulations to achieve all male or all female progenies. In male heterogametics, estrogen treatment of embryos, alevins, fry or juveniles (e.g., *Betta splendens*, Kavumpurath and Pandian, 1993a) or gravid female (e.g., *Poecilia reticulata*, Kavumpurath and Pandian, 1993b) yields F_1 XX neomales: When crossed with normal XX females, XX neomales produce all-female progenies. However, the production of all-male progenies is slightly complicated and may require an additional generation for progeny testing. The androgen treatment heterogametic male produces F_1 25% XX♀ : 50 XY♂ : 25 XY♂ progenies. The progeny testing indentifies YY♂, which when crossed with a normal XX♀ produces F_3 XY male progenies. Induction of gynogenesis of F_1 estrogen treated neofemales yields F_1 50% YY males, and the protocol eliminates the need for progeny testing at F_2. The manipulation at every generation has proven to be laborious, cumbersome and costlier. Hence manipulations have been aimed at producing a broodstock of males and females, which when crossed yields all male, all female or all sterile progenies alone.

In some fishes like *Oreochromis mossambicus* (Varadaraj and Pandian, 1989 see also Pandian, 1993) and *Oncorhynchus tshawytscha* (Devlin et al., 2001), Y^1Y^2 males do produce 2–8% unexpected female progenies. Using a DNA specific molecular marker, Kirankumar et al. (2003) showed that these unexpected female phenotypes are indeed genetic males. To eliminate the occurrence of such unexpected progenies, Muller-Belecke and Horstgen-Schwark (2007) generated Y^1Y^2 superfemale and crossed them with Y^1Y^2 supermale *O. niloticus* (Fig. 50). The protocol to produce absolutely cent percent male progenies requires seven generations. To produce all female or all male progenies in female-heterogametic fishes, the reverse protocols have to be employed. Figures 50, 51 illustrate the protocols that are to be employed to produce all male or all-female progenies in male and female hererogametic fishes. It may be indicated that protocol 2 shown by broken lines in Fig. 50 and 51 remain experimentally proved, as the number surviving mature YY supermale and WW superfemale are too few.

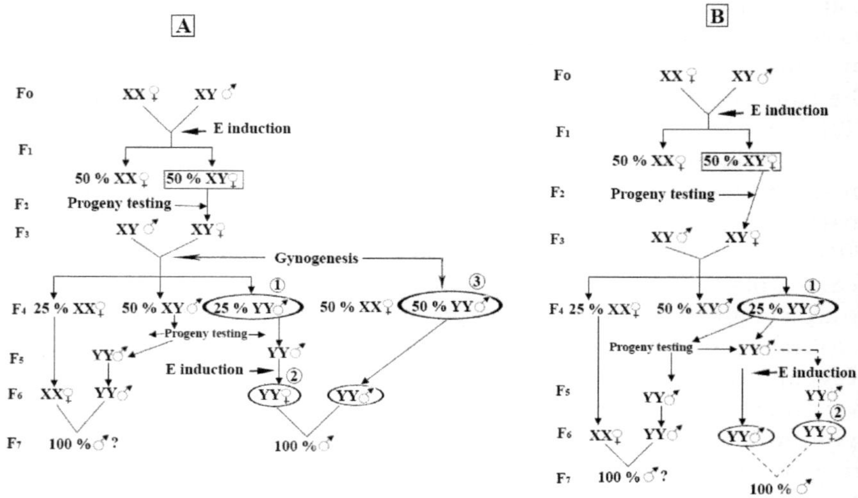

Fig. 50. Experimented and suggested (broken lines in B) protocols for 100% male production in oviparous (A, left panel) and viviparous (B, right panel) fishes characterized by male heterogamety. With induction of gynogenesis (in protocol 3) the oviparous fishes are amenable for 50% YY male production at F_2. Due to overriding autosomes, the cross between YY male and XX female produces XY female phenotypes. Estrogen induction (E induction) and production of YY female and cross YY female and YY male produces 100% males with total elimination of autosomal effect (compiled from Varadaraj and Pandian, 1989; Kirankumar et al., 2003 and Muller Belecke and Horstgen-Schwark, 2007).

Table 53 lists the authors, who are the first to generate supermale and superfemale in fishes. Despite considerable efforts, it has not been possible to generate live supermale *B. splendens* (Kavumpurath and Pandian, 1994), as it does not survive (George et al., 1994). The same is true of WW female, although ZZ superfemales of *O. aureus* (Melard, 1995) and *P. sphenops* (George and Pandian, 1995) are known to survive and reproduce. A reason for the inviability of Y^1Y^2 supermales and W^1W^2 superfemales is traced to their homozygozity. Re-analysis of the data reported on surviving F_2 male progenies of the cross between XY♀ and XY♂ indicates the following $Y^1Y^2 : X^1Y^2$ ratios : 0.2 Y^1Y^2 ♂ : 0.8 X^1Y^2 ♂ and 0.33 Y^1Y^2♂ : 0.67 X^1Y^2 ♂ for the EE_2 and E_2-induced *P. reticulata,* respectively (Kavumpurath and Pandian, 1992, 1993a), 0.3 Y^1Y^2 ♂ : 0.7 X^1Y^2 ♂ for the E_2 induced *O. mossambicus* (Varadaraj and Pandian, 1989), and 0.33 Y^1Y^2 ♂ : 0.67 X^1Y^2 ♂ for the E_2 induced *Puntius conchonius* (Kirankumar et al., 2003), against the expected ratio of 0.33 Y^1Y^2 ♂ : 0.67 X^1Y^2 ♂. Clearly, EE_2 seems to inflict heavier mortality in Y^1Y^2 males.

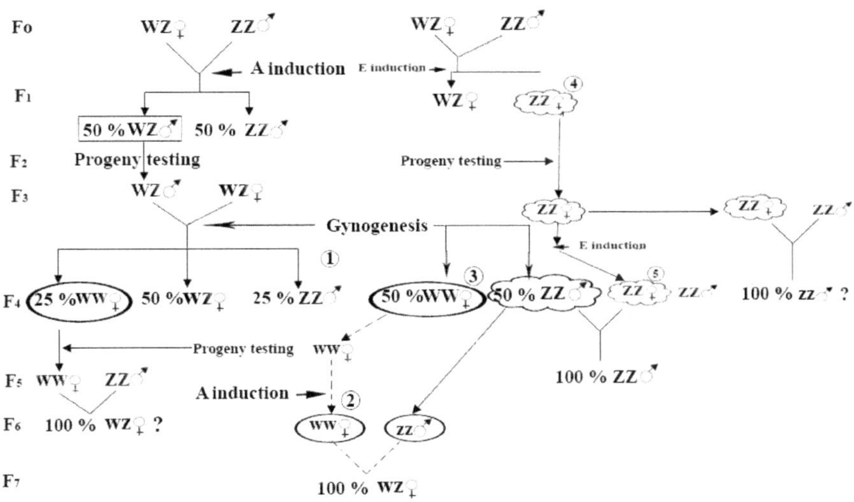

Fig. 51. Experimented (→) and suggested (broken arrow) protocols for 100% female or 100% male production in oviparous and viviparous fishes characterized by female heterogamety with induction of gynogenesis (in protocol 3), the oviparous fishes are able to produce 50 % WW females in F_2. Due to overriding autosomes, the cross between WW♀ and ZZ♂ may produce WZ female phenotypes. A (androgen) induction and production of WW male and cross between WW♂ and WW♀ produces 100% females with total elimination of autosomal effects. It is also possible to produce ZZ females in F_3 (protocol 4, George and Pandian, 1995) and pure ZZ female in F_5 (protocol 5, Melard, 1995).

Table 53. Generation of supermales and superfemales in fishes.

Species	Author(s)
Male heterogametics : Y^1 Y^2 supermale	
Oryzias latipes	Yamamota (1955)
Carassius auratus	Yamamota (1975)
Oreochromis mossambicus	Varadaraj and Pandian (1989)
O. niloticus	Scott et al. (1989)
Oncorhynchus tshawytscha	Devlin et al. (1991)
Poecilia reticulata	Kavumpurath and Pandian (1992)
Ictalurus puntatus	Davis et al. (1992)
Misgurnus anguillicaudatus	Arai et al. (1995)
M. mizolepis	Nam et al. (2001)
Puntius conchonius	Kirankumar et al. (2003)
Gymnocorymbus ternetzi	David and Pandian (2006a)
Hemigrammus caudovittatus	David and Pandian (2006b)
Male heterogametics : Y^1 Y^2 superfemale	
O. latipes	Fineman et al. (1975)
P. reticulata	Kavumpurath and Pandian (1993c)
I. punctatus	Davis et al. (1992)
O. niloticus	Noh et al. (2002)
Female heterogametics : Z^1 Z^2 superfemale	
O. aureus	Melard (1995)
P. sphenops	Geroge and Pandian (1995)

7.9 Residues and elimination

Natural and synthetic steroids have been used in animal husbandry to enhance feed conversion. Dietary administration of anabolic steroids in aquaculture has been practiced since 1970s (e.g., Nirmala and Pandian, 1983). Subsequently, hormones have been applied to induce sex reversal. The safety of this technique is of concern, as consumption of the steroid may lead to health hazards. There is a paucity of publications on steroid uptake and elimination in the hormone-treated fish. A reason for this paucity can be traced to the fact that the residues present in the treated juveniles and marketable adults are too low (e.g., 2 ng MT in *Oreochromis niloticus*, Phelps and Okoko, 2011) to be hazardous to human health (Pandian and Kirankumar, 2003).

The species subjected to residual analysis are limited to a few salmonids, cichlids and cyprinids and the hormones tested are also limited to T, MT and mibolerene (MB). Most investigations describe the pattern of uptake, organ-wise distribution and elimination of the administered labeled or unlabeled steroid. Three techniques have been used to evaluate residual levels of steroids by (i) high performance liquid chromatography (HPLC), (ii) indirect estimation of radioactivity, assumed to reflect the mass of steroid and bioassay of monitoring, (iii) induction of sexual activity in castrated rats fed on the carcass of steroid-treated fish. Rothbard et al. (1990) recognized that the HPLC technique cannot detect residual steroid of < 50 ng/g fish; likewise the bioassay technique of Das et al. (1990) is a useful tool in studying the effect of residual steroid but cannot quantify the residual level in the fish.

The uptake of labeled steroid, for example MT, is very fast (Table 54); within 2 hr after administration, 80–95% of the (MT) isotope is accumulated in gall bladder, pyloric caeca and posterior intestine (Fagerlund and Dye, 1979; Lone and Matty, 1981; Piferrer and Donaldson, 1994). The uptake and elimination are slower in embryos than in alevins of *Oncorhynchus kisutch* (Piferrer and Donaldson, 1994). However, 80–95% of the isotope is eliminated within a day after the withdrawal of steroid administration (Johnstone et al., 1983). A comparison of retention duration for T and M in *O. kisutch* (Fagerlund and Dye, 1979) and *Cyprinus carpio* (Lone and Matty, 1981) reveals that MT is retained at higher concentration in the gall bladder for a relatively longer period. Consequently, the residual steroid levels are at 0.1 ng MT/g *O. tshawytscha* on the 10th day (Fagerlund and Dye, 1979), 3.89 ng T/g *C. carpio* (Lone and Matty, 1981) and 5 ng MT/g *Oreochromis aureus* on the 21st day (Goudie et al., 1986) after the withdrawal of steroid administration (Table 54). A tilapia egg has as much as 3 ng T and an equal amount of E_2; hence a tilapia having 1 g ovarian tissue with 200 eggs may contain more than 600 ng of these steroids in its gonad (Rothbard et al.,

Table 54. Estimation of uptake, organ-wise disposition, and elimination of steroids administered to induce sex reversal in fish (T-Testosterone; MB-Mibolerone; MT-17α-Methyltestosterone).

Species, age/size, reference	Steroid and treatment protocol	Reported observations
Oncorhynchus kistuch 30 g Fagerlund and McBride (1978)	^{3}H T; 3-day isotope feeding followed by 10-d analysis	> 1% residue on the 4th day of depuration
O. kisutch 30 g Fagerlund and Dye (1979)	^{3}H MT; 3-d isotope followed by 10-d observation	Identification and quantification of isotope in 16 organs within 2 hours of feeding; 86% residue in gall bladder-pyloric caeca-intestine complex; 4.5% in muscles; on the 10th day 0.1 ng MT residue/g fish
O. mykiss alevins Johnstone et al. (1983)	fed 30 mg MT/kg feed for 21 d; dosed with ^{3}H MT for 1 d; analysis for 400 hr following the dosing	> 99% ^{3}H MT is eliminated within 4 d
O. mykiss Cravedi et al. (1989)	^{14}C MT; dosed for a day	95% of ingested isotope is absorbed within 24 hr; 67% excreted via gills, 22% via feces and 0.5% via urine. Bile accumulates 200–2,000 times more isotope than other organs
Oreochromis mossambicus 8 mg fry Johnstone et al. (1983)	MT fed 37 d after hatching; dosed with ^{3}H MT for 1 d; analysis for 400 hr following the dosing	100% elimination from tilapia carcass within 50 hr
O. aureus fry Goudie et al. (1986)	^{3}H MT and ^{14}C MT; dosed at 30 µg MT/g feed for 21 d followed by 21 d analysis	Identification and quantification of the isotope in 12 organs within 1 hr of feeding; 90% activity in viscera; 5 ng isotope residue/g fish on the 20th day
O. niloticus 10-day-old fry Curtis et al. (1991)	^{3}H MT feeding followed by 1-d closing and analysis until 10th day	Identification and quantification 97% residues as polar metabolites especially in bile; 100% elimination on 10 d but parent from of ^{3}H MT detectable even on the 10th day
Cyprinus carpio 3-day-old fry Das et al. (1990)	15 mg MB/kg diet fed for 30 d; bioassay of residue by feeding treated fish to castrated rats	No detectable sexual activity in the castrated rats
C. carpio 10-day-old fry Rao et al. (1990)	50 mg MT/kg diet fed for 45 d. T assayed in liquid scintillation counter	No detectable residual activity
C. carpio 60 g Lone and Matty (1981)	^{3}H T; 10 mg ^{3}H T fed for 12 d followed by 12-d analysis	Identification and quantification of ^{3}H T in 14 organs within 1 hr of feeding; after isotope feeding 91.5% in lower kidney and spleen; 3.8 ^{3}H T residue/g fish on the 20th day

1987). Hence, the estimated residual steroids of less than 5 ng/g in juvenile fishes is too low to cause any concern or hazard to humans, especially as the level is likely to decrease, as the fish grow to marketable size.

Steroid elimination is rapid initially but gradually gets stabilized. The rate at which the residues are eliminated depends on species, steroid potency, organ and treatment protocol (see Pandian and Kirankumar, 2003). In *O. mossambicus* and *O. mykiss,* the residue level is decreased to 9% within 50 hr of administration. In fact 99% ingested MT is eliminated within 4 d in *O. mykiss* (Johnstone et al., 1983). However, 90–99% of the administered hormone is eliminated within a day of administration, which must be taken more seriously, especially in the light of findings by Gomelsky et al. (1994) that metabolites of MT in the culture water are more potent to induce sex reversal than dietary MT. The calculated environmental input resulting from feminization of 100,000 fish with E_2 during 30 d can be 40 mg/d including the parent compound and its metabolites. For comparison, the daily release to the aquatic environment of the same hormone and its metabolite from one pregnant woman is in the range of 20–40 mg (see Piferrer, 2001). Hence the use of steroids in aquaculture system may pose no serious environmental problem. However the use of DES must be prohibited, especially in Asian countries, as it is proven to be carcinogenic for humans.

8

New Hypotheses

As indicated elsewhere, no other vertebrate taxa offers as much diversity and plasticity as teleost fishes for the study of sex determination and differentiation. Understandably, voluminous literature is available on this theme. From the point of sex determination and differentiation, this treatise series has continuously re-examined the results reported on I hermaphroditism and II allogenesis/xenogenesis including (i) gonadectomy, (ii) endocrine sex reversal and (iii) morphotypes (Pandian, 2010, 2011, 2012, this volume). The following hypotheses are the outcome of the analysis of the relevant literature:

1. *Hermaphroditic fishes are indeed functional bisexuals:* Among fishes, postponement of dimorphic sexual differentiation is not uncommon. For example, the differentiation occurs prior to hatching in *Oryzias latipes* but is postponed to post-hatching in *Oreochromis niloticus* and to as far as puberty in *Mugil cephalus* (Fig. 44). The trend to commence as all-females and 'sex change' by presumptive males either at alevin/fry stage, as in *Danio rerio,* or at puberty, as in *Dascyllus flavicaudus* is typical of secondary gonochores (Pandian, 2012). Hermaphrodites are shown to have originated from secondary gonochores (Pandian, 2012) and have arisen independently in more than 10 lineages (Warner, 1978) each lineage may be at a different geological time scales (Frisch, 2004). The following is the suggested course of origin and evolution of some of these lineages. Sub-optimal and super-optimal steroid doses are known to induce higher proportion of testicular and ovarian tissues, respectively (see p. 192). It is likely that some of these gonads became functional in hermaphrodites, with either male-active or female-active hermaphrodites. Hence the Okinawans and

Marians (for explanation, see Pandian, 2010) seem to have been the earliest among the hermaphrodites. They have different proportions of both ovarian and testicular tissues in their gonads. Being solitaries, the Marian hermaphrodites act either as male or female; but there is a single male or only a male-active hermaphrodite in a colony of Okinawan hermaphrodites. The need to ensure recombination to generate and sustain genetic diversity has been the 'driving force' to generate males and females. Selection pressure on this driving force has been intense that the hermaphrodites includes 68% (recalculated values) protogynics, 20% protandrics and 4% serials. Notably, these hermaphrodites act as either male or female at a specific ontogenetic stage (Pandian, 2012). Among 8% or 14 simultaneous hermaphrodites, 13 exchange gametes. In the self-fertilizing hermaphroditic *Kryptolebias marmoratus* too, 5% primary males (Harrington, 1967) and 25% secondary males (Tataronkov et al., 2009) are generated. In this hermaphrodite, 2–3% cross fertilization is made by the secondary males. Among the gamete exchanging simultaneous hermaphrodites, adult hermaphrodites like *Serranus baldwini* and *S. fasciatus* drop the ovarian tissues, as it happens at the alevin/fry and puberty stages in *D. rerio* and *D. flavicaudus*, respectively and these serranid males maintain harems each with a half dozen hermaphrodites. Indeed all the hermaphroditic fishes are potential hermaphrodites but are functionally gonochores. Figure 52 shows that the secondary gonochores, Okinawan hermaphrodites, simultaneous and self-fertilizing hermaphrodites, monandric protogynic hermaphrodites and serials like *Gobiodon histrio* commence as females/female-active hermaphrodites and undergo natural sex change to males/male active hermaphrodites; the frequency of males increases from 5–25% in *K. marmoratus* to 22–31% in protogynics (Pandian, 2010). The intense need for the male to generate genetic diversity through recombination has led protogynics to generate primary and secondary males, and serials to generate bisexual in a single individual by cyclically changing sex. In fact, the need to ensure recombination to generate and sustain genetic diversity is also true of unisexuals. The strategy of adopting, paternal leakage, paternal genome replacement by hybridogenesis and/or genome addition, as in *Poecilia formosa* and *Poeciliopsis* spp is inadequate. Males do appear in them but rarely. Some unisexuals have sterile males (e.g., *Carassius auratus*). Some like *Squalius alburnoides* and *Cobitis granoei taenia* do have functional males.

2. *All hermaphrodites undergo natural sex change*: the fact that all hermaphrodites are potential hermaphrodites but are functional gomochores implies that all hermaphrodites have to undergo natural sex change. However, simultaneous hermaphrodites were earlier considered not to undergo natural sex change. Sequentials

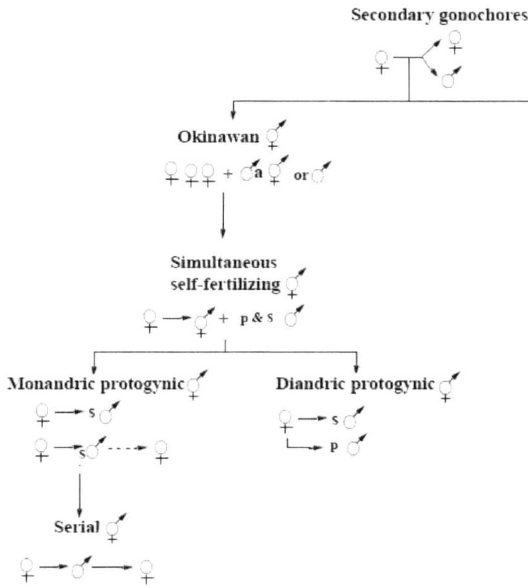

Fig. 52. The suggested course of origin and evolution of a hermaphroditic lineage. Note: In all of them sex differentiation commences from an ovarian gonad. p♂ = primary ♂, s♂ = secondary ♂, ♂a ⚥ = male active hermaphrodite, ♀a ⚥ = female active hermaphrodite, ⚥ female = hermaphrodite functioning as female, ⚥ male = hermaphrodite functioning as male. Broken lines indicate new report.

earlier were also considered to undergo natural sex change only once in one direction (i.e., from female to male and male to female) during their life time. This book has brought to light that (i) self-fertilizing *K. marmoratus* are hatched as females, undergo natural sex change to hermaphrodites and are capable of sex reversal to males, depending upon the habitat's temperature, (ii) among the dozen gamete exchanging simultaneous hermaphrodites, a couple of them undergo natural sex change to males and maintain about half a dozen hermaphrodites in the harem, and (iii) many sequentials are reported to undergo sex change more than once in either direction (Table 35, Fig. 37). Essentially, all hermaphroditic fishes are capable of undergoing natural sex change perhaps in either direction; only the number of times the hermaphrodite undergoes sex change is not frequent among simultaneous hermaphrodites, is minimum among sequentials and is more frequent among serials. Hence grouping the hermaphrodites under the three recognized patterns namely (i) simultaneous, (ii) sequential and (iii) serial may be required for a basic understanding but in terms of direction and times of sex change, the groupings may be fallacious.

3. *PGCs are required for sex change in sequential hermaphrodites*: Table 55 summarizes fragmentary information available on the presence of PGCs in sequential hermaphrodites. Brusle (1983) is perhaps the first to publish useful information on the appearance, features and potential functions of PGCs in the protogynic *Serranus hepatus*. The accumulation of PGCs in the ovarian zone and their possible migration to the testicular zone in delimited type of ovotestis is notable (Fig. 35). It is still to be known whether the failure of migration of the PGCs into the testicular zone causes fractions of *Pagrus pagrus* and *Epinephelus itajara* population to permanently remain as females (see p. 151). Apparently, PGCs with their bisexual potency are required to facilitate sex change in sequential hermaphrodites.
   More interesting is the castrated *Acanthopagrus schlegeli*, still left with the ovarian zone of the delimited ovotestis (Fig. 35) is able to successfully differentiate and develop its ovary. Following the removal of sex change-inhibiting testicular part of the ovotestis, the ovarian

Table 55. Reported observations on the appearance of PGCs in the gonads of hermaphroditic fishes.

Species and Reference	Reported observation
	Protogynics
Serranus hepatus Brusle (1983)	PGCs appear among oogonia in ovary and spermatogonia in testis. They are detected at the junction of ovary and testis and some of them migrate from the ovarian zone to the testicular zone. They exhibit the same ultrastructural features in both ovary and testis suggesting that they both retain bisexual potency. The SSCs and OSCs originate from the PGCs but the origin of PGCs is not traceable. The importance of ovarian tissue is reflected in its earlier maturation, larger size and transfer of PGCs to form SSCs in testis
Coris julis Brusle (1988)	Presence of a single type of bipotential PGCs is detected in the differentiated testis
Reinboth and Brusle-Sigard (1997)	PGCs play an active role in sex change
Epinephelus microdon Brusle-Sigard et al. (1992)	Presence of single type of bipotential PGCs is detected
E. coioides Yao et al. (2007)	PGCs are detected in the germinal epithelium of the transitional gonad
Halichoeres maculatus Kojima et al. (2008)	SGCs are detected in the transitional IP gonad
	Protandrics
Amphiprion frenatus Brusle-Sigard et al. (1994)	Simultaneous presence of PGCs, SSCs and OSCs are detected in the transitional gonad
A. polymnus Rattaunayuvakorn et al. (2006)	Presence of PGCs is detected

zone possessing PGCs successfully undergoes gonadal sex change (Wu et al., 2008a). Indeed the removal of testis from the ovotestis is analogous to the removal of the dominant inhibitor male from a harem! Contrastingly, the gonadectomized *Thalassoma bifasciatum,* which has suffered the total loss of PGCs from its undelimited gonad, is unable to regenerate either testis or ovary. Hence the changes observed by Semsar and Godwin (2003, see also Godwin, 2009) are limited to behavioral sex change alone. Apparently, the storage of PGCs is restricted to the gonads alone in the brain non-sexualized hermaphrodites. This is quite a contrast to the brain sexualized primary gonochore *Ctenopharyngodon idella,* which is able to regenerate the gonads of its primary sex, even after the removal of its entire gonad and surrounding mesentery (Table 38, see also Le Page et al., 2010; Pandian, 2012). Briefly, the presence of PGCs is obligately required to transdifferentiate the ovary or testis in the sex changing hermaphrodites.

As described elsewhere, with the loss of the ovary, *aromatase* and PGCs due to apoptosis in the presumptive male zebrafish, the pre-disposed germ cells supporting somatic cells (GCSSCs) induce masculinization. Hence estrogenic, i.e., protogynic sex reversal is possible but not androgenic, i.e., protandric sex reversal in this secondary gonochore (see p. 139). While confirming this conclusion, experimental evidence from a series of ovariectomized secondary gonochores including *Carassius auratus* indicates the possible brain sexualization and retention of PGCs/SSCs in organs other than the ovary/gonad in these fishes (Table 12, 22, 27). Thus there seems to be a continuum from complete non sexualization of brain and retention of PGCs in gonads alone in zebrafish to brain sexualization and retention of PGCs in organs other than gonads in primary gonochores.

4. *Primary gonochores retain PGCs in organs other than gonads and their brain is sexualized*: From their neuroendocrine studies mostly on the secondary gonochore *Danio rerio,* Le Page et al. (2010) proposed that non-sexualization of the brain is the root cause for the plasticity in sex differentiation in fishes. Considering their proposal, Pandian (2012) gathered evidence from seemingly unrelated areas to show that the proposal of Le Page et al. may hold true for secondary gonochores and hermaphrodites but not for primary gonochores. In primary gonochores, the brain is sexualized prior to puberty. This new hypothesis is supported by the following evidences: (i) The ontogenetic transformations from one phenotype like the primary male to female or female to male among hermaphrodites are intersexual, clearly indicating that the brain of hermaphrodites remain non-sexualized. In contrast, the morphotypes of gonochores that display alternative mating strategy (AMS), undergo intrasexual ontogenetic changes only,

suggesting brain sexualization in them. (ii) Gonadectomized adult fishes like *Oreochromis niloticus* and *Oncorhynchus nerka* (Table 38) regenerate the testis alone after castration or the ovary only following ovariectomy. That the regeneration of the testis and ovary following surgical removal of the entire gonads along with mesentery in *C. idella* following castration and ovariectomy, respectively, clearly reveals that the PGCs and/or their derivatives SSCs or OSCs are stored also in organs other than gonads and the entire organ sytems including the brain of adult primary gonochores are sexualized. Fragmented evidences have also become available in support of the hypothesis; the adult organ systems including the brain of gonochore are sexualized. For example, *Dmy* is expressed not only in the testis but also in the entire body of *O. latipes* (see Fig. 3B of Kobayashi et al., 2004). There is a clear sexual dimorphism in the muscles of adult *Poecilia reticulata* (Shen et al., 2007). At least one skeletal trait, the size of the opercular bones is mapped to the sex chromosomes of *G. aculeatus* (Kimmel et al., 2005). These evidences suggest that the entire body in *O. latipes*, muscles in *P. reticulata* and opercular bones of *G. aculeatus* are sexualized.

5. *Loss of bisexual potency by germ cells supporting somatic cells (GCSSCs) in gonochores:* Shinomiya et al. (2002) generated chimeras of medaka *O. latipes* by transplanting male (XY)- and female (XX)- specific PGCs into blastulae committed to differentiate into XX and XY genotypes, respectively. From these experiments, they concluded that it is the GCSSCs that determine sex of medaka progenies. Based on their studies on endocrine sex reversal in juvenile fishes, Nakamura et al. (2003) concluded that both PGCs and GCSSCs retain their bisexual plasticity. However, the view of Nakamura et al. is not consistent with the experimental study showing that the presence of adequate number of PGCs is obligately required for the maintenance of sexual dimorphism and in the absence of PGCs, the pre-disposed GCSSCs induces production of male progenies alone (Kurokawa et al., 2007). Understandably, sex is initially determined by the intimate 'cross talk' and interplay between PGCs and GCSSCs (Pandian, 2011).

   The generation of embryonic stem cells (ESCs) is limited to embryos in mammals. But it is extended upto the adult stage in fishes, or at least prior to adults attaining reproductive senescence, especially in females (Pandian, 2012). Not surprisingly, mature adult fishes are reported to retain PGCs (e.g., *Liza auratus*, Brusle and Brusle, 1978, *Xiphophorus* sp. Flores and Burns, 1993, *Fusconimus mus*, Moser, 2007). The histological investigation by Shibata and Hamaguchi (1988) in *O. latipes* suggests the loss of bisexual potency of OSCs. However, it is not clear whether this is related to sex determination by *DMY* alone prior to hatching. For OSCs of *O. mykiss* (Yoshizaki et al., 2010) and *D. rerio* (Wong et al., 2010), on transplantation to alevins/blastulae, are able to produce both male and female progenies (see Pandian, 2012), clearly indicating that the OSCs have not lost their bisexual potency. This is also true of SSCs. However, the transplantation of SSCs into sterilized adult testis of *O. niloticus* (Lacerdo et al., 2006, 2010) and *Odontesthes bonariensis* (Majhi et al., 2009) produces male progenies alone, as the GCSSCs in these primary gonochores have lost the bisexual potency. Another evidence is reported from transplantation of ovarian follicles. Csenki et al. (2010) transplanted ovarian follicles from YEP strain of zebrafish, which are known to retain bisexual potency (Knaut et al., 2002), into the differentiated ovary of foster females of AB strain. The surviving follicles in the foster females developed into fertilizable eggs but not into sperm (see Pandian, 2012), clearly indicating that GCSSCs in the differentiated ovary of the foster females have already lost by sexual potency.

Dietary administration of high doses of androgen to the gravid female having differentiated ovary and gestated embryos with undifferentiated gonad provides a unique opportunity to test whether or not the gonadal differentiation induces the loss of bisexual potency of GCSSCs. Having administered selected doses of natural and synthetic androgens to the gravid female *P. reticulata* during the labile period, Kavumpurath (1992), Kavumpurath and Pandian (1992, 1993b) found the masculinization of progenies but not the gravid females. Attempts to hormonally sex reverse the adults of *Clarias gariepinus* (Swapna and Senthilkumaran, 2009) and *Gobiocypris rarus* (Liao et al., 2009) all met with failure. In a sensationally titled publication, Ogawa et al. (2008) also came to the conclusion that the bisexual potency in *Cypris carpio* as in *O. niloticus* is retained only upto puberty. Similarly, Knapp et al. (2012) also reported the development of gonopodium only but not the gonadal sex reversal in *P. reticulata*. Receiving a daily injection of T for over nine weeks, the female *Betta splendens* is reported to display morphological features and aggressive behavior typical of a male and, ovarian degeneration. However, the development of a functional testis was not ascertained by Badura and Friedman (1988). Treatment with high doses of androgen for an extended duration induces sterility in males of salmonids and cyprinids. Efforts to regenerate testis by milder doses of androgens in sterile or sterility-induced fishes have not succeeded (Shelbourn et al., 1992) indicating that once the PGCs are lost, there is no way of regaining them. Taken together, all these evidences from hormonal administration experiments clearly indicate that GCSSCs of primary gonochores have lost their bisexual potency prior to the fish attaining sexual maturity.

From the foregone comprehensive analysis on endocrine and social mediated sex change, and experimented evidence from surgical removal of the testis or ovary as well as those from allogenesis and genogenesis emerges a picture of sex differentiation in fish (Fig. 53). Based on the evidence on the endocrine sex reversal, it is clear that all fishes amenable for endocrine induced sex reversal during pre-pubertal stages only. Following gonadal sex differentiation, adult gonochores are not readily amenable to sex change, while the hermaphrodites have retained that capacity. In primary gonochores, the PGCs retain bisexual potency but not the GCSSSCs. Consequently, ovariectomy or castration results in the regeneration of original sex namely the ovary or testis in adults of primary gonochores. However, the secondary gonochores seem to have lost the bisexual potency of both PGCs and GCSSCs; consequently, social or endocrine mediated sex change in the adults and surgical removal of the ovary or testis results in sex change from female to male but not female to male (see p. 138–140). On the other hand, hermaphrodites have retained the bisexual potency of PGCs and GCSSCs (see p. 147-148). However, the difference in structural organization of the gonad in the form of ovotestis or ovary/testis has endowed them with different bisexual potency. Surgical removal of the testicular zone from the ovotestis of *A. schlegeli* facilitates the precocious development of the ovary and it is likely that the reverse may also be possible. The loss of the entire basket of PGCs from the ovary or testis in *T. bifasciatum,* following castration or ovariectomy stands as a clear evidence for the loss of PGCs and the obligate requirement of PGCs to regenerate testis or ovary.

6. *Multiple pathways and multiple control*: The entire process of sex differentiation inclusive of sex change is regulated through multiple pathways and is also controlled at multiple levels. Considering the expression of the ubiquitous aggressive behavior in fishes, as an example, aggression is regulated through seven pathways (Filby et al., 2010, see Fig. 30) including the HPG-axis, which involves the presence of the testis. In the absence of testis, females of *Centropyge ferrugata* inhabiting branching harems are aggressive against each other (Sakai and Kohda, 1997). Obviously, they employ other than the HPG-axis. Apart from this, castration in males of many gonochoric species decreases the expression levels of aggression (see Table 24), clearly showing an impotant role for the HPG-axis. On the other hand, the castration in another half a dozen species evokes no negative effect indicating no role for the HPG-axis in regulation of aggression in these species. It is known that KT is not synthesized in the testis but it is in the liver of *C. gariepinus* and blood cells of *G. aculeatus* (see p. 133). It is likely that in species, in which castration does not invoke a negative

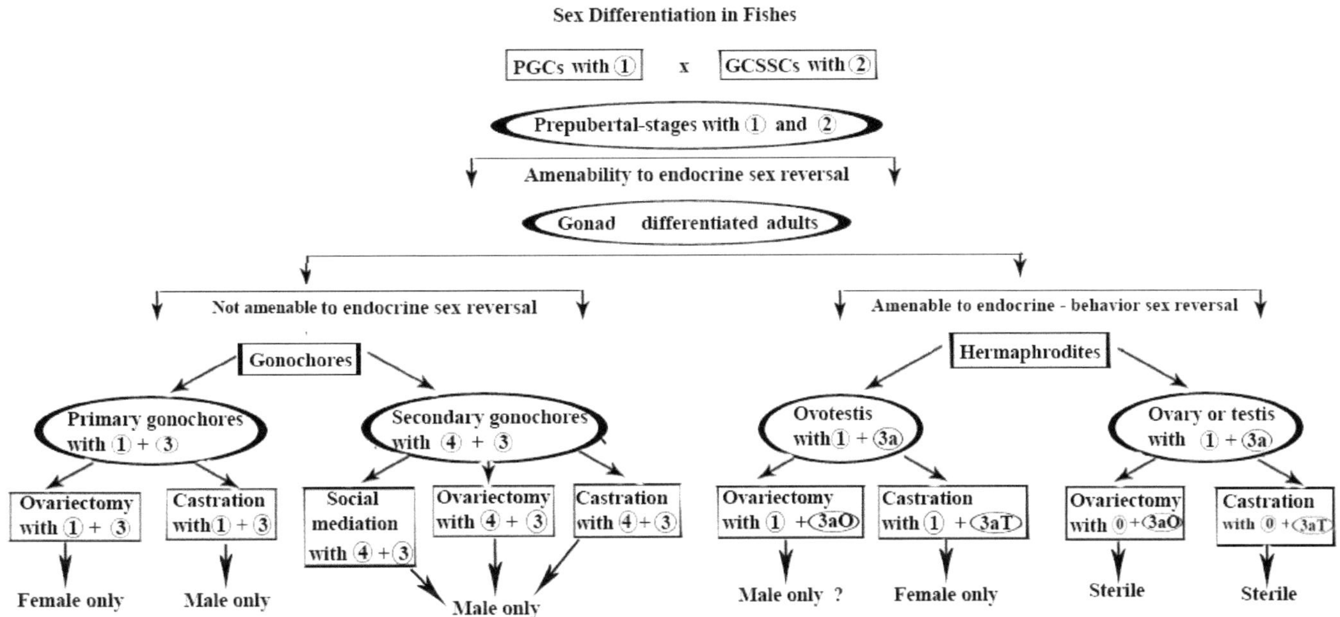

Fig. 53. Proposed pathways of sex differentiation in fishes. (1) = PGCs with bisexual potency, (2) = GCSSCs with bisexual potency, (3) = GCSSCs with unisexual potency, (3a) = GCSSCs with bisexual potency, (3aT) = GCSSCs with unisexual potency without testis of ovotestis, (3aO) = GCSSCs with unisexual potency without ovary of ovotestis, (4) = PGCs with unisexual potency (?), (0) = without PGCs.

effect on aggressive behavior, KT is synthesized in organs other than the testis. The pathways through which aggression is expressed may also be routed through KT synthesized in different organs.

7. *Genetic control of sex differentiation:* There are three hypotheses, the central ideas of which are interrelated: (i) Yamamoto's postulation that the sex steroids are the inducers of natural sex differentiation of male and female, (ii) Nakamura's conviction that *aromatase* and its enzyme cause feminization and their absence results in masculinization and (iii) A waging debate whether sex steroids are the 'cause' or 'result' of a specific event in the process of sex differentiation. In view of its implication to the other two hypotheses, Section 2.5 considered Yamamoto's hypothesis in the light of experimental evidence, especially those of Iwamatsu and Rougeot and their respective colleagues. From their investigations on *O. niloticus* (Nakamura and Nagahama, 1985), *O. rhodurus* (Nakamura and Nagahama, 1993), *O. mykiss* (Hines et al., 2009; Govoroun et al., 2001), many authors found that steroidogenic enzymes are present prior to histological sex diffentiation. These findings support those of Yamamoto's hypothesis. On the other hand, a series of experiments undertaken by others, especially Iwamatsu and his team on *O. latipes*, the fish with which Yamamoto experimented to propose his hypothesis, have brought strong evidence against the hypothesis. Having traced the maternal origin of E_2 in the eggs from steroidogenic follicular cells, Iwamatsu et al. (2006b) indicated the presence of 'a sex determinant' in the oocyte, which is sensitive or labile to sex steroid and manipulability of sex prior to fertilization. For example, injection of 5 ng EE_2/egg of *O. latipes* induces sex reversal in genetic male (Hano et al., 2005). From their studies on *Perca fluviatalis*, Rougeot et al. (2007) have suggested that the maternally derived endogenous T, KT and E_2 of the perch eggs may serve as precursors to induce initial sex differentiation. To initiate the cascade of sex differentiation, these natural steroid precursors await the trigger from 'a sex determinant', i.e., one or more genes. Clearly, sex differentiation is induced by the sex determining gene(s) and the maternally-derived one or another endogenous steroid may accordingly prime and initiate the cascade of sex differentiation.

Triggered by the corresponding gene(s) towards the ovarian differentiation, the brain is the first aromatization site in fishes, as it develops earlier than gonads (see Piferrer and Guiguen, 2008). In fact *cyp19a1b* is one of the earliest transcribed genes in a developing fish embryo (Sawyer et al., 2006). The potential for whole embryos to convert androgen to estrogen has been detected in *O. mykiss* (Yeoh et al., 1996a,b) and *Salvelinus alpinus* (Khan et al., 1997). Many studies have shown that estrogens are the major endocrine factors in deciding the destiny of sex (e.g., Bhandari et al., 2003, 2004, 2005), as synthesis of estrogens is required not only to initiate but also to maintain the ovarian differentiation. Hence it is the considered view of Nakamura et al. (2003) that just the depletion of E_2 alone is sufficient to initiate testicular differentiation. The E_2 depletion can be induced by aromatase inhibitors (AI) like fadrozole, or by anti-estrogenic chemicals (e.g., tamoxifen), which inhibit aromatase before or after the appearance of steroid producing cells (Fig. 54). Suzuki et al. (2004) reported for the first time that the endogenous estrogen is obligately required for the formation of ovarian cavity in *O. latipes*; this is strong evidence in support of the hypotheses of Yamamoto and Nakamura. However, the initial ovarian ontogeny varies with species, for example, the initial ovarian differentiation is indicated by the oogonial entry into meiosis in *O. latipes, Cichlasoma dimerus* and *Anguilla anguilla* but by the appearance of ovarian lumen in *Paralichthys olivaceus, P. lethostigma, O. niloticus* and *O. aureus* (Luckenbach et al., 2003).

Notably, the very first sign of sexually dimorphic differentiation in *O. latipes* occurs at the time point of proliferation of PGCs and their entry into meiosis (see Pandian, 2012); these events are not controlled by E_2 (see Table 40). Secondly, examining the role played by sex steroids in sex changing *Amphiprion melanopus*, Godwin and Thomas (1993) concluded that E_2 is the 'result' rather than a 'cause', as an increase in E_2 is not seen until the transition to female is completed. Hence an event in the sex differentiation

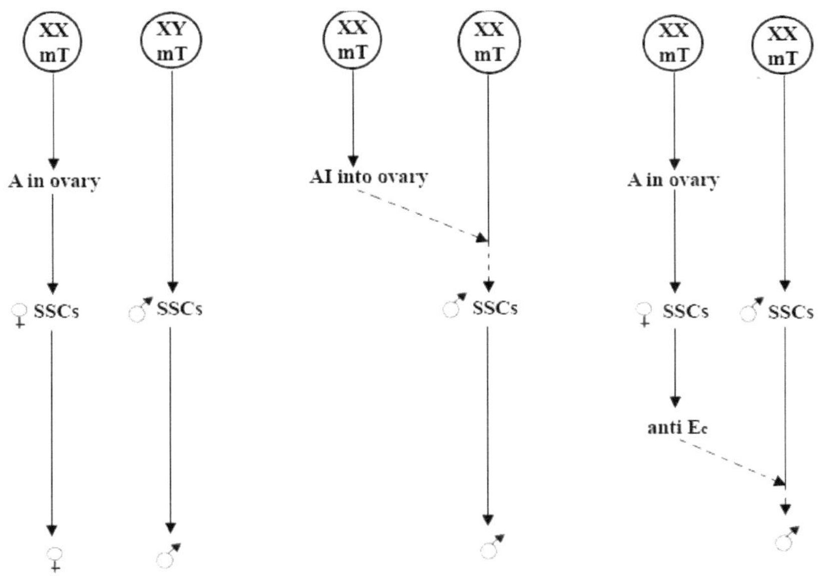

Fig. 54. Suggested cascades of sex differentiation and time point of interventions by aromatase inhibitors (AI) and anti-estrogenic chemicals (anti-E_c) during the cascade of ovarian and testicular differentiation. mT = maternal testosterone, SSCs = steroidogenic cells (compilation based on Nakamura et al., 2003 and others).

process, such as, oogonial entry into meiosis and formation ovarian lumen/cavity are controlled by different sets of genes; the one controlling oogonial entry into meiosis may be expressed prior to the formation of the ovarian cavity, as in *O. latipes*, or after the formation of the cavity, as in *P. olivaceus*. This explanation may resolve the debate on whether sex steroids are the 'cause' and 'result' of a specific event in sex differentiation. The proposed hypothesis of genetic control of all events in sex differentiation including (i) sex steroids as natural sex inducers, (ii) automatic testicular differentiation on E_2 depletion and (iii) steroids: 'cause' or 'result' may prove more useful.

Another good example is the control of dominant behavior through androgen level or androgen level through dominant behavior (see Fig. 55). It implies that the pathways of gene expression on dominant behavior may be routed through the androgen level or on the androgen level through dominant behavior. Removal of a harem master/mistress or housing two males/females of hermaphroditic species initiates a cascade of events commencing from behavioral to endocrine changes. Similarly, gene(s) responsible for a specific event in sex differentiation may be expressed on morphological differentitation through steroids or on steroids through morphological differentiation.

Fig. 55. Suggested pathways through which genes may act on dominant behavior/morphological differentiation through androgens/steroids and the reverse of the same.

References

Adams, B.A., E.D. Vickers, C. Warby et al. 2002. Three forms of gonadotropin-releasing hormone including a novel form in a salmonid *Coregonus clupeoformis*. Biol Reprod, 67: 232–239.

Agulleiro, M.J., V. Anguis, J.P. Canavate et al. 2006. Induction of spawning of captive-reared Senegal sole (*Solea senegalensis*) using different administration methods for gonadotropin-releasing hormone agonist. Aquaculture, 257: 511–524.

Aizen, J., I. Meiri, I. Tzchori et al. 2005. Enhancing spawning in the grey mullet (*Mugil cephalus*) by removal of dopaminergic inhibition. Gen Comp Endocrinol, 142: 212–221.

Ajithkumar, T.T. and T. Balasubramanian. 2009. Broodstock development, spawning and larval rearing of the false clownfish *Amphiprion ocellaris* in captivity using estuarine water. Curr Sci, 97: 1–4.

Ajithkumar, T.T., S.K. Setu, P. Murugesan and T. Balasubramanian. 2010. Studies on captive breeding and larval rearing of clownfish, *Amphiprion sebae* (Bleeker, 1853) using estuarine water. Indian J Mar Sci, 39: 114–119.

Ajithkumar, T.T., K.V. Dhaneesh, R. Vinoth et al. 2012. Hatchery production of the clownfish *Amphiprion nigripes* at Agati Island, Lakshadweep, India. J Environ Biol, 33: 623–628.

Akhtar, N. 1984. Anesthesia, abdominal surgery, efficacy of ovariectomy and subsequent androgen treatments in inducement of sex reversal in *Tillapia nilotica*. Ph.D. Thesis, Auburn University, Auburn, USA.

Alavi, S.M.H., M. Psenicka, J. Policar et al. 2009. Sperm quality in male *Burbus barbus* L fed on different diets during spawning season. Fish Physiol Biochem, 35: 683–693.

Alfonso, L.O., G.J. Wassermann and R.T. De Oliveira. 2001. Sex reversal in Nile tilapia (*Oreochromis niloticus*) using a nonsteroidal aromatase inhibitor. J Exp Zool, 290: 177–181.

Allen, G.R., J. Drew and D. Fenner. 2010. *Amphiprion pacificus*, a new species of anemonefish (Pomacentridae) from Fiji, Samoa and Wallis Island. Internatl J Ichthyol, 16: 129–138.

Allsop, D.J. and S.A. West. 2003. Constant relative age and size at sex change in sequentially hermaphroditic fish. J Evol Biol, 16: 921–929.

Allyn, M.L., R.J. Sheelan and C.C. Kohler. 2001. The effects of capture and transportation stress on white bass semen osmolarity and their alleviation via sodium chloride. Trans Am Fish Soc, 130: 706–711.

Almeida, F.F.L., C. Kristoffersen, G.L. Taranger and R.W. Schulz. 2008. Spermatogenesis in Atlantic cod (*Gadus morhua*): a novel model of cystic germ cell development. Biol Reprod, 78: 27–34.

Alves, M.J., M.M. Coelho and M.J. Collares-Pereira. 1998. Diversity in the reproduction modes of females of the *Rutilus alburnoides* complex (Teleostei, Cyprinidae): A way to avoid the genetic constraints of uniparentalism. Mol Biol Evol, 15: 1233–1242.

Alves, M.J., M.M. Coelho, M.I. Prospero and M.J. Collares-Pereiro. 1999. Production of fertile unreduced hybrid sperm by hybrid males of the *Rutilus alburnoides* complex (Teleostei, Cyprinidae): An alternative route to genome tetraploidization in unisexuals. Genetics, 151: 277–283.

Alves, M.J., M.J. Collares-Pereira, T.E. Dowling and M.M. Coelho. 2002. The genetics of maintenance of an all male lineage in *Squalius alburnoides* complex. J Fish Biol, 60: 649–662.

Alves, M.J., M. Gromicho, M.J. Collares-Pereira et al. 2004. Simultaneous production of triploid and haploid eggs by triploid *Squalius alburnoides* (Teleostei: Cyprinidae). J Exp Zool, 301A: 552–558.

Amer, M.A., T. Miura, C. Miura and K. Yamaguchi. 2001. Involvement of sex steroid hormones in the early stages of spermatogenesis in Japanese huchen (*Hucho perryi*). Biol Reprod, 65: 1057–1066.

Amiri, B.M., M. Macbayashi, S. Adachi and K. Yamaguchi. 1996. Testicular development and serum sex steroid profiles during annual sexual cycle of the male sturgeon hybrid the bester. J Fish Biol, 48: 1039–1050.

Amole, N. and S. Unniappan. 2009. Fasting induces preproghrelin mRNA expression in the brain and gut of zebrafish, *Danio rerio.* Gen Comp Endocrinol, 161: 133–137.

Appelt, C.W. and P.W. Sorensen. 2007. Female goldfish signal spawning readiness by altering, when and where they release a uninary pheromone, Anim Behav, 74: 1329–1338.

Arabaci, M., I. Diler and M. Sari. 2004. Induction and synchronization of ovulation in rainbow trout *Oncorhynchus mykiss* by administration of emulsified busserelin (GnRHs) and its effects on egg quality. Aquaculture, 237: 475–484.

Arai, K., M. Ikeno and R. Suzuki. 1995. Production of androgenetic diploid loach *Misgurnus anguillicaudatus* using sperm of natural tetraploids. Aquaculture, 137: 131–138.

Arai, K., K. Taniura and Q. Zhang. 1999. Production of second generation progeny of hexaploid loach. Fish Sci, 65: 186–192.

Argue, B.J. and R.A. Dunham. 1999. Hybrid fertility, introgression and backcrossing in fish. Rev Fish Sci, 7: 137–195.

Arslan, T. and R.P. Phelps. 2004. Production of monosex male black crappie *Pomoxis nigromaculatus* populations by multiple androgen immersion. Aquaculture, 234: 561–573.

Arukwe, A. 2008. Steroidogenic acute regulatory (StAR) protein and cholesterol side-chain cleavage (P450sec)-regulated steroidogenesis as an organ-specific molecular and cellular target for endocrine disrupting chemicals in fish. Cell Biol Taxicol, 24: 527–540.

Arukwe, A. and A. Goksoyr. 2003. Egg shell and egg yolk proteins in fish: hepatic proteins for the next generation: oogenetic, population, and evolutionary implications of endocrine disruption. Comp Hepatol, 2: 4 doi:10.1186/1476-5926-2-4.

Astaurino, J.F., L.A. Sorbera, M. Carrillo et al. 2001. Reproductive performance in male European sea bass (*Dicentrarchus labrax* L.) fed two PUFA-enriched experimental diets : a comparison with two males fed a wet diet. Aquaculture, 194: 173–190.

Atz, J.W. 1964. Intersexuality in fishes. pp. 145–232. *In:* C.N. Armstrong and A.J. Marshall [eds.]. Intersexuality in Vertebrates including Man. Academic Press, London, UK.

Aubin-Horth, N., C.R. Landry, B.H. Letcher and H.A. Hofmann. 2005. Alternative life histories shape brain gene expression profiles in males of the same population. Proc R Soc Lond B, 272: 1655–1662.

Babin, P.J., O. Carnevali, E. Lubzens and W.J. Schneider. 2007. Molecular aspects of oocyte vitellogenesis in fish. pp. 39–76. *In:* P.J. Babin, J. Cerda and E. Lubzens [eds.]. The Fish Oocyte: From Basic Studies to Biotechnological Applications. Springer, Dordrecht, The Netherlands.

Badura, L.L. and H. Friedman. 1988. Sex reversal in female *Betta splendens* as a function of testosterone manipulation and social influence. J Comp Physchol, 102: 202–268.

Baker, I.J., I.I. Solar and E.M. Donaldson. 1998. Masculinisation of Chinook salmon (*Oncorhynchus tshawytscha*) by immersion treatment using 17α-methyltestosterone around the time of hatching. Aquaculture, 72: 359–367.

Balamurugan, K. and S. Haider. 1998. Partial purification of maturation-promoting factor from catfish, *Clarias batrachus*: identification as the histone H1 kinase and its periodic activation. Comp Biochem Physiol, 120C: 329–342.

Balasubramani, T. and T.J. Pandian. 2008a. Endosulfan suppresses growth and reproduction in zebrafish. Curr Sci, 94: 883–890.
Balasubramani, T. and T.J. Pandian. 2008b. Effect of norethindrone on masculinisation of fighting fish. Curr Sci, 95: 1446–1453.
Balinsky, D.L., W.V. King and T.I.J. Smith. 2005. The use of luteinizing hormone releasing hormone analogue for ovulation induction in black sea bass (*Centropristis striata*). Aquaculture, 250: 813–822.
Balment, R.J., W. Lu, E. Weybourne and J.M. Warne. 2006. Arginine vasotocin a key hormone in fish physiology and behaviour: a review with insights from mammalian models. Gen Comp Endocrinol, 147: 9–16.
Balsano, J.S., E.J. Randle, E.M. Rasch and P.J. Monaco. 1985. Reproductive behaviour and maintenance of all female *Poecilia*. Env Biol Fish, 12: 257–263.
Balsano, J.S., E.M. Rasch and P.J. Monaco. 1989. The evolutionary ecology of *Poecilia formosa* and its triploid associate. pp. 277–299. *In:* G.K. Meffe and F.F. Snelson Jr. [eds.]. Ecology and Evolution of Live Bearing Fishes (Poeciliidae). Prentice Hall Englewood Cliffs, New Jersey, USA.
Banet, A.I. and D.N. Reznick. 2008. Do placental species abort offspring? Testing an assumption of the Trexler-De Angelis model. Funct Ecol, 22: 323–331.
Barbosa, M. and A.E. Magurran. 2010. Guppies control offspring size at birth in response to differences in population sex ratio. Biol J Lin Soc, 100: 414–419.
Baroiller, J.F. and A. Toguyeni. 1996. Comparative effects of a natural androgen 11β-hydroxy androstenedione and a synthetic androgen 17α-methyltestosterone on the sex ratios of *Oreochromis niloticus*. Third Internatl Symp Tilapia in Aquaculture, ICLARM Conf Proc 41: 238–245.
Baron, D., J. Montfort, R. Houlgatte et al. 2007. Androgen-induced masculinisation in rainbow trout results in a marked dysregulation of early gonadal gene expression profiles. BMC Genomics, 8: 357–doi:10.1186/1471-2164-8-357.
Baron, D., R. Houlgatte, A. Fostier and Y. Guiguen. 2008. Expression profile of candidate genes during ovary-to-testis trans-differentiation in rainbow trout masculinised by androgens. Gen Comp Endocrinol, 156: 369–378.
Basavaraja, N., B.U. Chandrashekara and R.H. Ahamed. 2012. Production of all-population of guppy *Poecilia reticulata* Schneider through application of an aromatase inhibitor. (MS reviewed).
Bass, A. 1992. Dimorphic male brains and alternative reproductive tactics in a vocalizing fish. Trends Neurosci, 15: 139–145.
Bass, A.H. 1993. From brain to behaviour: Hormonal cascades and alternative mating tactics in teleost fishes. Rev Fish Biol Fisher, 3: 181–186.
Basu, D., A.K. Navneet, S. Dasgupta and S. Bhattacharya. 2004. Cdc2-Cyclin B-induced G2 to M transition in perch oocytes is dependent on Cdc25. Biol Reprod, 71: 894–900.
Bell, A.M. 2001. Effects of endocrine disrupter on courtship and aggressive behavior of male three spined stickleback *Gasterosteus aculeatus*. Anim Behav, 62: 775–790.
Bender, N., D. Heg, I.M. Hamilton et al. 2006. The relationship between social status, behavior, growth and steroids in male helpers and breeders of a cooperatively breeding cichlid. Horm Behav, 50: 173–182.
Bender, N., Z. Heg-Bacher, R.F. Oliveira et al. 2008. Hormonal control of brood care and social status in a cichlid fish with brood care helpers. Physiol Behav, 94: 349–358.
Bentivegna, F. and F. Benedetto. 1989. Gonochorism and seasonal variations in the gonads of the labrid *Symphodus (Grenilabrus) ocellatus* (Forsskal). J Fish Biol, 34: 343–348.
Berglund, A., G. Rosenqvist and I. Svensson. 1986. Reversed sex roles and parental energy investment in zygotes of two pipefish (Syngnathidae) species. Mar Ecol Prog Ser, 29: 209–215.
Bergmuller, R. and M. Taborsky. 2005. Experimental manipulation of helping in a cooperative breeder-helpers "pay to stay" by pre-emptive appeasement. Anim Behav, 69: 19–28.

Bergmuller, R., D. Heg and M. Taborsky. 2005. Helpers in a cooperatively breeding cichlid stay and pay or disperse and breed, depending on ecological constraints. Proc R Soc Lond B, 272: 325–331.

Berishivili, G., H. D'Cotta, J.F. Baroiller et al. 2006. Differential expression of IGF-1 mRNA and peptide in the male and female gonad during early development of a bony fish, the tilapia *Oreochromis niloticus*. Gen Comp Endocrinol, 146: 204–210.

Berkowitz, P. 1937. Effect of oestrogenic substances in *Lebistes reticulatus* (Guppy). Proc Soc Exp Med, 36: 416–418.

Bernet, D., T. Wahli, S. Kipfer et al. 2009. Macroscopic gonadal deviations and intersex in developing whitefish *Coregonus lavaretus*. Aquat Biol, 6: 1–13.

Bernstein, V.J., J.R. Woldman and W.E. Benis. 1997. Sturgeon Biodiversity and Conservation. Kluwer Academic Publishers, Dordrecht, The Netherlands.

Beulbens, K., E.H. Eding, P. Gilson et al. 1997. Gonadal differentiation, intersexuality and sex ratios of European eel (*Anguilla anguilla* L.) maintained in captivity. Aquaculture, 153: 135–150.

Bhandari, R.K., H. Komuro, S. Nakamura et al. 2003. Gonadal restructuring and correlative steroid hormone profiles during natural sex change in protogynous honeycomb grouper (*Epinephelus merra*). Zool Sci, 20: 1399–1404.

Bhandari, R.K., H. Komuro, M. Higa and M. Nakamura. 2004. Sex inversion of sexually immature honeycomb grouper (*Epinephelus merra*) by aromatase inhibitor. Zool Sci, 21: 305–310.

Bhandari, R.K., M.A. Alam, M. Higa et al. 2005. Evidence that estrogen regulates the sex change of honeycomb grouper (*Epinephelus merra*), a protogynous hermaphrodite fish. J Exp Zool, 303A: 497–503.

Billard, R. 1969. La spermatogenese de *Poecilia reticulata*. 1. Estimation du nombre de generations goniales et rendement de la spermatogenese. Ann Biol Anim Bioch Biophys, 9: 251–271.

Billard, R. 1992. Reproduction in rainbow trout: sex differentiation, dynamics of gametogenesis, biology and preservation of gametes. Aquaculture, 100: 263–298.

Biran, J., S. Ben-Dor and B. Levavi-Sivan. 2008. Molecular identification and functional characterization of the kisspeptin/kisspeptin receptor system in lower vertebrates. Biol Reprod, 79: 776–786.

Black, M.P., J. Battnazart, M. Baillien and M.S. Grober. 2005a. Socially induced and rapid increases in aggression is inversely related to brain aromatase activity in a sex changing fish *Lythrypnus dalli*. Proc R Soc Lond B, 272: 2435–2440.

Black, M.P., B. Moore, A.V.M. Canario et al. 2005b. Reproduction in context: Field testing a laboratory model of socially controlled sex change in *Lythrypnus dalli* (Gilbert). J Exp Mar Biol Ecol, 318: 127–143.

Black, M.P., J. Balthazarat, M. Baillien and M.S. Grober. 2011. Rapid increase in aggressive behaviour preceeds the decrease in brain aromatase activity during socially mediated sex change in *Lythrypnus dalli*. Gen Comp Endocrinol, 170: 119–124.

Blazquez, M., P.T. Bosma, E.J. Fraser et al. 1998. Fish as models for the neuroendocrine regulation of reproduction and growth. Comp Biochem Physiol, 119C: 345–364.

Blazguez, M., M. Carillo, S. Zanury and F. Piferrer. 1999. Sex ratios of offspring of sex reversed sea bass and the relationship between growth and phenotypic sex differentiation. J Fish Biol, 55: 916–930.

Bobe, J., T. Nguyen and B. Jalabert. 2004. Targeted gene expression profiling in the rainbow trout (*Oncorhynchus mykiss*) ovary during maturational competence acquisition and oocyte maturation. Biol Reprod, 71: 73–82.

Bobe, J., J. Montfort, T. Nguyen and A. Fostier. 2006. Identification of new participants in the rainbow trout (*Oncorhynchus mykiss*) oocyte maturation and ovulation processes using cDNA microarrays. Reprod Biol Endocrinol, 4: 39.

Bobe, J., S. Mahe, T. Nguyen et al. 2008. A novel functional and highly divergent sex hormone-binding globulin that may participate in the local control of ovarian functions in salmonids. Endocrinology, 149: 2980–2989.

Bogart, M.H. 1987. Sex determination: A hypothesis based on steroid ratios. J Theor Biol, 128: 349–357.

Bogerd, J., M. Blomenrohr, E. Andersson et al. 2001. Discrepancy between molecular structure and ligand selectivity of a testicular follicle-stimulating hormone receptor of the African catfish (*Clarias gariepinus*). Biol Reprod, 64: 1633–1643.

Bonnet, E., A. Fostier and J. Bobe. 2007. Microarray-based analysis of fish egg quality after natural or controlled ovulation. BMC Genomics, 8: 55 doi:10.1186/1471-2164-8-55.

Borg, B. 1987. Stimulation of reproductive behavior by aromatizable and non-aromatizable androgens in the male three-spine stickleback *Gasterosteus aculeatus*. pp. 269–271. *In:* S.O.K. Kullander and B. Fernholm [eds.]. Proc 5th Cong Europe Icthylogists. Swedish Museum of Natural History, Stockholm, Sweden.

Borg, B. 1994. Androgens in teleost fish. Comp Biochem Physiol, 109C: 219–245.

Borowsky, R. and K.D. Kallman. 1976. Patterns of mating in natural populations of *Xiphophorus* (Pisces: Poeciliidae). 1. *X. maculatus* from Beleize and Mexico. Evolution, 30: 693–706.

Bownds, C., R. Wilson and D.J. Marshall. 2010. Why do colder mother produce larger eggs? An optimality approach. J Exp Biol, 213: 3796–3801.

Bradford, C.S. and M.H. Taylor. 1987. Semilunar changes in estradiol and cortisol coincident with gonadal maturation and spawning in the killifish *Fundulus heteroclitus*. Gen Comp Endocrinol, 66: 71–78.

Bradley, K.M., J.P. Breyer, D.B. Melville et al. 2011. An SNP based linkage map for zebrafish reveals sex determination loci. Genes/Genomes/Genetics, 1: doi 10.1534/93.111.000190.

Brantley, R.K., J.C. Wingfield and A.H. Bass. 1993. Sex steroid levels in *Porichthys notatus* a fish with alternative reproductive tactics and a review of the hormonal bases of male dimorphism among teleost fish. Horm Behav, 27: 332–347.

Brantley, R.K. and A.H. Bass. 1994. Alternative male spawning tactics and acoustic signals in plainfin midshipman fish *Porichthys notatus* Girad (Teleostei: Batrachoididae). Ethology, 96: 213–232.

Brouwer, S.L. and M.H. Griffiths. 2005. Reproductive biology of carpenter seabream (*Argyrozona argyrozona*) (Pisces: Sparidae). Fish Bull, 103: 258–269.

Brusle, S. 1983. Contribution to the sexuality of a hermaphroditic teleost *Serranus hepatus* L. J Fish Biol, 22: 283–292.

Brusle, S. 1987. Sex-inversion of the hermaphroditic protogynous teleost *Coris julis* L. (Labridae). J Fish Biol, 30: 605–616.

Brusle, S. 1988. Sex differentiation in teleosts: primordial germ cells as stem cells. Reproduction in fish: basic and applied aspects of in endocrinology and genetics. Colloq Inst Natl Rech Agron, 44: 21–24.

Brusle, S. and J. Brusle. 1978. An ultrastructural study of early germ cells in *Mugil (Liza) auratus* Risso, 1810 (Teleostei: Mugilidae). Ann Biol Anim Biochim Biophys, 18: 1141–1457.

Brusle-Sigard, S., L. Debas, S. Fourcault and J. Fuchs. 1992. Ultrastructural study of sex inversion in a protogynous hermaphrodite *Epinephelus microdon* (Teleostei: Serranidae). Reprod Nutr Dev, 32: 393–406.

Brusle-Sigard, S., R. Reinboth and B. Fourcault. 1994. Germinal potentialities during sexual state changes in a protandric hermaphrodite *Amphiprion frenatus* (Teleostei: Pomacentridae), J Fish Biol, 45: 597–611.

Bullock, L.H., M.D. Murphy, M.F. Godcharles and M.E. Mitchell. 1992. Age, growth and reproduction of jewfish *Epinephelus itajara* in the eastern Gulf of Mexico. Fish Bull, 90: 243–249.

Burgus, R., M. Butcher, M. Amoss et al. 1972. Primary structure of the ovine hypothalamic luteinizing hormone-releasing factor (LRF) (LH hpothalmus-LRF gas chromatography-mass spectrometry-decapeptide Edman degradation). Proc Natl Acad Sci USA, 69: 278–282.

Burnard, D., R.E. Gozlan and S.W. Griffiths. 2008. The role of pheromones in freshwater fishes. J Fish Biol, 73: 1–16.

Burness, G., S. Casselman, A.I. Schulte-Hostedde et al. 2004. Sperm swimming speed and energetics vary with sperm competition risk in bluegill (*Lepomis macrochirus*). Behav Ecol Sociobiol, 56: 65–70.

Burness, G., C.D. Moyes and R. Montgomerie. 2005. Motility, ATP levels and enzyme activity of sperm from bluegill (*Lepomis macrochirus*). Comp Biochem Physiol, Part A Mol Integ Physiol, 140: 11–17.

Burns, J.R. and S.H. Weitzman. 2004. Insemination in Ostariophysan fish. pp. 107–135. *In:* H.J. Grier and M.C. Uribe [eds.]. Viviparous Fishes. New Life Publications, Homestead, Florida, USA.

Burns, J.R., S.H. Weitzman, H.J. Grier and N.A. Menezes. 1995. Internal fertilization, testis and sperm morphology in glandulocaudine fishes (Teleostei: Characidae: Glandulocaudinae). J Morphol, 224: 131–145.

Burns, J.R., S.H. Weitzman, K.R. Lange and L.R. Malabarba. 1998. Sperm ultrastructure in characid fishes (Teleostei: Ostariophysi). pp. 235–244. *In:* Z.M. Malabarba, R.E. Reis, R.P. Vari, Z.M. Lucena and C.A.S. Lucena [eds.]. Phylogeny and Classification of Neotropical Fishes. Porto Alegre, Brazil.

Buxton, C.D. and J.R. Clarke. 1991. The biology of the white mussel cracker *Sparodon durbanensis* (Pisces: Sparidae) on the eastern Cape coast, South Africa. S Afr J Mar Sci, 10: 285–296.

Campbell, B., J.T. Dickey and P. Swanson. 2003. Endocrine changes during onset of puberty in male spring Chinook salmon (*Oncorhynchus tshawytscha*). Boil Reprod, 69: 2109–2117.

Campbell, P.M., T.G. Pottinger and J.P. Sumpter. 1992. Stress reduces the quality of gametes produced by rainbow trout. Biol Reprod, 85: 371–378.

Campbell, P.M., T.G. Pottinger and J.P. Sumpter. 1994. Preliminary evidence that chronic confinement stress reduces the quality of gametes produced by brown and rainbow trout. Aquaculture, 120: 151–169.

Cardwell, H.K., H.J. Lee, A.H. Macbeth and W.S. Young. 2008. Vasopression: behavioural roles of an 'original' neuropeptide. Prog Neurobiol, 84: 1–24.

Cardwell, J.R. and N.R. Liley. 1991a. Hormonal control of sex and color-change in the spotlight parrotfish *Sparisoma viride.* Gen Com Endocrinol, 81: 7–20.

Cardwell, J.R. and N.R. Liley. 1991b. Androgen control of social-status in males of a wild population of spotlight parrotfish *Sparisoma viride* (Scaridae). Horm Behav, 25: 1–18.

Cardwell, J.R., P.W. Sorensen, G.J. van der Kraak and N.R. Liley. 1996. Effect of dominance status on sex hormone levels in laboratory and wild spawning male trout. Gen Comp Endocrinol, 101: 333–341.

Carlisle, S.L., S.K. Marxer-Miller, A.V.M. Canario et al. 2000. Effect of 11-ketotestosterone on genital papilla morphology in the sex changing fish *Lythrypnus dalli.* J Fish Biol, 57: 445–456.

Carmona, J.A., O.I. Sanjur, I. Dadrio et al. 1997. Hybridogenic reproduction and maternal ancestry of polyploidy Iberian fish: The *Tropidophoxinellus albroinoides* complex. Genetics, 146: 983–993.

Carnevali, O. 2007. Reproductive endocrinology and gamete quality. Gen Comp Endocrinol, 153: 273–274.

Carnevali, O., R. Carletta, A. Cambi et al. 1999. Yolk formation and degradation during oocyte maturation in seabream *Sparus aurata* involvement of two lysosomal proteinases. Boil Reprod, 60: 140–146.

Carolsfeld, J., J.F. Powell, M. Park et al. 2000. Primary structure and function of three gonadotropin releasing hormones including a novel form from an ancient teleost herring. Endocrinology, 141: 505–512.

Carpenter, R.E., W.J. Korzan, C. Bockholt et al. 2009. Corticotropin releasing factor influence territorial aggression in free-living female mountain spinny lizards. Horm Behav, 35: 205–214.

Carragher, J.F. and J.P. Sumpter. 1990. The effect of cortisol on the secretion of sex steroids from cultured ovarian follicles of rainbow trout. Gen Comp Endocrinol, 77: 403–407.

Carragher, J.F., J.P. Sumpter, T.G. Pottiger and A.D. Pickering. 1989. The deleterious effects of cortisol implantation on reproductive function in two species of trout *Salmo trutta* L. and *Salmo gaidneri* Richardson. Gen Comp Endocrinol, 76: 310–321.

Carruth, L.L. 2000. Freshwater cichlid *Crenicara punctulata* is a protogynous sequential hermaphrodite. Copeia, 2000: 71–82.

Casadevall, M., E. Delgado, O. Colleye et al. 2009. Histological study of the sex change in the skunk clownfish *Amphiprion akallopisos*. Open Fish Sci J, 2: 55–58.

Castelnuovo, G. 1937. Effectidi alcuni ormoni sulla maturazione della carpe. Riv Biol, 23: 363–372.

Castro, N., A.F.H. Ros, K. Becker and R.F. Oliveira. 2006. Metabolic costs of aggressive behavior in the Siamese fighting fish *Betta splendens*. Agg Behav, 32: 474–480.

Cavaco, J.E.B., J.G.D. Lambert, R.W. Schulz and H.J.T.H. Goos. 1996. Pubertal development of male African catfish *Clarias gariepinus*. *In vitro* steroidogenesis by testis and interregnal tissue and plasma levels of sexual steroids. Fish Physiol Biochem, 16: 129–138.

Cavaco, J.E.B., H.F. Vischer and J.G.D. Lambert. 1997. Mismatch between patterns of circulating and testicular androgens in African catfish *Clarias gariepinus*. Fish Physiol Biochem, 17: 155–162.

Cavaco, J.E.B., C. Vilrokx, V.L. Trudean et al. 1998. Sex steroid and the initiation of puberty in male African catfish *Clarias gariepinus*. Am J Physiol, 44: R1793–R1802.

Cavaco, J.E.B., J. Bogerd, H. Goos and R.W. Schulz. 2001. Testosterone inhibits 11-Ketotestosterone-induced spermatogenesis in African catfish *Clarias gariepinus*. Biol Reprod, 65: 1807–1812.

Ceinos, R.M., S. Polakof, A.R. Illamola et al. 2008. Food deprivation and re-feeding effects on pineal indoles metabolism and melatonin synthesis in the rainbow trout *Oncorhynchus mykiss*. Gen Comp Endocrinol, 156: 410–417.

Cek, S., N. Bromage, C. Randall et al. 1998. Reproductive biology of the rosy barb (*Puntius conchonius*). pp. 167–176. *In:* M.S. Celikkale, E. Duzgunes and I. Okumus [eds.]. Sex Differentiation and Gametogenesis. Proc Internatl Symp Fish Ecol, Trabzon, Turkey.

Cek, S., F. Twan and E. Atik. 2007a. Masculinisation of convict cichlid (*Cichlasoma nigrofasciatum*) by immersion in *Tribulus terrestris* extract. Aquacult Int, 15: 109–119.

Cek, S., F. Twan and E. Atik. 2007b. The effect of gokshura *Tribulus terrestris* on sex reversal of guppy *Poecilia reticulatus*. Pak J Biol Sci, 10: 718–725.

Cerda, J., M. Fabra and D. Raldua. 2007. Physiological and molecular basis of fish oocyte hydration. pp. 349–396. *In:* P.J. Babin, J. Cerda and E. Lubzens [eds.]. The Fish Oocyte: From Basic Studies to Biotechnological Applications. Springer, Dordrecht, The Netherlands.

Chang, C.F. and B.Y. Lin. 1998. Estradiol-17β stimulates aromatase activity and reversible sex change in protandrous block porgy *Acanthopagrus schlegeli*. J Exp Zool, 280: 165–173.

Chang, C.F., E.L. Lau and B.Y. Lin. 1995a. Stimulation of spermatogenesis or of sex reversal according to the dose of exogenous estradial-17β in juvenile males of protandrous black porgy *Acanthopagrus schlegeli*. Gen Comp Endocrinol, 100: 355–367.

Chang, C.F., E.L. Lau and B.Y. Lin. 1995b. Estradiol 17-β suppresses testicular development and stimulates sex reversal in protandrous black porgy *Acanthopagrus schlegeli*. Fish Physiol Biochem, 14: 481–488.

Chang, C.F., S.C. Lan and B.S. Pan. 1995c. Feed administration of estradiol-17β stimulates female differentiation in juvenile grey mullet *Mugil cephalus*. Zool Stud, 34: 257–264.

Chang, C.F., B.Y. Lin, E.L. Lau et al. 1997. The endocrine mechanism of sex reversal in the protandrous black porgy *Acanthopagrus schlegeli*: a Review. Chinese J Physiol, 40: 197–205.

Chang, J.P. and R.E. Peter. 1983. Effects of dopamine on gonadotropin release in female goldfish *Carassius auratus.* Neuroendocrinology, 36: 351–357.

Chang, X., R. Patino, P. Thomas and G. Yoshizaki. 1999. Developmental and protein kinase dependent regulation of ovarian connexin mRNA and oocyte maturatunal competence in Atlantic croaker. Gen Comp Endocrinol, 114: 330–339.

Chang, X.T., T. Kobayashi, H. Kajiura et al. 1997. Isolation and characterization of the DNA encoding tilapia (*Oreochromis niloticus*) cytochrome p450 Aromatase (p450 arom): changes in p450 arom mRNA, protein and enzyme activity in the ovarian follicle during oogenesis. J Mol Endocrinol, 18: 57–66.

Chang, X., T. Kobayashi, B. Senthilkumaran et al. 2005. Two types of aromatase with different encoding genes, tissue distribution and developmental expression in Nile tilapia (*Oreochromis niloticus*). Gen Comp Endocrinol, 141: 101–115.

Chaube, R. and K.P. Joy. 2003. Thyroid hormone modulation of brain *in vivo* tyrosine hydroxylase activity and kinetics in female catfish *Heteropneustes fossilis.* J Endocrinol, 179: 205–215.

Cheek, A.O., P. Thomas and C.V. Sullivan. 2000. Sex steroids relative to alternative mating behaviors in the simultaneous hermaphrodite *Serranus subligarius* (Perciformes: Serranidae). Horm Behav, 37: 198–211.

Chen, F.Y., T. Chow, M. Chao and R. Lim. 1977. Artificial spawning and larval rearing of the grouper *Epinephelus tauvina* in Singapore. Singapore J Prim Ind, 5: 11–21.

Chen, W. and W.L. Ge. 2011. Ontogenic expression profiles of gonadotropins (*fshb* and *lhb*) and growth hormone (*gh*) during sexual differentiation and puberty onset in female zebrafish. Biol Reprod, DOI:10.1095/biolreprod.111.094730.

Chiang, E.F., Y.L. Yan, Y. Guiguen et al. 2001. Two Cyp19 (P450 aromatase) genes on duplicated zebrafish chromosomes are expressed in ovary or brain. Mol Biol Evol, 18: 542–550.

Chiba, H., T. Miura, M. Nakamura and K. Yamaguchi. 1997. Differentiation and development of Leydig cell and induction of spermatogenesis during testicular differentiation in the Japanese eel *Anguilla japonica.* Fish Physiol Biochem, 16: 187–195.

Chopelet, J., R.S. Waples and S. Mariani. 2009. Sex change and genetic structure of marine fish. Fish Fish, 10: 329–343.

Chowdhury, I. and K.P. Joy. 2000. Annual cyclic variations in some biochemical constituents of seminal vesicle and testis of the catfish *Heteropneustes fossilis* (Bloch): a study correlating plasma testosterone level. Acta Biol Hung, 51: 53–62.

Chowdhury, I. and K.P. Joy. 2001a. Seminal vesicle and testis secretions in *Heteropneustes fossilis* (Bloch): composition and effects on sperm motility and fertilization. Aquaculture, 193: 355–371.

Chowdhury, I. and K.P. Joy. 2001b. Effects of castration and cyproterone acetate on some biochemical constituents of the seminal vesicles in the catfish *Heteropneustes fossilis* (Bloch). Netherland J Zool, 51: 51–69.

Chowdhury, I. and K.P. Joy. 2001c. Responses of seminal vesicle and testicular β glucuronidase and β–N-acetylglucosaminidase to testosterone and some metabolites in *Heteropneustes fossilis* (Bloch). J Exp Zool, 290: 777–782.

Chowdhury, I. and K.P. Joy. 2007. Seminal vesicle and its role in the reproduction of teleosts. Fish Physiol Biochem, 33: 383–398.

Ciereszko, A. and K. Dabrowski. 1995. Sperm quality and ascorbic acid concentration in rainbow trout semen are affected by dietary Vitamin C: an across-season study. Biol Reprod, 52: 982–988.

Cimino, M.C. 1972a. Egg production, polyploidization evolution in a diploid all female fish of the genus *Poeciliopsis.* Evolution, 26: 294–306.

Cimino, M.C. 1972b. Meiosis in triploid all female fish (*Poeciliopsis* Poeciliidae), Science, 175: 1484–1486.

Clearwater, S.J. and N.W. Pankhurst. 1997. The response to capture and confinement stress of plasma cortisol, plasma sex steroids and vitellogenic oocytes in the marine teleost red gunard. J Fish Biol, 50: 429–441.

Clemens, H.P. and T. Inslee. 1968. The production of normal male and androgen sex reversed *Tilapia aurea* and *T. nilotica* fed a commercial catfish diet in ponds. pp. 34–42. *In:* R.O. Smitherman, W.L. Shelton and J.M. Grover [eds.]. Proc Cult Exotic Fishes. American Fishery Society, Auburn Univ, Alabama, USA.

Clemens, H.P., C. McDermitt and T. Inslee. 1996. The effects of feeding methyltestosterone to guppies for 60 days after birth. Copeia, 1966: 280–284.

Clotfelter, E.D. and A.C. Rodriquez. 2006. Behavioral changes in fish exposed to phytoestrogens. Env Pollu, 144: 833–839.

Coburn, M.M. 1986. Egg diameter variation in eastern North American minnows (Pisces: Cyprinidae): Correlation with vertebral number, habitat and spawning behavior. Ohio J Sci, 86: 110–120.

Cochran, R.C. 1987. Serum androgens during the annual reproductive cycle of the male mummichog *Fundulus heteroclitus.* Gen Comp Endocrinol, 65: 141–148.

Cole, K.S. and D.L.G. Noakes. 1997. Gonadal development and sexual allocation in mangrove killifish *Rivulus marmoratus* (Pisces, Atherinomorph). Copeia, 1997: 596–600.

Colombo, G. and G. Grandi. 1990. Gonadal sex differentiation of *Anguilla anguilla* by sex steroids. Int Rev Hydrodiol, 76: 763–773.

Colombo, G. and G. Grandi. 1996. Histological study of the development and sex differentiation on the gonad in the European eel. J Fish Biol, 48: 493–512.

Condeca, J.A.B. and A.V.M. Canario. 1995. Steroidogenesis during estrogen-induced sex inversion in the seabream *Sparus aurata.* Fifth Internatl Symp Reprod Physiol Fish, Austin, Texas. USA. p. 306.

Consten, D., J. Bogerd, J. Komen et al. 2001a. Long-term cortisol treatment inhibits pubertal development in male common carp *Cyprinus carpio* L. Biol Reprod, 64: 1063–1071.

Consten, D., E.D. Keuning, M. Terlou et al. 2001b. Cortisol effects on the testicular androgen synthesizing capacity in common carp *Cyprinus carpio* L. Fish Physiol Biochem, 25: 91–98.

Consten, D., J.G.D. Lambert and H.J.T. Goos. 2001c. Cortisol affects testicular developmet in male common carp *Cyprinus carpio* L but not via an effect on LH secretion. Comp Biochem Physiol, B: Biochem Mol Biol, 129: 671–677.

Consten, D., J.G.D. Lambert, H. Komen and H.J.P. Goos. 2002a. Corticosteroids affect the testicular androgen production in male common carp (*Cyprinus carpio* L.). Biol Reprod, 6: 106–111.

Consten, D., E.D. Keuning, J. Bogerd et al. 2002b. Sex steroids and their involvement in the cortisol induced inhibition of pubertal development in male common carp *Cyprinus carpio* L. Biol Reprod, 67: 465–472.

Contreras-Sanchez, W.M., C.B. Schreck, M.S. Fitzpatrick and C.B. Pereira. 1998. Effects of stress on the reproductive performance of rainbow trout (*Oncorhynchus mykiss*). Biol Reprod, 58: 439–447.

Coward, K. and N.R. Bromage. 1999. Spawning periodicity, fecundity and egg size in laboratory held stocks of a subtrate-spawning tilapiine *Tilapia zilli.* Aquaculture, 171: 251–267.

Coward, K., N.R. Bromage and D.C. Little. 1998. Inhibition of spawning and associated suppression of sex steroid levels during confinement in the substrate spawning *Tilapia zilli.* J Fish Biol, 52: 152–165.

Cravedi, J.P., G. Delons and D. Rao. 1989. Disposition and elimination routes of 17α-methyltestosterone in rainbow trout (*Salmo gairdneu*). Can J Fish Aquat Sci, 46: 159–165.

Csenki, Z., A. Zaucker, B. Kovacs et al. 2010. Intraovarian transplantation of stage I-II follicles results in viable zebrafish embryos. Int J Dev Biol, 54: 585–589.

Cunha, C., I. Doadrio and M.M. Coerho. 2008. Speciation towards teraploidization after intermediate processes on non-sexual reproduction. Phil Trans R Soc B, 363: 2921–2929.

Curtis, L.R., F.T. Diren, M.D. Hurley et al. 1991. Disposition and elimination of 17α-methyltestosterone in Nile tilapia (*Oreochromis niloticus*). Aquaculture, 99: 193–201.

Cyr, D.G., N.R. Bromage, J. Dusto and J.G. Eales. 1988. Seasonal patterns in serum levels of thyroid hormones and sex steroids in relation to photoperiod-induced changes in spawning time in rainbow trout *Salmo gaidneri*. Gen Comp Endocrinol, 69: 217–225.

Dadzie, S. and M. Hyder. 1976. Compensatory hypertrophy of the remaining ovary and the effects of methallibure in the unilaterally ovariectomized *Tilapia aurea*. Gen Comp Endocrinol, 29: 433–440.

Darwin, C. 1859. On the Origin of Species by Means of Natural Selection. Murray, London, UK.

Das, P., M.K. Mukhopadhyay, K.M. Das and P.K. Pandit. 1987. Gonadal sex manipulation of *Oreochromis mossambicus* (Peters). *In:* K. Tiews [ed.]. *Proc World Symp Selection Hybridization and Genetic Engg in Aquacult*. Bordeaux, Berlin, Germany.

Das, S.K., H.P.C. Shetty, M.C. Narayana et al. 1990. Androgenic bioassay for the residual effect of administered mibolerone in common carp *Cyprinus carpio* (Linn). Proc Sec Indian Fish Forum, Mangalore, India. pp. 85–86.

Dasgupta, S. 2000. Hormonal induction of maturation and breeding in an Indian carp and perch. Ph. D. Thesis, Visva Bharathi University, Shantiniketan, India.

Dasgupta, S., S.K. Sarkar, N. Sarangi and S. Bhattacharya. 2009. Variation in spawning responses, egg and larvae productions from induced rohu (*Labeo rohita)* during pre-monsoon and monsoon seasons. Relationships with hormonal changes and oocyte responsiveness during final maturation. Aquaculture, 290: 320–326.

David, C.J. 2004. Experimental sperm cryopseservation and genetic studies in selected fish. Ph. D. Thesis, Madurai Kamaraj University, Madurai, India.

David, C.J. and T.J. Pandian. 2006a. GFP reporter gene confirms paternity in the androgenote Buenos Aires tetra *Hemigrammus caudovittatus*. J Exp Zool, 305 A: 83–95.

David, C.J. and T.J. Pandian. 2006b. Cadaveric sperm induces intergeneric androgenesis in the fish *Hemigrammus caudovittatus*. Theriogenology, 65: 1048–1070.

Davidson, B., M. Soodak, H.V. Strout et al. 1979. Thiurea and cyanamide as inhibitors of thyroid peroxide: the role of iodide. Endocrinology, 69: 217–225.

Davis, K.B. and G.M. Ludwig. 2004. Hormonal effects on sex differentiation and growth in sunshine bass *Merone chrysops x Merone saxatalis*. Aquaculture, 3: 587–597.

Davis, K.B., B.A. Simco, C.A. Goudie et al. 1990. Hormonal sex manipulation and evidence for female heterogamety in channel catfish. Gen Comp Endocrinol, 78: 218–323.

Davis, K.B., C.A. Goudie, B.A. Simco et al. 1992. Influence of dihydrotestosterone on sex determination and in channel catfish and blue catfish: period of development sensitivity. Gen Comp Endocrinol, 86: 107–115.

Dawley, R.M. 1987. Hybridization and polyploidy and a community of three sunfish species (Pisces, Centrarchidae). Copeia, 1987: 326–335.

Dawley, R.M. 1989. An introduction to unisexual vertebrates. Bulletin, 466: 1–18. *In:* Evolution and Ecology of Unisexual vertebrates. New York State University Museum, Albany, New York, USA. Bulletin, 466.

Dawley, R.M. 1992. Clonal hybrids of the common laboratory fish *Fundulus heteroclitus*. Proc Natl Acad Sci USA, 89: 2485–248.

Dawley, R.M. and K.A. Goddard. 1988. Diploid triploid mosaics *Phoxinus eos* and *Phoxinus neogaeus*. Evolution, 42: 649–659.

Defraipont, M. and P.W. Sorensen. 1993. Exposure of pheromone 17α, 20β dihydroxy-4-pregnen-3-one enhances the behavioural spawning success, sperm production and sperm motility of male goldfish. Anim Behav, 46: 245–256.

Degani, G. 1993. Effect of social behavior on oocyte development and steroid changes in *Trichogaster trichopterus* (Pallas). Copeia, 4: 1091–1096.

Desprez, D., C. Melard and J.C. Philippart. 1995. Production of a high percentage of male offspring with 17α-ethylnylestradiol sex-reverssed *Oreochromis aureus*. II Comparative reproductive biology of females and F_2 pseudofemales and large scale production of male progeny. Aquaculture, 30: 35–41.

de Vlaming, V.L. and B.I. Sundararaj. 1972. Endocrine influences on seminal vesicles of the estuarine gobiid fish *Gillichthys mirabilis*. Biol Bull, 142: 243–250.

Devlin, R.H. and Y. Nagahama. 2002. Sex determination and sex differentiation in fish: an overview of genetic, physiological and environmental influences. Aquaculture, 208: 191–364.

Devlin, R.H., K.B. McNeil, T.D.D. Groves and E.M. Donaldson. 1991. A rapid PCR-based test for Y chromosomal DNA probe capable of determining genetic sex in Chinook salmon (*Oncorhynchus tshawytscha*). Can J Fish Aquat Sci, 48: 1606–1612.

Devlin, R.H., C.A. Biagi and D.E. Smailus. 2001. Genetic mapping of Y-chromosomal DNA markers in Pacific salmon. Genetica, 111: 43–58.

DeWoody, J.A. and A.C. Avise. 2001. Genetic perspectives on the natural history of fish mating systems. J Hered, 92: 167–172.

Dhaneesh, K.V., K. Nanthini Devi, T.T. Ajithkumar et al. 2011. Breeding, embryonic development and salinity tolerance of skunk clownfish *Amphiprion akallopisos*. J King Saud Univ Sci, doi:10.1016/y.jsus.2011.03.005.

Diana, J.S. 1983. Growth, maturation and production of northern pike in three Michigan lakes. Fish Bull, 105: 582–587.

Dierkes, P., M. Taborsky and U. Kohler. 1999. Relative parasitism of brood care helpers in a cooperatively breeding fish. Behav Ecol, 10: 510–515.

Dierkes, P., D. Heg, M. Taborsky et al. 2005. Genetic relatedness in groups is sex-specific and declines with age of helpers in a cooperatively breeding cichlid. Ecol Lett, 8: 968–975.

Diotel, N., J.-L. Do Rego, I. Anglade et al. 2011. Activity and expression of steroidogenic enzymes in the brain of adult zebrafish. Eur J Neurosci, doi:10.1111/ j.1460-9568. 2011.07731.x.

Down, N.E., E.M. Donaldson, H.M. Dye et al. 1988. Recombinant bovine somatotropin more than doubles the growth of coho salmon (*Oncorhynchus kisutch*) acclimated to sea water and ambient winter conditions. Aquaculture, 68: 141–155.

Downing, A.L. Burns. 1995. Testis morphology and spermatozeugma formation in the three genera of viviparous halfbeaks: *Normorhamphus, Dermogenys* and *Hemirhamphodon* (Teleostei: Hemirhamphidae). J Morphol, 225: 329–343.

Dreanno, C., J. Cosson, M. Suquet et al. 1999. Nucleotides content oxidative phosphorylation, morphology and fertilizing capacity of turbot (*Psetta maxima*) spermatozoa during the motility period. Mol Reprod Dev, 53: 230–243.

Duan, C. 1998. Nutritional and developmental regulation of insulin-like growth factors in fish. J Nutr, 128: 3065–3145.

Dufour, S., M.-E. Sebert, F.-A. Weltzien et al. 2010. Neuroendocrine control by dopamine of teleost reproduction. J Fish Biol, 76: 129–160.

Duftner, N., K.M. Sefe, S. Koblmuller et al. 2007. Parallel evolution of facial stripe patterns in the *Neolamprologus brichardi/pulcher* species complex endemic to Lake Tanganyika. Mol Phylogenet Evol, 45: 706–715.

Duncan, N.J., G.A. Rodriguez, M. deO et al. 2003. Effects of controlled delivery and acute injections of LHRHa on bullseye pufferfish (*Sphoeroides annulatus*) spawning. Aquaculture, 218: 625–635.

Dunhan, R.A. 1990. Production and use of monosex or sterile fishes in aquaculture. Rev Aquat Sci, 2: 1–17.

Duponchelle, F., P. Cecchi, D. Corbin et al. 2000. Variations in fecundity and egg size of female Nile tilapia, *Oreochromis niloticus,* from man-made lakes of Cote d'Ivoire. Env Biol Fish, 57: 155–170.

Ebisawa, A. 1997. Some aspects of reproduction and sexuality in the spotcheek emperor *Lethrinus rubrioperculatus* in waters off the Ryuku Islands. Icthyol Res, 44: 201–212.

Echelle, A.A., A.F. Echelle and D.P. Middaugh. 1989. Evolutionary biology of the *Menidia clarkhubbsi* complex of unisexual fishes (Athernidae): origins, clonal diversity and mode of reproduction. pp. 144–152. *In:* R.M. Dawley and J.P. Bogart [eds.]. Evolution and Ecology of Unisexual Vertebrates. New York State Museum, Albany, USA. Bulletin, 466.

Egami, N. 1955. Production of testis-ova in adult male of *Oryzias latipes*. I. Testis-ova in the fish receiving estrogens. Jap J Zool, 11: 353–365.

Ekstrom, P. and H. Meissl. 2003. Evolution of photosensory pineal organs in new light: the fate of neuroendocrine receptors. Phil Trans R Soc, B 358: 1679–1700.

Elder, J.F. and I.J. Schlosser. 1995. Extreme clonal uniformity of *Phoxinus eos naegaeus* gynogens (Pisces, Cyprinidae) among variable habitats in northern Minnesota beaver ponds. Proc Natl Acad Sci, USA, 92: 5001–5005.

Elofsson, H., B.G. McAllister, D.E. Kime et al. 2003. Long lasting sticklebacks sperm; is ovarian fluid a key to success in freshwater? J Fish Biol, 63: 240–253.

Elofsson, U.O.E., S. Winberg and G.E. Nilsson. 1999. Relationship between sex and the size and number of forebrain gonadotropin-release hormone-immunoreactive neurones in the ballan wrasse (*Labrus berggylta*) a protogynous hermaphrodite. J Comp Neurol, 410: 158–170.

Elofsson, U., I. Mayer, B. Damsgrad and S. Winberg. 2000. Intermale competition in sexually mature Arctic charr: effects on brain monoamines, endocrine stress responses, sex hormone levels and behavior. Gen Comp Endocrinol, 118: 450–469.

Epler, P., J. Gales and S. Stoklosowa. 1997. Steroidogenic activity of carp ovarian follicular and interstitial cells at the pre-spawning and resting-time: a tissue culture approach. Comp Biochem Physiol, 116C: 167–170.

Eriksen, M.S., M. Bakken, A. Espmark et al. 2006. Prespawning stress in farmed Atlantic salmon *Salmo salar*: maternal cortisol exposure and hyperthermia during embryonic development affects offspring survival growth and incidence of malformations. J Fish Biol, 69: 114–129.

Essenberg, J.M. 1926. Complex sex reversal in viviparous teleost *Xiphophorus helleri*. Biol Bull, 51: 98–111.

Fabra, M., D. Raldua, D.A. Power et al. 2005. Marine fish egg hydration is aquaporin-mediated. Science, 307: 545.

Fabra, M., D. Raldua, M.G. Bozzo et al. 2006. Yolk proteolysis and aquaporin-1o play essential roles to regulate fish oocyte hydration during meiosis resumption. Dev Biol, 295: 50–262.

Fagerlund, U.H.M. and J.R. McBride. 1978. Distribution and disappearance of radio-activity in blood and tissues of coho salmon (*Oncorhynchus kisutch*) after oral administration of 3H-testosterone. J Fish Res Bd, Can, 35: 893–900.

Fagerlund, U.H.M. and H.M. Dye. 1979. Depletion of radioactivity from yearling coho salmon (*Oncorhynchus kisutch*) after extended ingestion of anabolically effective doses of 17α-methyltestosterone-1, 2-3h. Aquaculture, 18: 303–315.

Fagundes, T., D.M. Goncalves and R.F. Oliveira. 2007. Female mate choice and mate search tactics in a sex role reversed population of the peacock blenny *Salaria pavo* (Risso, 1810). J Fish Biol, 71: 77–89.

Falcon, J., L. Besseau, S. Sauzet and G. Boeuf. 2007. Melatonin effects on the hypothalamo-pituitary axis in fish. Trend Endocrinol Metabol, 18: 81–88.

Falcon, J., H. Migaud, J.A. Munoz-Cueto and M. Carrillo. 2009. Current knowledge on the melatonin system in teleost fish. Gen Comp Endocrinol, doi:10.1016/ j.ygcen. 2009.04.026.

Feist, G., C.B. Schreck, M.S. Fitzpatrick and J.M. Redding. 1990. Sex steroid profiles of coho salmon (*Oncorhynchus kisutch*) during early development and sexual differentiation. Gen Comp Endocrinol, 80: 299–313.

Fenske, M. and H. Segner. 2004. Aromatase modulation alters gonadal differentiation in developing zebrafish (*Danio rerio*). Aquat Toxicol, 67: 105–126.

Fernald, R.D. and R.B. White. 1999. Gonadotropin releasing hormone gene: phylogeny, structure and functions. Front Neuroendocrinol, 20: 224–230.

Filby, A.L., C.P. Gregory, F.A. Tamsin et al. 2010. Unraveling the neurophysiological basis of aggression in a fish model. BMG Genomics, 11: 498: 1–7.
Filipenko, M., O.V. Alekseyenko, A.G. Beilina et al. 2001. Increase of tyrosine hydroxylase and dopamine transporter mRNA levels in ventral tegmental area of male mice under influence of repeated aggression experience. Brain Res Mol Brain Res, 96: 77–81.
Fineman, R., J. Hamilton and G. Chase. 1975. Reproductive performance of male and female phenotypes in three sex chromosomal genotypes (XX, XY, YY) in the killifish *Orizias latipes.* J Exp Zool, 192: 349–354.
Finn, R.N., G.C. Ostby, B. Norberg and H.J. Fyhn. 2002. *In vivo* oocyte hydration in Atlantic halibut (*Hippoglossus hippoglossus*): proteolytic liberation of free amino acids, and ion transport are driving forces for osmotic water influx. J Exp Biol, 205: 211–224.
Fishelson, L. 1992. Comparative gonad morphology and sexuality of the Muraenidae (Pisces: Teleostei). Copeia, 197–209.
Fishelson, L. and O. Gon. 2009. Comparison of the ovaries and oogenesis of some Australian and South African viviparid clinid fishes (Clinidae, Blennioidei, Perciformes). Env Biol Fish, 86: 527–540.
Fishelson, L., J.H.J. Vanvuren and A. Tyran. 1994. Ontogeny and ultrastructure of seminal vesicles of the catfish *Clarias gariepinus.* J Morphol, 219: 59–71.
Fitzpatrick, J.L., J.K. Desjardins, N. Milligan et al. 2007. Reproductive-tactic-specific variation in sperm swimming speeds in a shell brooding cichlid. Boil Reprod, 77: 280–284.
Flores, J.A. and J.R. Burns. 1993. Ultrastructural study of embryonic and early adult germ cells and their support cells in both sexes of *Xiphophorus,* Teleostei, Poeciliidae. Cell Tissue Res, 271: 263–270.
Foran, C.M. and A.H. Bass. 1999. Pre-optic GnRH and AVT: Axes for sexual plasticity in teleost fish. Gen Comp Endocrinol, 116: 141–152.
Forrester, G.E. 1991. Social rank, individual size and group composition as determinants of food consumption by humbug damselfish *Dascyllus aruanas.* Anim Behav, 42: 701–711.
Francis, C.R. 1992. Sexual lability in teleosts: developmental factors. Quart Rev Biol, 67: 1–17.
Fraser, P. and N.E. Stacey. 2002. Isolation increases milt production in goldfish. J Exp Zool, 293: 511–524.
Fricke, H. and S. Fricke. 1977. Social control of sex reversal in a coral-reef fish. Nature, 266: 829–832.
Frisch, A. 2004. Sex-change and gonadal steroids in sequentially-hermaphroditic teleost fish. Rev Fish Biol Fish, 14: 481–499.
Froiland, E., K. Murashita, E.H. Jorgensen and T. Kurokawa. 2010. Leptin and ghrelin in anadromous Arctic charr: cloning and change in expressions during a seasonal feeding cycle. Gen Comp Endocrinol, 165: 136–143.
Fujiwara, A., S. Abe, E. Yamaha et al. 1997. Uniparental chromosome elimination in the early embryogenesis of the inviable salmonid hybrids between masu salmon female and rainbow trout male. Chromosoma, 106: 44–52.
Garber, A.F., S.E. Fordham, J.E. Symonds et al. 2009. Hormonal induction of ovulation and spermiation in Atlantic cod (*Gadus morhua*). Aquaculture, 296: 179–183.
Garcia-Diaz, M., J.A. Gonzalea, M.J. Lorente and V.M. Tuset. 2006. Spawning season maturity size and fecundity in blacktail comber (*Serranus atricauda*) (Serranidae) from the eastern central Atlantic. Fish Bull, 104: 159–166.
Garcia-Lopez, A., G. Martinez-Rodriques and C. Sarasquete. 2005. Male reproductive system in Senegalese sole *Solea senegalensis* (Kaup): anatomy, histology and histochemistry. Hist Histopathol, 20: 1179–1189.
Garcia-Lopez, A., J. Bogerd, J.C. Granneman et al. 2008. Leydig cells express FSH receptors in African catfish. Endocrinology, 150: 357–365.
Garcia-Lopez, A., M.I. Sanchez-Amaya, C.R. Tyler and F. Prat. 2011a. Mechanisms of oocyte development in European sea bass (*Dicentrarchus labrax* L.): investigations via application of unilateral ovariectomy. Reprod Res, DOI:10.1530/REP-11-0040.

Garcia-Lopez, A., M.I. Sanchez, S. Halm et al. 2011b. Bone morphogenetic protein 15 and growth differentiation factor 9 expression in the ovary of European sea bass (*Dicentrarchus labrax*): Cellular localization, developmental profiles and response to unilateral ovariectomy. Gen Comp Endocrinol, 174: 326–334.

Gardiner, D.M. 1978. Fine structure of the spermatozoon of the viviparous teleost *Cymatogaster agregata.* J Fish Biol, 13: 435–438.

Garrett, G.P. 1982. Variation in the reproductive traits of the Pecos pupfish, *Cyprinodon pecosensis.* Amer Midl Nat, 108: 355–363.

Garrett, G.P. 1989. Hormonal sex control of largemouth bass. Prog Fish Cult, 51: 146–148.

Ge, W. 2005. Activin and its receptors in fish reproduction. pp. 128–154. *In:* P. Melamed and N. Sherwood [eds.]. Hormones and their Receptors in Fish Reproduction. World Scientific Publishing, Hackensack NJ, USA.

George,T. 1995. Endocrine sex reversal and hybridization in selected ornamental fish. Ph.D. Thesis, Madurai Kamaraj University, Madurai, India.

George, T. and T.J. Pandian. 1995. Production of ZZ females in the female-heterogametic black molly *Poecilia sphenops* by endocrine sex reversal and progeny testing. Aquaculture, 136: 81–90.

George, T. and T.J. Pandian. 1996. Hormonal induction of sex reversal and progeny testing in the zebra cichlid *Cichlasoma nigrofasciatum.* J Exp Zool, 275: 374–382.

George, T. and T.J. Pandian. 1997. Interspecific hybridization in poeciliids. Indian J Exp Biol, 35: 628–637.

George, T. and T.J. Pandian. 1998. Dietary administration of androgens induces sterility in the female-heterogametic block molly *Poecilia sphenops* (Cuvier & Valenciennes, 1984). Aquacult Res, 29: 167–175.

George, T., T.J. Pandian and S. Kavumpurath. 1994. Inviability of the YY zygole of the fishting fish *Betta splendens.* Isr J Aquacult, 46: 3–8.

Ghalambor, C.K., D.N. Reznick and J.A. Walker. 2004. Constraints on adaptive evolution: the functional trade-off between reproduction and fast-start swimming performance in the Trinidadian guppy (*Poecilia reticulata*). Am Nat, 164: 38–50.

Ghosh, S., T.T. Ajithkumar, K. Nanthinidevi and T. Balasubramanian. 2012. Reef fish breeding and hatchery production using brackish water, a sustainable technology with special reference to clark's clownfish *Ampriprion clarkii* (Bennett, 1830). Internatl J Env Sci Dev, 3: 56–60.

Giesing, E.R., C.D. Suski, R.E. Warner and A.M. Bell. 2011. Female sticklebacks transfer information via eggs: effects of maternal experience with predators on offspring. Proc R Soc Lond B, 278: 1753–1759.

Gilling, C.J., D.O.F. Skibinski and J.A. Beardmore. 1996. Sex reversal of tilapia fry by immersion in water containing oestrogens. pp. 314–319. *In:* R.S.V. Pullin, J. Lazard, M. Legendre et al. [eds.]. The Third Int Symp Tilapia Aquacult. ICLARM, Manila, The Philippines.

Gnani, A. and B. Levavi-Sivan. 2009. Sexual development in fish, practical applications for aquaculture. Sex Dev, 3: 164–175.

Goddard, K.A, and R.M. Dawley. 1990. Clonal inheritance of a diploid nuclear genome by a hybrid freshwater minnow (*Phoxinus eos neogacus* pisces: Cyprinidae). Evolution, 44: 1052–1065.

Godwin, J. 2009. Social determination of sex in reef fishes. Sem Cell Dev Biol, 20: 264–270.

Godwin, J.R. and P. Thomas. 1993. Sex change and steroid profiles in the protandrous anemonefish *Amphiprion melanopus* (Pomacentridae, Teleostei). Gen Comp Endocrinol, 91: 144–157.

Godwin, J., D. Crews and R.R. Warner. 1996. Behavioural sex change in the absence of gonads in a coral reef fish. Proc R Soc Lond, B, 263: 1683–1688.

Godwin, J., R. Sawby, R.R. Warner et al. 2000. Hypothalamic orginine vasotocin mRNA abundance variation across sexes and with sex change in a coral reef fish. Brain Behav Evol, 55: 77–84.

Godwin, J., J.A. Luckenbach and R.J. Borski. 2003. Ecology meets endocrinology: environmental sex determination in fishes. Evol Dev, 5: 40–49.

Goetz, F.W. and M. Garczynski. 1997. The ovarian regulation of ovulation in teleost fish. Fish Physiol Biochem, 17: 33–38.

Goetz, F.W., E.M. Donaldson and G.A. Hunter. 1979. Effects of estradiol-17β methyltestosterone on gonadal differentiation in the coho salmon *Oncorhynchus kisutch*. Aquaculture, 17: 207–278.

Goffeau, A., B.G. Barrell, H. Bussey et al. 1996. Life with 6000 genes. Science, 274: 547–563.

Goldstein, M.E., A.W. Trank, L.H. Fossom and R.W. Hamill. 1992. Molecular aspects of the regulation of tyrosine hydroxylase by testosterone. Mol Brain Res, 14: 79–86.

Gomelsky, B., N.B. Cherfas, N. BenDon and G. Hulata. 1994. Hormonal sex inversion in the common carp *Cyprinus carpio* L. Aquaculture, 126: 265–270.

Gomelsky, B., N.B. Cherfas, A. Gissis and G. Hulata. 1999. Hormonal sex inversion in striped bass and whitebass x striped bass hybrids. North Am J Aquacult, 61: 199–205.

Goncalves, D.M. and R.F. Oliveira. 2011. Hormones and sexual behavior of teleost fishes. pp. 119–147. *In:* D.O. Norris and K.H. Lopez [eds.]. Hormones and Reproduction of Vertebrates: Fishes. Elsevier, Amsterdam, The Netherlands.

Goncalves, D.M., T. Fagundes and R.F. Oliveira. 2003. Reproductive behaviour of sneaker males of the peacock blenny. J Fish Biol, 63: 528–532.

Goncalves, D., R. Matos, T. Fagundes and R. Oliveira. 2005. Bourgeois males of the peacock blenny *Salaria pavo* discriminate female mimics from females? Ethology, 111: 559–572.

Goncalves, D.M., J. Alpedrenha, M. Teles and R.F. Oliveira. 2007. Endocrine control of sexual behavior of sneaker males of the peacock blenny *Salaria pavo:* effects of castration aromatase inhibition, testosterone and estradiol. Horm Behav, 51: 534–541.

Goncalves, D., T.A. Joao and R.F. Oliveira. 2008. Brain and gonadal aromatase activity and steroid hormone levels in females and polymorphic males of the peacock *Salario pavo*. Horm Behav, 54: 717–725.

Gonzalez-Martinez, D., T. Madigou, N. Zmora et al. 2001. Differential expression of three different prepro-GnRH (gonadotrophin-releasing hormone) messengers in the brain of the European sea bass (*Dicentrarchus labrax*). J Comp Neurol, 429: 144–155.

Gonzalez-Martinez, D., N. Zmora, E. Mananos et al. 2002. Immunohistochemical localization of three different prepro-GnRHs in the brain and pituitary of the European sea bass (*Dicentrarchus labrax*) using antibodies to the corresponding GnRH-associated peptides. J Comp Neurol, 446: 95–113.

Goodwin, N.B., N.K. Dulvy and J.D. Reynolds. 2002. Life-history correlates of the evolution of live bearing in fishes. Phil Trans R Soc Lond B, 357: 259–267.

Goswami, S.V. and B.I. Sundararaj. 1968. Compensatory hypertrophy of the remaining ovary after unilateral ovariectomy at various phases of the reproductive cycle of catfish *Heteropneustes fessilis* (Bloch). Gen Comp Endocrinol, 11: 401–413.

Goudie, C.A., B.D. Redner, B.A. Simco and K.B. Davis. 1983. Feminization of channel catfish by oral administration of steroid sex hormones. Trans Am Fish Soc, 112: 670–672.

Goudie, C.A., W.L. Shelton and N.C. Parker. 1986. Tissue distribution and elimination of radiolabeled methyltestosterone fed to sexually undifferentiated blue tilapia. Aquaculture, 58: 215–226.

Govoroun, M., O.M. McMeel, H. D'Cotta et al. 2001. Steroid enzyme gene expression during natural and androgen induced gonadal differentiation in the rainbow trout *Oncorhynchus mykiss*. J Exp Zool, 290: 558–566.

Govourn, M.S., M. Pannetier, E. Pailhoux et al. 2004. Isolation of chicken homolog of the Foxl2 gene and comparison of its expression patterns with those of aromatase during ovarian development. Dev Dyn, 231: 859–870.

Grandi, G. and G. Colombo. 1997. Development and early differentiation of gonad in the European eel (*Anguilla anguilla* L. Anguilliformes, Teleostei): a cytological and ultrastructural study. J Morphol, 231: 195–216.

Grandi, G., M. Chicca, E. Tagliati et al. 2010. Immunolocalization of steroidogenic enzymes in gonads of European eel *Anguilla anguilla* (L.). J Appl Ichthyol, 26: 40–48.
Greeley, Jr., M.S., D.R. Calder and R.A. Wallace. 1986. Changes in teleost yolk proteins during oocyte maturation: correlation of yolk proteolysis with oocyte hydration. Comp Biochem Physiol, B 84: 1–9.
Greenwood, A.K., A.R. Wark, R.D. Fernald and H.A. Hoffmann. 2008. Expression of arginine vasotocin in distinct preoptic regions is associated with dominant and subordinate behavior in an African cichlid fish. Proc Biol Sci, 275: 2393–2340.
Grier, H.J. 1981. Cellular organization of the testis and spermatogenesis in fishes. Am Zool, 21: 345–357.
Grier, H.J. 1993. Comparative organization of Sertoli cells including the Sertoli cells barrier. pp. 703–739. *In*: L.D. Russel and M.D. Grisworld [eds.]. The Sertolli Cells. Chace River Press, Clearwater, FL., USA.
Grier, H.J. and B.B. Collette. 1987. Unique spermatozeugmata in testes of halfbeaks of the genus *Zenarchopterus* (Teleostei, Hemiramphidae). Copeia 1987: 300–311.
Grier, H.J. and L.R. Parenti. 1994. Reproductive biology and systematic of phallostethid fishes as revealed by gonad structure. Env Biol Fish, 41: 287–299.
Grier, H.J. and R.G. Taylor. 1998. Testicular maturation and regression in the common snook. J Fish Biol, 53: 521–542.
Grier, H.J. and M.C. Uribe. 2005. Gonads and Gametogenesis of Viviparous Fishes. New Life Publications, Homestead, Florida, USA.
Grier, H.J., M.C. Uribe, L.R. Parenti, Rosa-Cruz, G. Dela. 2005. Fecundity, the germinal epithelium, and folliculogenesis in viviparous fishes. pp. 193–217. *In:* H.J. Grier and M.C. Uribe [eds.]. Viviparous Fishes. New Life Publications, Homestead, Florida, USA.
Grober, M.S. 1998. Socially controlled sex change: intergrating ultimate and proximate levels of analysis. Acta Ethol, 1: 3–17.
Grober, M.S. and A.H. Bass. 1991. Neuronal correlates of sex role changes in labrid fishes: LHRH-like immune reactivity. Brain Behav Ecol, 38: 302–312.
Grober, M.S. and T. Sunobe. 1996. Serial adult sex change involves rapid and reversible changes in forebrain neurochemistry. Neuro Rep, 7: 2945–2949.
Grober, M.S., A.A. George, K.K. Watkins et al. 2002. Forebrain AVT and courtship in a fish with male alternative reproductive tactics. Brain Res Bull, 57: 423–425.
Gross, M.R. 1982. Sneakers satellites and parentals. Polymorphic mating strategies in North American sunfishes. Z Tierpsychol, 60: 1–26.
Gross, M.R. 1985. Disruptive selection of alternative life histories in salmon. Nature, 313: 47–48.
Gross, M.R. 1991. Evolution of alternative reproductive strategies, frequency-dependent sexual selection in male bluegill sunfish. Phil Trans R Soc Lond, B, 332: 59–66.
Gross, M.R. 1996. Alternative reproductive strategies and tactics: diversity within sexes. Trends Ecol Evol, 11: 92–98.
Guerrero, R.D. 1975. Use of androgens for the production of all-male *Tilapia aurea* (Steindachner). Trans Am Fish Soc, 2: 342–348.
Guiguen, Y., M.J. Ricordel and A. Fostier. 1998. Involvement of estrogens in the process of sex differentiation in rainbow trout (*Oncorhynchus mykiss*): *in vivo* treatments, aromatase gene expression. J Exp Zool, 281: 506.
Guiguen, Y., J.F. Baroiller, M.J. Ricordel et al. 1999. Involvement of estrogens in the process of sex differentiation in two fish species: the rainbow trout (*Oncorhynchus mykiss*) and a tilapia (*Oreochromis niloticus*). Mol Reprod Dev, 54: 154–162.
Guin, S., A. Bandyopadhyay, N.R. Jana and S. Bhattacharya. 1993. Thyroid hormone stimulates progesterone release from the ovary of a fish *Anabas testudines.* Curr Sci, 64: 327–329.
Guzman, O.M., J. Ramos, C.C. Mylonas and E.L. Mananos. 2009. Spawning performance and plasma levels of GnRHa and sex steroids in cultured Senegalese sole (*Solea senegalensis*) treatd with GnRHa-delivery systems. Aquaculture, 291: 200–209.

Haddy, J.A. and N.W. Pankhurst. 1999. Stress induced changes in concentrations of plasma sex steroids in black bream. J Fish Biol, 55: 1304–1316.
Haffray, P., V. Petit, Y. Guiguen et al. 2009. Successful production of monosex female brook trout *Salvelinus fontinalis* using gynogenetic sex reversal males by a combination of methyltestosterone immersion and oral treatment. Aquaculture, 290: 47–52.
Haider, S. 2003. Cyclic AMP level and phosphodiesterase activity during 17alpha, 20beta-dihydroxy-4-pregnen-3-one induction and theophylline inhibition of oocyte maturation in the catfish, *Clarias batrachus*. Comp Biochem Physiol, 134A: 267–274.
Hano, T., Y. Oshima, T. Oe et al. 2005. Quantitative bio-imaging analysis for evaluation of sexual differentiation in germ cells of olvas-GFP/ST-II YI medaka (*Oryzias latipes*) nanoinjected in sex change in ova with ethynylestradiol. Environ Toxicol Chem, 24: 70–71.
Harrington, Jr., R.W. 1967. Environmentally controlled induction of primary male gonochorists from the eggs of self-fertilizing hermaphroditic fish *Rivulus marmoratus* Poey. Biol Bull, 132: 174–179.
Harrington, J., J.W. 1971. How ecological and genetic factors interact to determine, when self-fertilizing hermaphrodites of *Rivulus marmoratus* change into functional males with a reappraisal of the modes of intersexuality among fishes. Copeia, 1971: 389–342.
Harrington, Jr., J.W. 1975. Sex determination and differentiation among uniparental homozygotes of the hermaphroditic fish *Rivulus marmoratus* (Cyprinodontidae, Atheriniformes). pp. 249–262. *In*: R. Reinboth [ed.]. Intersexuality in the Animal Kingdom. Springer Verlag, Berlin, Germany.
Harris, P.J. and J.C. McGovern. 1997. Changes in the life history of red porgy *Pagrus pagrus* from the southeastern United States. Fish Bull, 95: 732–747.
Hattori, A. and Y. Yanagisawa. 1991. Life history pathways in relation to gonad sex differentiation in the anemonefish *Amphiprion clarkii* in temperate waters of Japan. Env Biol Fish, 31: 139–155.
Hayakawa, Y., H. Munehara and A. Komaru. 2002. Obstructive role of the dimorphic sperm in a non-copulatory marine sculpin *Hemilepidotus gilberti* to prevent other male's sperm from fertilization. Env Biol Fish, 64: 419–427.
Heath, D.D., J.W. Heath and G.K. Iwama. 1991. Maturation of chinook salmon *Oncorhynchus tshawytscha* (Walbaum) early identification based on the development of a bimodal weight frequency distribution. J Fish Biol, 39: 565–575.
Heath, D.D., L. Rankin, C.A. Bryden et al. 2002. Heritability and Y-chromosome influence in the jack male life history of chinook salmon (*Oncorhynchus tshawytscha*). Heredity, 89: 311–317.
Heg, D., L. Brouwer, Z. Bachar and M. Taborsky. 2005. Large group size yields group stability in the cooperatively breeding cichlid *Neolamprologus pulcher.* Behaviour, 142: 1615–1641.
Hendry, C.I., D.J. Martin-Robichand and T.J. Benfey. 2003. Hormonal sex reversal of Atlantic habibut (*Hippoglossus hippoglossus* L.). Aquaculture, 219: 769–781.
Henrich, S. 1988. Variation in offspring sizes of the poeciliid fish *Heterandria formosa* in relation to fitness. Oikos, 51: 13–18.
Herpin, A. and M. Schartl. 2011. Dmrt1 genes at the crossroads: a widespread and central class of sexual development factors in fish. FEBS J, 278: 1010–1019.
Hess, R.A. and L.R. Franca. 2007. Spermatogenesis and cycle of the seminiferous epithelium. pp. 1–5. *In:* C.Y. Cheng [ed.]. Molecular Mechanisms of Spermatogenesis. Landes Bioscience, Austin, Texas, USA.
Higa, M., K. Ogasawara, A. Sakaguchi et al. 2003. Role of steroid hormones in sex change of protogynous wrasse. Fish Physiol Biochem, 28: 149–150.
Higgs, D.A. and E.M. Donaldson. 1975. A preliminary investigation of the effect of bovine growth hormone on growth and muscle composition of coho salmon (*Oncorhynchus kisutch*). Gen Comp Endocrinol, 27: 240–253.
Hines, G.A., L.R. Boots, T. Wibbels and S.A. Watts. 1999. Steroid levels and steroid metabolism in relation to early gonadal development in the tilapia *Oreochromis niloticus* (Teleostei: Cyprinoidei). Gen Comp Endocrinol, 114: 235–248.

Hioki, S. and K. Suzuki. 1996. Sex changing from male to female on the way of protogynous process in three *Centropyge* angelfishes (Pomacanthidae, Teleostei). Bull Inst Oceanic Res Develop, Tokai Univ, 17: 27–34.

Hirai, N., A. Nanba, M. Koshio et al. 2006. Feminization of Japanese medaka (*Oryzias latipes*) exposed to 17β-estradiol: Formation of testis-ova and sex-transformation during early-ontogeny. Aquat Toxicol, 77: 78–86.

Hiramatsu, N., T. Matsubara, G.M. Weber et al. 2002. Vitellogenesis in aquatic animals. Fish Sci, 68 (suppl.1): 694–699.

Hirschenhauser, K. and R.F. Oliveira. 2006. Social modulation of androgens in male vertebrates: meta-analyses of the challenge hypothesis. Anim Behav, 71: 265–277.

Hishida, T. 1965. Accumulation of estrone–16-C^{14} and diethylstilbestrol (Monoethyl-1-C^{14}) in larval gonads of the medaka *Oryzias latipes* and determination of the minimum dosage of estrogen for sex reversal. Gen Comp Endocrinol, 5: 137–144.

Hishida, T. 1969. Hydroxysteroid dehydrogenase in larval gonad of the medaka *Oryzias latipes*. Zool Mag, Tokyo, 78: 25.

Hobbs, J.P.A., P.L. Munday and G.P. Jones. 2004. Social induction of maturation and sex determination in a coral reef fish. Proc R Soc Lond B, 271: 2109–2114.

Hofmann, H.A. 2006. Gonadotropin-releasing hormone signaling in behavioural plasticity. Curr Opinion Neurobiol, 16: 343–350.

Holmgren, K. 1996. Effect of water temperature and growth variation on the sex ratio of experimentally reared eels. Ecol Freshwat Fish, 5: 203–212.

Hong, Y., T. Liu, H. Zhao et al. 2004. Establishment of a normal spermatogonial germ cell line capable of sperm production *in vitro*. Proc Natl Aced Sci USA, 101: 8011–8016.

Hornung, M.W., K.M. Jensen, J.J. Korte et al. 2004. Mechanistic basis for estrogenic effects in fathead minnow (*Pimephales promelas*) following exposure to the androgen 17α-methyltestrosterone: conversion of 17α-methyltestrosterone to 17α methylestradiol. Aquat Toxicol, 66: 15–23.

Hourigan, T.F., M. Nakamura, Y. Nagahama et al. 1991. Histology, ultrastructure and *in vitro* steroidogenesis of the testis of two male phenotypes of the protogynous fish *Thalassoma duperrey* (Labridae). Gen Comp Endocrinol, 83: 183–217.

Hsu, C.-C., M.-F. Hou, J.-R. Hong et al. 2010. Inducible male sterility by targeted cell ablation in zebrafish testis. Mar Biotechnol, 12: 466–478.

Hsu, H.J., N. Hsu, M.C. Hu and B.C. Chung. 2006. Steroidogenesis in zebrafish and mouse models. Mol Cell Endocrinol, 248: 160–163.

Hulak, M., M. Psenicka, D. Gela et al. 2010. Morphological change upon treatment by endocrine modulates in meiogenetic tench (*Tinca tinca* L.). Aquacult Res, 41: 233–239.

Hunter, G.A. and E.M. Donaldson. 1983. Hormonal sex control and its application to fish culture. 9: 223–303. *In:* W.S. Hoar, D.J. Randall and E.M. Donaldson [eds.]. Fish Physiology: Reproduction, Part B. Behavior and Fertility Control. Academic Press, New York, USA.

Hunter, G.A., I.I. Solar, I.J. Baker and E.M. Donaldson. 1980. Feminization of coho salmon (*Oncorhynchus kisutch*) and Chinook salmon (*Oncorhynchus tshawytscha*) by immersion of alevins in a solution of estradiol-17β. Aquaculture, 53: 295–302.

Idler, D.R., U.H.M. Fagerlund and A.P. Ronald. 1960. Isolation of pregnen-4-one-17α-20β-diol-3-one from plasma of Pacific salmon (*Oncorhynchus nerka*). Biochem Biophy Res Commun, 2: 133–137.

Ijiri, S., H. Kaneko, T. Kobayashi et al. 2008. Sexual dimorphic expression of genes in gonads during early differentiation of a teleost fish the Nile tilapia *Oreochromis niloticus*. Biol Reprod, 78: 333–341.

Immler, S., C. Mazzoldi and M.B. Rasotto. 2004. From sneaker to parental male: change of reproductive traits in the black goby. *Gobius niger* (Teleostei: Gobiidae). J Exp Zool, 301A: 177–185.

Iwamatsu, T., H. Kobayashi, S. Hamaguchi et al. 2005. Estradiol-17β content in developing eggs and induced sex reversal of the medaka (*Oryzias latipes*). J Exp Zool, 303A: 161–167.

Iwamatsu, T., H. Kobayashi, R. Sagegami and T. Shuo. 2006a. Testosterone content of developing eggs and sex reversal in the medaka (*Oryzias latipes*). Gen Comp Endocrinol, 145: 67–74.

Iwamatsu, T., H. Kobayashi and M. Yamashita. 2006b. Sex reversal in medaka treated *in vitro* with 17α-methyldihydrotestosterone during oocyte maturation. Dev Growth Differ, 48: 59–64.

Iwamoto, R.N., B.A. Alexander and W.K. Herschberger. 1984. Genotypic and environmental effects on the incidence of sexual precocity in coho salmon (*Oncorhynchus kisutch*). Aquaculture, 43: 105–121.

Iwata, E., Y. Nagai and H. Sasaki. 2008. Social rank modulates brain arginine vasotocin immunoreactivity in false clown anemonefish (*Amphiprion ocellaris*). Fish Physiol Biochem, DOI 10.1007/s10695-008-9298-y.

Jamieson, B.G.M. 1991. Fish Evolution and Systematic Evidence from Spermatozoa. Cambridge University Press, Cambridge, UK.

Janz, D.M. and G. van Der Kraak. 1997. Suppression of apoptosis by gonadotropin 17β estradiol and epidermal growth factor in rainbow trout preovulatory ovarian follicle. Gen Comp Endocrinol, 105: 186–193.

Jasra, S.K., W.J. Arbuckle, L.D. Corkum et al. 2007. The seminal vesicle synthesizes steroids in the round goby *Neogobius melanostonus.* Comp Biochem Physiol, Part 1 Mol Integr Physiol, 148: 117–123.

Jeng, S.-R., G.-R. Chen, J.-Y. Lai et al. 2002. Regulation of pituitary gonadotropin II and growth hormone content by sex steroids and pituitary extract in the aquacultured Japanese eel *Anguilla japonica*. Aquaculture, 209: 319–332.

Jeng, S.-R., S. Dufour and C.-F. Chang. 2005. Differential expression of neural and gonadal aromatase enzymatic activities in relation to gonadal development in Japanese eel *Anguilla japonica.* J Exp Zool, 303A: 802–812.

Jeng, S.-R., J. Pasquier, W.-S. Yueh et al. 2011. Differential regulation of the expression of cytochrome P450 aromatase, estrogen and androgen receptor subtypes in the brain-pituitary-ovarian axis of the Japanese eel (*Anguilla japonica*) reveals steroid dependent and independent mechanisms. Gen Comp Endocrinol, 175: 163–172.

Jia, Z.Y., L.Y. Shi, X.W. Sun and Q.Q. Lei. 2008. Inheritance of microsatellite DNA in by sexual gynogenesis complex of Fangzheng silver crucian carp *Carassius auratus gibelio* (Bloch). J Fish Biol, 73: 1161–1169.

Johnstone, R., T.H. Simpson and A.F. Youngson. 1978. Sex reversal in salmonid culture. Aquaculture, 13: 115–134.

Johnstone, R., T.H. Simpson and A.F. Youngson. 1979. Sex reversal in salmonid culture. Part II. The progeny of sex reversed rainbow trout. Aquaculture, 18: 13–19.

Johnstone, R., D.J. MacIntosh and R.S. Wright. 1983. Elimination of orally administered 17α-methyltestosterone by *Oreochromis mossambicus* (tilapia) and *Salmo gairdneri* (rainbow trout) juveniles. Aquaculture, 35: 249–259.

Jones, A.G. and J.C. Avise. 2001. Mating systems and sexual selection in male pregnant pipefishes and seahorses: insights from microsatellite-based studies on maternity. J Hered, 92: 150–158.

Jones, A.G., E. Walker, C. Kvarnemo et al. 2001. How cuckoldry can decrease the opportunity for sexual selection: Data and theory from a genetic parentage analysis of the sand goby *Pomatoschistus minutus.* Proc Natl Acad Sci, USA, 98: 9151–9156.

Joy, K.P. and M. Singh. 1998. A comparative study on the distribution, morphology and some biochemical constituents of seminal vesicles in catfishes. J Endocrinol Reprod, 2: 26–33.

Joy, K.P., B. Senthilkumaran and C.C. Sudhakumari. 1998. Periovulatory changes in hypothalamic and pituitary monoamines following GnRH analogue treatment in the catfish *Heteropneustes fossilis*: a study correlating changes in plasma hormone profiles. J Endocrinol, 156: 365–372.

Kagawa, H., H. Tanaka, K. Okuzawa and M. Kobayashi. 1998. GTH II but not GTH I induces final maturation and the development of maturational competence of oocytes of red sea bream *in vitro*. Gen Comp Endocrinol, 112: 80–88.

Kah, O. and S. Dufour. 2011. Conserved and divergent features of reproductive neuroendocrinology in teleost fishes. 1: 15–42. *In:* D.O. Norris and K.H. Lopez [eds.]. Hormones and Reproduction of Vertebrates: Fishes. Elsevier, Amsterdam, The Netherlands.

Kah, O., C. Lethimonier, G. Somoza et al. 2007. GnRH and GnRH receptors in metazoan: a historical, comparative, and evolutive perspective. Gen Comp Endocrinol, 153: 346–364.

Kanamori, A., Y. Nagahama and N. Egami. 1985. Development of the tissue architecture in the gonads of the medaka *Oryzias latipes*. Zool Sci, 2: 695–706.

Kanda, S., Y. Akazome, T. Matsunaga et al. 2008. Identification of KiSS-1 product kiss-peptin and steroid-sensitive sexually dimorphic kisspeptin neurons in medaka (*Oryzias latipes*). Endocrinology, 149: 2467–2476.

Kanoh, Y. 2000. Reproductive success associated with territoriality, sneaking and grouping in male rose bitterlings *Rhodeus ocellatus* (Pisces: Cyprinidae). Env Biol Fish, 57: 143–154.

Katano, O. 1983. Spawning of the dark chub *Zacco temmincki* (Temmineket Schlegel) (Pisces: Cyprinidae) with special reference to male social status. Phy Biol Ecol Japan, 20: 33–52.

Katano, O. 1992. Spawning tactics of paired males of the dark chub *Zacco temmincki* reflect potential fitness costs of satellites. Env Biol Fish, 35: 343–350.

Kavanaugh, S.I., M. Nozaki and S.A. Sower. 2008. Origins of gonadotropin-releasing hormone (GnRH) in vertebrates: identification of a novel GnRH in a basal vertebrate the sea lamprey. Endocrinology, 149: 3860–3869.

Kavitha, P. and P. Subramanian. 2011a. Influence of *Tribulus terrestris* on testicular enzyme in freshwater ornamental fish *Poecilia latipinna*. Fish Physiol Biochem, 37: 801–807.

Kavitha, P. and P. Subramanian. 2011b. Effect of *Tribulus terrestris* on monosex production of *Poecilia latipinna*. Curr Sci, 101: 100–104.

Kavumpurath, S. 1992. Ploidy induction in ornamental fish. Ph.D. Thesis, Madurai Kamaraj University, Madurai, India.

Kavumpurath, S. and T.J. Pandian. 1992. Production of YY male guppy (*Poecilia reticulata*) by endocrine sex reversal and selective breeding. Asian Fish Sci, 57: 265–276.

Kavumpurath, S. and T.J. Pandian. 1993a. Determination of labile period and critical dose for sex reversal by oral administration of estrogens in *Betta splendens* (Regan). Indian J Exp Biol, 31: 16–21.

Kavumpurath, S. and T.J. Pandian. 1993b. Masculinisation of *Poecilia reticulata* by dietary administration of synthetic or natural androgens to gravid females. Aquaculture, 116: 83–89.

Kavumpurath, S. and T.J. Pandian. 1993c. Production of YY female guppy *Poecilia reticulata* by endocrine sex reversal and progeny testis. Aquaculture, 116: 183–189.

Kavumpurath, S. and T.J. Pandian. 1994. Masculinising of fighting fish *Betta splendens* using synthetic or natural androgens. Aquacult Fish Mgmt. 25: 389–393.

Kawahara, T. and I. Yamashita. 2000. Estrogen-independent ovary formation in the medaka fis, *Oryzias latipes*. Zool Sci, 17: 56–68.

Kazeto, Y., M. Kohara, T. Miura et al. 2008. Japanese eel follicle-stimulating hormone (Fsh) and luteinizing hormone (Lh): production and biologically active recombinant FSH and Lh by *Drosophila* S2 cells and their differential actions on the reproductive biology. Boil Reprod, 79: 938–946.

Kersten, C.A., M. Krisfabusi, J.E. Parsons et al. 2001. Gonadal regeneration in masculinised female or steroid treated rainbow trout (*Oncorhynchus mykiss*). J Exp Zool, 290: 396–401.

Khan, I.A., E. Lopez and J. Leloup-hatey. 1986. Effects of hypophysectomy on the testis of the European ell (*Anguilla anguilla* L.). Gen Comp Endocrinol, 62: 411–418.

Khan, M.N., R.L. Ronaud and J.F. Leatherland. 1997. Steroid metabolism by embryonic tissues Arctic charr *Salvelinus alpinus*. Gen Comp Endocrinol, 105: 344–357.

Kim, D.S., H.J. Cho, I.C. Bang et al. 2001. Effects of immersion of larvae in oestradiol-17β on feminization, structural changes of gonad and growth performance in the far eastern catfish *Silurus asotus*. Aquacult Res, 32: 323–328.

Kime, D.E. 1993. 'Classical' and 'non-classical' reproductive steroids in fish. Rev Fish Biol Fish, 3: 160–180.

Kime, D.E. and H. Tiveiten. 2002. Unusual motility characteristics of the spotted wolfish. J Fish Biol, 61: 1549–1559.

Kime, D.E., K.J.W. Van Look, B.G. McAllister et al. 2001. Computer-assisted sperm analysis (CASA) as a tool for monitoring sperm quality in fish. Comp Biochem Physiol, 130: 425–433.

Kimmel, C.B., B. Ullman, C. Walker et al. 2005. Evolution and development of facial bone morphology in three spine stickleback. Proc Natl Acad Sci, USA, 102: 5791–5796.

Kindler, P., D.P. Philipp, M.R. Gross and J.M. Bahr. 1989. Serum 11-ketotestosterone and testosterone concentrations associated with reproduction in male bluegill (*Lepomis macrochirus*: Centrarchidae). Gen Comp Endocrinol, 75: 446–453.

Kindler, P.M., K.M. Bahr and D.P. Philipp. 1991a. The effects of exogenous 11-ketotestosterone, testosterone and cyproterone acetate on pre-spawning and parental care behaviors of male bluegill. Horm Behav, 25: 410–423.

Kindler, P.M., J.M. Bahr, M.R. Gross and D.P. Philipp. 1991b. Hormonal regulation of parental care behavior in nesting male bluegills: do the effects of bromocriptone suggest a role for prolactin? Physiol Zool, 64: 310–332.

King, L.S., D. Kozono and P. Agre. 2004. From structure to disease: the evolving tale of aquaporin biology. Nat Rev Mol Biol, 5: 687–698.

Kirankumar, S. 2003. Induction of intra and interspecific androgenesis in fish. Ph.D. Thesis, Madurai Kamaraj University, Madurai, India.

Kirankumar, S. and T.J. Pandian. 2002. Effect on growth and reproduction of hormone immersed and masculinised fighting fish *Betta splendens*. J Exp Zool, 293: 606–616.

Kirankumar, S. and T.J. Pandian. 2004. Interspecific androgenetic restoration of rosy barb using cadaveric sperm. Genome, 47: 66–73.

Kirankumar, S., V. Anathy and T.J. Pandian. 2003. Hormonal induction of supermale golden rosy barb and isolation of Y chromosome specific markers. Gen Comp Endocrinol, 134: 62–71.

Kishida, M. and J.L. Specker. 2000. Paternal mouthbrooding in the black-chinned tilapia *Sarotherodon melanotheron* (Pisces: Cichlidae): change in gonadal steroids and potential for vitellogenin transfer to larvae. Horm Behav, 37: 40–48.

Kitahashi, T., S. Ogawa and I.S. Parhar. 2009. Cloning and expression of kiss in the zebrafish and medaka. Endocrinology, 150: 821–831.

Kitano, T., N. Yoshinaga, E. Shiraishi et al. 2007. Tamoxifen induces masculinisation of genetic females and regulates p450 aromatase and Mullerian inhabiting substances mRNA expression in Japanese flounder (*Paralichthys olivaceus*). Mol Reprod Dev, 74: 1171–1177.

Kjesbu, O.S., H. Murua, F. Saborido-Rey and P.R. Witthames. 2010. Method development and evaluation of stock reproductive potential of marine fish. Fish Res, 104: 1–7.

Kline, R.J., I.A. Khan and G.J. Holf. 2011. Behavior, color change and time for sexual inversion in the protogynous grouper *Ephinephelus adscensionis*. PLoS One, 6: e.19576 doi:10.1371/journalpone00119576.

Knapp, R. 2004. Endocrine mediation of vertebrate male alternative reproductive tactics: The next generation of studies. Integ Comp Biol, 43: 658–668.

Knapp, R. and S.L. Carlisle. 2011. Testicular function and hormonal regulation in fishes. 1: 43–61. *In*: D.O. Norris and K.H. Lopez [eds.]. Hormones and Reproduction of Vertebrates: Fishes. Elsevier, Amsterdam, The Netherlands.

Knapp, R., J.C. Wingfield and A.H. Bass. 1999. Steroid hormones and parental care in the plainfin midshipmen fish (*Porichthys notatus*). Horm Behav, 35: 81–89.

Knapp, R., E. Marsh-Mathew, L. Vo and S. Rosencrans. 2012. Stress hormone masculinizes female morphology and behaviour. Biol Lett, doi:10.1098/rsbl.0514.

Knaut, H., F. Pelgri, K. Bohmann et al. 2002. Zebrafish *vasa* RNA but not its protein component of the germplasm and segregates asymmetrically before germ line specification. J Cell Biol, 149: 875–888.

Kobayashi, K. and K. Suzuki. 1992. Hermaphroditism and sexual function in *Cirrhitichthys aureus* and the other Japanese hawkfishes (Cirrhitidae, Teleostei). Jap J Ichthyol, 38: 397–410.

Kobayashi, K., K. Suzuki and Y. Shibara. 1993a. Reproduction and hermaphroditism in *Parapersis snyderi* (Teleost: Paraercidae) in Suruga Bay, Central Japan. J Fac Mar Sci Technol, Tokai Univ, 35: 149–168.

Kobayashi, K., K. Suzuki and Y. Shibara. 1993b. Studies on the gonadal sex succession in *Parapercis snyderi* cross examined in captivity. Bull Inst Occanic Res Develop, Tokai Univ, 14: 83–91.

Kobayashi, M. and N. Stacey. 1993. Prostaglandins-induced female spawning behavior in goldfish (*Carassius auratus*) appears independent of ovarian influence. Horm Behav, 27: 38–55.

Kobayashi, M., K. Aida and N.E. Stacey. 1991. Induction of testis development by implantation of 11-Ketotestosterone in female goldfish. Zool Sci, 8: 389–393.

Kobyashi, M., P.W. Sorensen and N.E. Stacey. 2002. Hormonal and pheromonal control of spawning behaviour in the goldfish. Fish Physiol Biochem, 26: 71–84.

Kobayashi, T., X.-T. Chang, M. Nakamura et al. 1996. Fish 3β-hydroxysteroid dehydrogenase/ $\Delta^5\Delta^4$ isomerase: antibody production and their use for the immune histochemical detection of fish steroidogenic tissue. Zool Sci, 13: 909–914.

Kobayashi, T., H. Kajiura-Kobayashi and Y. Nagahama. 2000. Differential expression of *vasa* homologue gene in the germ cells during oogenesis and spermatogenesis in teleost fish *Oreochromis niloticus.* Mech Dev, 99: 134–142.

Kobayashi, T., M. Matsuda, H. Kajiura-Kobayashi et al. 2004. Two Dm-domain genes *Dmy* and *Dmrt1* involved in testicular differentiation and development in medaka *Oryzias latipes.* Dev Dyn, 231: 518–526.

Kobayashi, T., P. Pakarinen, J. Torgersen et al. 2008a. The gonadotropin receptors FSH-R and LH-R of Atlantic halibut (*Hippoglossus hippoglossus*)—2. Differential follicle expression and asynchronous oogenesis. Gen Comp Endocrinol, 156: 595–602.

Kobayshi, T., H. Kajura-Kobayashi, G. Guan et al. 2008b. Sexual dimorphic expression and development in medaka DmRT1 and sox9a during gonadal differentiation and hormone induced sex reversal in the teleost fish (*Oreochromis niloticus*). Dev Dyn, 237: 297–306.

Kobayashi, Y., T. Sunobe, T. Kobayashi et al. 2005. Gonadal structure of the serial sex changing gobiid fish *Trimma okinawae.* Dev Growth Differ, 47: 7–13.

Kobayashi, Y., M. Nakamura, T. Sunobe et al. 2009. Sex change in the gobiid fish is mediated through rapid switching gonadotropin receptors from ovarian to testicular portion and vice versa. Endocrinology, 150: 1503–1511.

Kobayashi, Y., R. Horiguchi, S. Miura and M. Nakumura. 2010a. Sex-and tissue-specific expression of p450 aromatase (*cyp19a1a*) in the yellowtail clownfish *Amphiprion clarkia.* Comp Biochem Physiol Part A, 155: 237–244.

Kobayashi, Y., M.A. Alam, R. Horiguchi et al. 2010b. Sexually dimorphic expression of gonadotropin subunits in the pituitary of protogynous honeycomb grouper (*Epinephelus merra*): Evidence that follicle stimulating hormone (FSH) induces gonadal sex change. Biol Reprod, 82: 1030–1036.

Kobayashi, Y., R. Horiguchi, R. Nozu, M. Nakamura. 2010c. Expression and localization of forkhead transcriptional factor 2 (FoxL2) in the gonads of protogynous *Halichoeres trimaculatus*. Biol Sex Diff, 1.3doi:10.7186/2042-6410-1-3.

Kobayashi, Y., R. Nozu and M. Nakamura. 2011. Role of estrogen in spermatogenesis in initial phase males in the three spotted wrasse (*Halichoeres trimaculatus*). Effect of aromatase inhibitor on the testis. Dev Dyn, 240: 116–121.

Kochakpour, N. and P.B. Moens. 2008. Sex specific crossover patterns in zebrafish (*Danio rerio*). Heredity, 100: 489–495.

Kohda, M., M. Tanimura, M. Kikue-Nakamura and S. Yamagishi. 1995. Sperm drinking by female catfish: a novel mode of insemination. Env Biol Fish, 42: 1–16.

Kojima, Y., R.K. Bhandari, Y. Kobayashi and M. Nakamura. 2008. Sex change of adult initial-phase wrasse *Halichoeres trimaculatus* by estradiol-17β treatment. Gen Comp Endocrinol, 156: 628–632.

Kokokris, L., S. Brusle, M. Kentouri and A. Fostier. 1999. Sexual maturity and hermaphroditsm of the red porgy *Pagrus pagrus* (Teleostei: Sparidae). Mar Biol, 134: 621–629.

Komatsu, T., S. Nakamura and M. Nakamura. 2006. Masculinization of female golden rabbitfish *Siganus guttatus* using an aromatase inhibitor treatment during sex differentiation. Comp Biochem Physiol, Part C, 143: 402–409.

Komen J., J.G.D. Lambert, C.J.J. Richter and H.J.T.H. Goos. 1995. Endocrine control of sex differentiation in XX female, and in XY and XX male common carp (*Cyprinus carpio* L.). p. 383. *In:* F. Goetz and P. Thoms [eds.]. Fifth Internatl Symp Reprod Physiol Fish, University of Texas, Austin, USA.

Kondo, M., I. Nanda, U. Hornung et al. 2003. Absence of the candidate male determining gene *dmrt1(Y)* of medaka from other fish species: Cur Biol, 14: 1664–1669.

Korta, M., R. Dominguez-Petit, H. Murua and F. Saborido-Rey. 2010. Regional variability in reproductive traits of European hake *Merluccias meruluccias* L. populations. Fish Res, 104: 64–72.

Koteeswaran, R. and T.J. Pandian. 2002. Live sperm from post-mortem preserved Indian catfish. Curr Sci, 82: 447–450.

Koulish, S. and C.R. Kramer. 1989. hCG induces sex reversal in *Thalassoma bifasciatum*, a protogynous fish. J Exp Zool, 252: 156–168.

Koulish, S., C.R. Kramer and H.J. Grier. 2002. Organization of male gonad in a protogynous *Thalassoma bifasciatum* (Teleostei: Labridae). J Morphol, 254: 292–311.

Kramer, C. and K.D. Kallman. 1985. Sex differentiation of somatic tissue in the unsexualized gonad primodia of the embryos of three species of poeciliid fish. J Anat, 140: 269–277.

Kramer, C.R., M.T. Cadell and L. Bubenheimer-Livolsi. 1993. sGnRH-A [(D-Arg6. Pro9, Net-) LHRH] in combination with domperidone induces gonad sex reversal in a protogynous fish, the bluehead wrasse *Thalassoma bifasciatum*. J Fish Biol, 42: 185–195.

Kroon, F.J. and N.R. Liley. 2000. The role of steroid hormones in protogynous sex change in blackeye goby *Coryphopterus nicholsii* (Teleostei: Gobiidae). Gen Comp Endocrinol, 118: 273–283.

Kroon, F.J., P.L. Munday, D.A. Westcott et al. 2005. Aromatase pathway mediates sex change in each direction. Proc R Soc Lond B, 272: 1399–1409.

Kumakura, N., K. Okuzawa, K. Gen and H. Kagawa. 2003. Effects of gonadotropin-releasing hormone agonist and dopamine antagonist on hypothalamus-pituitary-gonadal axis of pre-pubertal female red seabream (*Pagrus major*). Gen Comp Endocrinol, 131: 264–273.

Kumar, S.R., I. Shigeho, K. Katherine et al. 2000. Cloning and functional expression of a thyrotropin receptor from the gonads of a vertebrate (boney fish): potential thyroid-independent role for thyrotropin in reproduction. Mol Cell Endocrinol, 167: 1–9.

Kuo, C.M., Y.Y. Ting and S.L. Yeh. 1988. Induced sex reversal and spawning of blue spotted grouper *Epinephelus fario*. Aquaculture, 74: 113–126.

Kurokawa, H., D. Saito, S. Nakamura et al. 2007. Germ cells are essential for sexual dimorphism in the medaka gonad. Proc Natl Acad Sci USA, 104: 16958–16963.

Kusakabe, N., I. Nakamura, J. Evans et al. 2006. Changes in mRNAs encoding steroidogenic acute regulatory protein, steroidogenic enzyme and receptors for gonadotropins during spermatogenesis in rainbow trout testes. J Endocrinol, 189: 541–554.

Kuwamura, T. and Y. Nakashima. 1998. New aspects of sex change among reef fishes: recent studies in Japan. Env Biol Fish, 52: 125–135.

Kuwamura, T., Y. Nakashima and Y. Yogo. 1994. Sex change in either direction by growth rate advantage in the monogamous goby *Paragobiodon echinocephalus*. Behav Ecol, 5: 434–438.

Kuwamura, T., N. Tanaka, Y. Nagashima et al. 2002. Reversed sex change in the protogynous reef fish *Labroides dimidiatus*. Ethology, 108: 443–450.

Kuwamura, T., S. Suzuki, N. Tanaka et al. 2007. Sex change of primary males in a diandric labrid *Halichoeres trimaculatus*: coexistence of protandry and protogyny within a species. J Fish Biol, 70: 1898–1906.

Kuwamura, T., S. Suzuki and T. Kadota. 2011. Reversed sex change by widowed males in polygynous and protogynous fishes: female removal experiments in the field. Naturwissenschaften, 98: 1041–1048.

Kweon, H.-S., E.-H. Park and N. Peters. 1998. Spermatozoan ultrastructure in the internally self-fertilizing hermaphroditic teleost, *Rivulus marmoratus* (Cyprinodontiformes, Rivulidae). Copeia, 1101–1106.

Kwok, H., W. So, Y. Wang and W. Ge. 2005. Zebrafish gonadotropins and their receptors: 1. Cloning and characterization of zebrafish follicle-stimulating hormone and luteinizing hormone receptors—evidence for their distinct functions in follicle development. Biol Reprod, 72: 1370–1381.

Kwon, J.Y., J.M. Brendan and D.J. Penman. 2001a. Cloning of brain aromatase gene and expression of brain and ovarian aromatase genes during sexual differentiation in genetic male and female Nile tilapia *Oreochromis niloticus*. Mol Reprod Dev, 59: 359–370.

Kwon, J., B. McAndrew and D. Penman. 2001b. Cloning of brain aromatase gene during sexual differentiation in genetic male and female Nile tilapia *Oreochromis niloticus*. Mol Reprod Dev, 62: 456–469.

Lacatello, L., A. Pelastro, R. Diana et al. 2007. Variation pattern of sperm quality traits in two gobies with alternative mating tactics. Funct Eol, 21: 975–981.

Lacerda, S.M.S.N., S.R. Batlouni, G.M.J. Costa et al. 2006. Germ cells transplantation in fish: the Nile tilapia model. Anim Reprod, 3: 146–156.

Lacerda, S.M.S.N., S.R. Batlouni, S.B.G. Silva et al. 2010. A new and fast technique to generate offspring after germ cells transplantation in adult fish. The Nile tilapia (*Oreochromis niloticus*) model. PLoS ONE, 5: e10740.doi.10.1371/journal.pone.0010740.

Lahnsteiner, F. and R.A. Patzner. 1990. Spermiogenesis and structure of mature spermatozoa in bleniid fishes (Pisces, Bleniidae). J Submicr Cytol Pathol, 22: 565–576.

Lahnsteiner, F. and R.A. Patzner. 2009a. Male reproductive system: spermatic duct and accessory organs of the testis. Part A: 143–186. *In:* B.M.G. Jamieson [ed.]. Reproductive Biology and Phylogeny of Fishes (Agnathans and Bony Fishes). Science Publishers, Enfield, NH, USA.

Lahnsteiner, F. and R.A. Patzner. 2009b. Sperm morphology and ultrastructure in fish. pp. 1–6. *In:* S.M.H. Alavi, J.J. Cosson, K. Coward and G. Rafee [eds.]. Fish Spermatology. Alpha Science, Oxford, UK.

Lahnsteiner, F., R.A. Patzner and T. Weismann. 1993. The spermatic duct of salmonid fishes (Salmonidae, Teleostei). Morphology histology and composition of secretion of seminal vesicle. J Fish Biol, 42: 79–93.

Lahnsteiner, F., B. Berger and T. Weismann. 1999. Sperm metabolism of the telest fish *Chalcoburunus chalooids* and *Oncorhynchus mykiss* and its relation to mobility and viability. J Exp Zool, 284: 454–465.

Lamatsch, D.K. and M. Stock. 2009. Sperm-dependent parthenogenesis and hybridogenesis in teleost fish. pp. 399–431. *In:* I. Schon, K. Martens and P. van Dijik [eds.]. Lost

Sex: The Evolutionary Biology of Parthenogenesis. Springer Verlag, Dordrecht, The Netherlands.
Lamatsch, D.K., I. Nanda, J.T. Epplen et al. 2000. Unusual triploid males in a microchromosome carrying clone of the Amazon molly *Poecilia formosa*. Cytogenet Cell Genet, 9: 148–159.
Lamatsch, D.K., I. Nanda, I. Schlupp et al. 2004. Distribution and stability of supernumerary microchromosomes in natural populations of the Amazon molly *Poecilia formosa*. Cytogenet Genome Res, 106: 189–194.
Lampert, K.P., D.K. Lamatsch, P. Fischer and M. Schartl. 2008. A tetraploid Amazon molly *Poecilia formosa*. J Hered, doi:10.1098/hered/esm 102.
Lance, V.A. and M.H. Bogart. 1991. Tamoxifen "sex reverses" alligator embryos at male producing temperatures, but is an antiestrogen in female hatchlings. Cell Mol Life Sci, 47: 263–266.
Larsen, M.G. and E. Baatrup. 2010. Functional behavior and reproduction in androgenic sex reversed zebrafish *Danio rerio*. Env Toxicol Chem, 29: 1828–1833.
Larsen, M.G., K. Bilbery and E. Baatrup. 2009. Reversibility of estrogenic sex changes in zebrafish (*Danio rerio*). Environ Toxicol Chem, 28: 1783–1785.
Larson, E.T. 2011. Neuroendocrine regulation in sex-changing fishes. 1: 149–168. *In*: D.O. Norris and K.H. Lopez [eds.]. Hormones and Reproduction of Vertebrates: Fishes. Elsevier, Amsterdam, The Netherlands.
Larson, E.T., D.O. Norris and C.H. Summers. 2003a. Manoaminergic changes associated with socially induced sex reversal in the saddleback wrasse. Neuroscience, 119: 251–263.
Larson, E.T., D.O. Norris, E.G. Gran and C.H. Summers. 2003b. Monoamines stimulate sex reversal in the saddleback wrasse. Gen Comp Endocrinol, 130: 289–298.
Larsson, D.G., T. Sperry and P. Thomas. 2002. Regulation of androgen receptors in Atlantic croaker brain by testosterone and estradiol. Gen Comp Endocrinol, 128: 224–230.
Lebrun, C., R. Billard and B. Jalabart. 1982. Changes in the number of germ cells in the gonads of the rainbow trout (*Salmo gaidneri*) during the first 10 post-hatching weeks. Reprod Nutr Dev, 22: 405–412.
Lee, J.S.E. and A.H. Bass. 2004. Effects of 11-ketotestosterone on brain, sonic muscle and behaviour in type- II midshipman fish. Horm Behav, 46: 115–116.
Lee, J.S.F. and A.H. Bass. 2006. Dimorphic male midshipmanfish: reduced sexual selection or sexual selection for reduced charactors? Behav Ecol, 17: 670–675.
Lee, S.T.L., D.E. Kime, T.J. Lam and C.H. Tan. 1998. Synthesis of 17, 20α/β-dihydroxy-4-pregnen-3-one and 5β-pregnen in spermatozoa of primary and 17α-methyltestosterone-induced secondary male grouper (*Epinephelus coioides*). Gen Comp Endocrinol, 112: 1–9.
Lee, Y.D. and B.H. Kim. 1996. Reproductive biological studies on the seven band grouper *Epinephelus septemfasciatus*. Proc Third Internatl Symp Mar Sci, 55–60.
Lee, Y.H., F.Y. Lee, W.S. Yuch et al. 2000. Profiles of gonadal development sex steroids aromatase activity and gonadotropin II in the controlled sex change of protandrous black porgy *Acanthopagrus schlegeli* Blacker. Gen Comp Endocrinol, 119: 111–120.
Lee, Y.H., J.L. Du, W.S. Yuch et al. 2001. Sex change in protandrous black porgy *Acanthopagrus schlegeli*: A review in gonadal development, estradiol, estrogen receptor, aromatase activity and gonadotropin. J Exp Zool, 290: 715–726.
Lee, Y.H., W.S. Yuch, J.L. Du et al. 2002. Aromatase inhibitors block natural sex change and induce male function in the protandraus black porgy *Acanthopagrus schlegeli* Blacker: Possible mechanism of natural sex change. Biol Reprod, 66: 1749–1754.
Le Gac, F. and M. Loir. 1999. Male reproductive system in fish. 3: 20–30. *In:* E. Knobil and J.D. Neil [eds.]. Encyclopaedia of Reproduction. Academic Press, San Diego, USA.
Le Gac, F., M. Loir, P.Y. Le Bail and M. Ollitrault. 1996. Insulin-like growth factor (IGF-1) mRNA and IGF-1 receptor in trout testis and in isolated spermatogenic and Sertoli cells. Mol Reprod Dev, 44: 23–35.
Legendre, M.G., G. Teugels, C. Canty and B. Jalabert. 1992. A comparative study on morphology, growth rate and reproduction of *Clarias gariepinus* (Burchell, 1822), *Heterobranchus longifilis* (Valenciennes 1840) and reciprocal hybrids (Pisces, Clariidae). J Fish Biol, 40: 59–79.

Leitz, T. and R. Reinboth. 1987. The biosynthesis of 11-ketotestosterone by the testis of the Siamese fighting fish *Betta splendens* Regan (Anabantoidei, Belontiidae). Gen Comp Endocrinol, 66: 145–157.

Le Menn, F., J. Cerda and P.J. Babin. 2007. Ultrastructural aspects of the ontogeny and differentiation of ray—finned fish ovarian follicles. pp. 1–37. *In:* P.J. Babin [ed.]. The Fish Oocyte: From Basic Studies to Biotechnological Application. Springer, Dordrecht, The Netherlands.

Le Page, Y., N. Diotel, C. Vaillant et al. 2010. Aromatase, brain sexualization and plasticity: the fish paradigm. Eur J Neurosci, 32: 2105–2115.

Lethimonier, C., G. Flouriot, Y. Valotaire et al. 2000. Transcriptional interference between glucocorticoid receptor and estradiol receptor mediates the inhibitory effect of cortisol on fish vitellogenesis. Biol Reprod, 62: 1763–1771.

Lethimonier, C., G. Flouriot, O. Kah and B. DeCouret. 2002. The glucocorticoid receptor represses the positive autoregulation of the trout estrogen receptor gene by preventing the enhancer effect of a C/EBP beta-like protein. Endocrinology, 143: 2961–2974.

Levavi-Sivan, B., R. Vaiman, O. Sachs and I. Tzchori. 2004. Spawning induction and hormonal levels during final oocyte maturation in the silver perch *Bidyanus bidyanus*. Aquaculture, 229: 419–431.

Lewis, Z.R., C. McClellan, J.H. Postlethwait et al. 2008. Female specific increase in primordial germ cells marks sex differentiation in threespine stickleback (*Gasterostus aculeatus*). J Morphol, 269: 909–921.

Li, G.L., X.C. Liu, H.R. Lin et al. 2006. Effect of aromatizable and non aromatizable androgens on the sex inversion of red-spotted grouper (*Epinephelus akaara*). Fish Physiol Biochem, 32: 25–33.

Liao, T., Q.L. Guo, S.W. Jin et al. 2009. Comparative responses in rare minnow exposed to 17β estradiol during different life stages. Fish Physiol Biochem, 35: 341–349.

Liley, N.R. 1972. The effects of estrogens and other steroids on the sexual behavior of the female guppy *Poecilia reticulata*. Gen Comp Endocrinol, Suppl, 3: 542–552.

Lin, H.-R., G. Xie, L.-H. Zhang et al. 1998. Artificial induction of gonadal maturation and ovulation in the Japanese eel (*Anguilla japonica* T et S). Bull Fr Peche Piscie, 349: 163–176.

Linhart, O., M. Radina, M. Flajhans et al. 2006. Studies on sperm of diploid and triploid tench *Tinca tinca* L. Aquacult Int, 14: 9–25.

Liu, J.F., Y. Guiguen and S.J. Liu. 2009. Aromatase (*p450 arom*) and 11β-hydroxylase (*p450 11β*) genes are differentially expressed during sex change process of the protogynous rice field eel *Monopterus albus*. Fish Physiol Biochem, 35: 511–518.

Liu, M. and Y. Sadovy. 2004. Early gonadal development and primary males in the protogynous epinepheline *Cephalopholis boenak*. J Fish Biol, 65: 987–1002.

Liu, M. and Y. Sadovy de Mitcheson. 2009. Gonadal development during sexual differentiation in hatchery-produced orange-spotted grouper (*Epinephelus coioides*) and humpback grouper (*Cromileptes altivelis*) (Pisces: Serranidae, Epinephelinae). Aquaculture, 287: 191–202.

Liu, S.J., Q. Qin, J. Xiao et al. 2007. The formation of the polyploidy hybrids from different sub family fish crossings and evolutionary significance. Genetics, 176: 1023–1034.

Liu, X., B. Liang and S. Zhang. 2010. Sequence and expression of cytochrome p450 aromatase and FTZ—F1 genes in the protandrous black porgy (*Acanthopagrus schlegeli*). Gen Comp Endocrinol, 138: 247–254.

Liu, Z., F. Wu, B. Ziao et al. 2007. Molecular cloning of doublesex and mab 3 related transcribtion factor 1, forkhead transcription factor gene 2 and two types of cytochromone p450 aromatases in Southern catfish and their possible roles in sex differentiation. J Endocrinol, 194: 223–241.

Liu, Z.H., Y.G. Zhang and D.S. Wang. 2008. Studies on feminization, sex determination, and differentiation of the Southern catfish, *Silurus meridionalis*—a review. Fish Physiol Biochem, DOI 10.1007/s10695-008-9281-7.

Lodi, E. 1979. Instances of sex inversion in the domesticated swordtail *Xiphophorus helleri* Heckel (Pisces, Osteichthys). Experientia, 35: 1440–1441.

Loir, M., C. Caulty, P. Planquette and P.Y. LeBail. 1989. Comparative study of the male reproductive tract in seven families of South-American catfish. Aquat Living Resour, 2: 45–56.

Lone, K.P. and A.J. Matty. 1981. Uptake and disappearance of radioactivity in blood and tissue of carp (*Cyprinus carpio*) after feeding 3H-testosterone. Aquaculture, 24: 315–326.

Lopez-Olmeda, J.F., J.A. Madrid and F.J. Sanchez-Vasquez. 2006. Light and temperature cycles as zeitgebers of zebrafish (*Danio rerio*) circadian activity rhythms. Chronobiol Internat, 23: 537–550.

Lorenzi, V., R.L. Early, E.W. Rodgers et al. 2008. Diurnal patterns and sex differences in cortisol, 11-ketotestosterone, testosterone and 17-beta estradiol in the blue banded goby (*Lythrypnus dalli*). Gen Comp Endocrinol, 155: 438–446.

Love, M.S., M. Yoklarich and L. Thornstein. 2002. The rockfishes of the northeast Pacific. University of California Press, Berkeley, CA, USA.

Lovejoy, D.A., W.H. Fischer, S. Ngamvongchon et al. 1992. Distinct sequence of gonadotropin-releasing hormone (GnRH) in dogfish brain provides insight into GnRH evolution. Proc Natl Acad Sci USA, 89: 6373–6377.

Lowe, T.P. and J.R. Larkin. 1975. Sex reversal in *Betta splendens* Regan with emphasis on the problem of sex determination. J Exp Zool, 191: 25–32.

Lu, Y.S., W. Hu, X.C. Liu et al. 2010. Molecular cloning and mRNA expression pattern of sox9 during sex reversal in orange-spotted grouper (*Epinephelus coioides*). Aquaculture, 306: 322–328.

Lubzens, E., G. Young, J. Bobe and J. Cerda. 2010. Oogenesis in teleosts: How fish eggs are formed. Gen Comp Endocrinol, 165: 367–389.

Luckenbach, J.A., J. Godwin, H.V. Daniels et al. 2003. Gonadal differentiation and effects of temperature on sex differentiation in Southern flounder (*Paralichthys lethostigma*). Aquaculture, 216: 315–327.

Luckenbach, J.A., L.W. Early, A.H. Rowe et al. 2005. Aromatase cytochrome p450: Cloning, intron variation, and ontogeny of gene expression in Southern flounder (*Paralichthys lethostigma*). J Exp Zool, 303A: 643–656.

Luckenbach, J.A., D.B. Iliev, F.W. Goetz and P. Swanson. 2008. Identification of differentially expressed ovarian genes during primary and secondary oocyte growth in coho salmon, *Oncorhynchus kisutch*. Reprod Biol Endocrinol, 6: 2–16.

Luckenbach, J.A., J.T. Dickey and P. Swanson. 2010. Regulation of pituitary GnRH receptor and gonadotropin subunits by IGF1 and GnRH in prepubertal male coho salmon. Gen Comp Endocrinol, 167: 387–396.

Lynn, S.G., K.A. Powell, D.F. Westneat and B.S. Shepherd. 2009. Seasonal and sex specific mRNA level of key endocrine genes in adult yellow perch (*Perca fluviatalis*) from Lake Erie. Mar Biotechnol, 11: 210–222.

Ma, Y., O.S. Kjesbu and T. Jorgensen. 1998. Effects of ration on the maturation and fecundity in captivity of Atlantic herring (*Clupea harengus*). Can J Fish Aquat Sci, 55: 900–908.

Macchi, C.J. 1998. Preliminary estimate of spawning frequency and batch fecundity of striped weakfish *Cynoscion striatus* in coastal waters of Buenos Aires province. Fish Bull, 96: 375–381.

Macchi, G.J. and E.M. Acha. 2000. Spawning frequency and batch fecundity of Brazilian menhaden *Brevoortia aurea* in the Rio de la Plata estuary off Argentina and Uruguay. Fish Bull, 98: 283–289.

MacFarlane, R.B. and M.J. Bowers. 1995. Matrotrophic viviparity in the yellowtail rockfish *Sebastes flavidus*. J Exp Biol, 198: 1197–1206.

MacGregor, J.I. and V.C. Jordan. 1998. Basic guide to the mechanisms of antiestrogen action. Pharmocol Rev, 50: 151–196.

Mackie, M.C. 2003. Socially controlled sex change in the half-moon grouper *Ephinephelus rivulatus*, at Ningaloo Reef, Western Australia. Coral Reefs, 22: 132–142.

Madhu, K. and R. Madhu. 2005. Protandrous hermaphroditism in the clownfish *Amphiprion percula* from Andaman and Nicobar Islands. Indian J Fish, 53: 373–380.

Madhu, R., K. Madhu and K.M. Venugopalan. 2010. Sex change of hatchery produced *Amphiprion ocellaris*: Influence of mating system removal gonad maturation and nesting success. J Mar Biol Ass India, 52: 62–69.

Maitra, S.K., A. Chattoraj and S. Bhattacharyya. 2005. Implication of melatonin in oocyte maturation in Indian major carp *Catla catla*. Fish Physiol Biochem, 31: 201–207.

Majhi, S.K., R.S. Hattori, M. Yokota et al. 2009. Germ cell transplantation using sexually competent fish: An approach for rapid propagation of endangered and valuable germlines. PLoS ONE, 4: e6132. Doi:10.1371/journal.pone, 0006132.

Malavasi, S., P. Torricelli, M. Liigli et al. 2003. Male courtship in a teleost with alternative reproductive tactics, the grass goby *Zosterisessor ophiocephalus*. Env Biol Fish, 66: 231–236.

Malison, J.A., L.S. Procarione, T.P. Barry et al. 1994. Endocrine and gonadal changes during the annual reproductive cycle of the freshwater teleost *Stizostedion vitreum*. Fish Physiol Biochem, 13: 473–784.

Malison, J.A., T.B. Kayes, C.D. Best and C.H. Amudson. 1986. Sexual differentiation and use of hormones to control sex in yellow perch (*Perca flavesceus*). Can J Fish Aquat Sci, 43: 26–35.

Manickam, P. and K.P. Joy. 1989. Induction of maturation and ovulation by pimozide-LHRH analogue treatment and resulting in high quality egg production in the Asian catfish *Clarias batanchus* (L). Aquaculature, 83: 193–199.

Manning, A.J., H.M. Murray, J.W. Gallant et al. 2008. Ontogenetic and tissue-specific expression of preproghrelin in the Atlantic halibut, *Hippoglossus hippoglossus* L. J Endocrinol, 196: 181–192.

Manosroi, J., K. Petcrjul and A. Manosroi. 2003. Effect of fluoxymesterone fish feed granule on sex reversal of the hybrid, Thai Red tilapia (*Oreochromis niloticus* Linn x *Oreochromis mossambicus* Linn.). MS reviewed.

Mansour, N., F. Lahnsteiner and R. Patzner. 2002. The spermatozoan of the African catfish: fine structure, motility, viability and its behaviour in seminal vesicle secretion. J Fish Biol, 60: 545–560.

Marino, G., E. Panini, A. Longobardi et al. 2003. Induction of ovulation in captive-reared dusky grouper *Epinephelus marginatus* (Lowe, 1834) with a sustained release GnRHa implant. Aquaculture, 219: 841–858.

Marshall, D.J., S.S. Heppell, S.B. Munch and R.R. Warner. 2010. The relationship between maternal phenotype and offspring quality: Do older mothers really produce the best offspring? Ecology, 91: 2862–2873.

Martin, E. and M. Taborsky. 1997. Alternative male mating tactic in a cichlid, *Pelvicachromis pulcher:* a comparison of reproductive effort and success. Behav Ecol Sociobiol, 41: 311–319.

Martinez-Chavez, C.C., M. Minghetti and H. Migaud. 2008. GPCR54 and rGnRH I gene expression during the onset of puberty in Nile tilapia. Gen Comp Endocrinol, 156: 224–233.

Martin-Robichaud, D.J., R.H. Peterson, T.J. Benfey and L.W. Crim. 1994. Direct feminization of lumpfish (*Cyclopterus lumpus* L.) using 17β-oestrodiol-enriched *Artemia* as food. Aquaculture, 123: 137–151.

Maruska, K.P. and R.D. Fernald. 2011. Plasticity of the reproductive axis caused by social status change in an African cichlid fish: II Testicular gene expression and spermatogenesis. Endocrinology, 152: 291–302.

Maruska, K.P., B. Levavi-Sivan and R.D. Fernald. 2011. Plasticity of the reproductive axis caused by social status change in an African cichlid fish: I Pitutary gonadotropins. Endocrinology, 152: 281–290.

Mathew, G. 2011. Gonadal restructuring during sex transformation in the protogynous greasy grouper *Epinephelus tauvina.* (Forskal) (Perciformes: Serranidae). Indian J Fish, 58: 30–41.

Matsuda, M., Y. Nagahama, A. Shinomiya et al. 2002. *Dmy* is a Y specific *Dm*-domain gene required for male development in the medaka fish. Nature, 417: 553–563.

Matsuda, M., T. Sato, Y. Toyazaki et al. 2003. *Oryzias curvinotus* has *Dmy*, a gene that is required for male development in the medaka, *O. latipes*. Zool Sci, 20: 159–161.

Matsubara, K., K. Arai and R. Suzuki. 1995. Survival potential and chromosomes of progeny of triploid and pentaploid females in the loach *Misguramus angillicandatus*. Aquaculture, 131: 37–48.

Matsuyama, M., S. Onozato and M. Kashiwagi. 2002. Endocrine control of diurnal oocyte maturation in the Kyusen wrasse *Halichoeres poecilopterus.* Zool Sci, 19: 1045–1053.

Matta, S.L., D.A. Vilela, H.P. Godinho and L.R. Franca. 2002. The goitrogen 6-n-propyl-2-thiouracil (PTV) given during testis development increases Sertoli and germ cell number per cyst in fish: the tilapia (*Oreochromis niloticus*) model. Endocrinology, 143: 970–978.

Mattei, X. 1993. Peculiarities in the organization of testis of *Ophidion* sp. (Pisces: Teleostei). Evidence for two types of spermatogenesis in teleost fish. J Fish Biol, 43: 931–937.

Matty, A.J. 1985. Fish Endocrinology. Croom Hills, Kent, UK. pp. 272.

Maugars, G. and M. Schmitz. 2008. Gene expression profiling during spermatogenesis in early maturing Atlantic salmon parr testes. Gen Comp Endocrinol, 159: 178–187.

Mayer, I., B. Borg and R. Schultz. 1990a. Seasonal changes in and effects of castration/androgen replacement on the plasma levels of five androgens in the male three-spined stickleback *Gasterostcus aculeatus* L. Gen Comp Endocrinol, 79: 23–30.

Mayer, I., B. Borg and R. Schultz. 1990b. Conversion of 11-ketoandrostenedione to 11-ketotestosterone by blood cells of six fish species. Gen Comp Endocrinol, 77: 70–74.

Mayer, I., H. Lundqvist, I. Berglund et al. 1993. Plasma levels of sex steroids in three species of pipefish (Syngnathidae). Can J Zool, 71: 1903–1907.

Mazzoldi, C., M. Scagginate, E. Ambrosin and M.B. Rasolta. 2000. Mating system and alternative male mating tactics in the grass goby *Zosterisessor ophiocephalus* (Teleostei: Gobiidae). Mar Biol, 137: 1041–1048.

Mazzoldi, C., V. Lorenzi and M.B. Rasotto. 2007. Variation of male reproductive apparatus in relation to fertilization modalities in the catfish families Auchenipteridae and Callichthyidae (Teleostei: Siluriformes). J Fish Biol, 70: 243–256.

McCaffrey, K., M.B. Hawkins and J. Godwin. 2011. Sexual phenotype differences in zic2 mRNA abundance in the preoptic area of a protogynous teleost *Thalassoma bifasciatum.* PLoS ONE, 6: 23213.doi: 10.1371/journal.pone.0023213.

McCormick, M.I. 1998. Behaviourally induced maternal stress in a fish influences progeny quality by a hormonal mechanism. Ecology, 79: 1873–1883.

McCormick, M.I. 1999. Experimental test of the effect of maternal hormones on larval quality of a coral reef fish. Oecologia, 118: 412–422.

McIntosh, D.J., T.J. Varghese and G.P.S. Rao. 1985. Hormonal sex reversal of wild spawned tilapia in India. J Fish Biol, 26: 44–52.

Meghan, L.M.F., J.B. Nicholas and G. van der Kraak. 2011. Stress and reproduction. pp. 103–117. *In:* D.O. Norris and K.H. Lopez [eds.]. Hormones and Reproduction of Vertebrates: Fishes. Elsevier, Amsterdam, The Netherlands.

Mehault, S., R. Dominguez-Petit, S. Cervino and F. Saborido-Rey. 2010. Variability in total egg production and implications for management of the southern stock of European hake. Fish Res, 104: 111–122.

Melard, C. 1995. Production of a high percentage of male offspring with 17α-ethynylestradiol sex-reversed *Oreochromis aureus.* I Estrogen sex reversal and production of F_2 pseudofemales. Aquaculture, 130: 25–34.

Meredith, R.W., M.N. Pires, D.N. Reznick and M.S. Springer. 2009. Molecular phylogenetic relationships and the evolution of the placenta in *Poecilia* (*Micropoecilia*) (Poeciliidae: Cyprinodontiformes). Mol Phylogenet Evol, 55: 631–639.

Meredith, R.W., M.N. Pires, D.N. Reznick and M.S. Springer. 2011. Molecular phylogenetic relationships and the coevolution of placentotrophy and superfetation in *Poecilia* (Poeciliidae: Cyprinodontiformes). Mol Phylogenet Evol, doi:10.1016/j.ympev.2011.01.014.

Migaud, H., J.F. Taylor, G.L. Taranger et al. 2006. A comparative *ex vivo* and *in vivo* study of day and night perception in teleost species using the melatonin rhythm. J Pineal Res, 41: 42–52.

Migaud, H., A. Davie, C.C. Martonez-Chavez and S. Al-Khamees. 2007. Evidence for differential photic regulation of pineal melatonin synthesis in teleosts. J Pineal Res, 43: 327–335.

Migaud, H., A. Davie and J.F. Taylor. 2010. Current knowledge on the photoneuroendocrine regulation of reproduction in temperate fish species. J Fish Biol, 76: 27–68.

Milla, S., B. Jalabart, H. Rime et al. 2006. Hydration of rainbow trout oocyte during meiotic maturation and *in vitro* regulation by 17, 20β-dihydroxy-4-pregnen-3-one and cortisol. J Exp Biol, 209: 1147–1156.

Milla, S., X. Terrien, A. Strum et al. 2008. Plasma 11-deoxycorticosterone (DOC) and mineral corticoid receptor testicular expression during rainbow trout *Oncorhynchus mykiss* spermiation: implication with 17α, 20β-dihydroxyprogestrone on the milt fluidity. Reprod Biol Endocrinol, 19: 6–19.

Milla, S., N. Wang, S.N.M. Madiki and P. Kestemont. 2009. Corticosteroids: friends or foes of teleost fish reproduction. Comp Biochem Physiol, 153A: 242–251.

Minamimoto, M., Y. Sakakura, K. Soyano et al. 2006. Plasma sex steroid levels and steroidogenesis in the gonad of the self-fertilizing fish *Rivulus marmoratus*. Env Biol Fish, 75: 159–166.

Mingist, M., T. Kitani, N. Koide and H. Ueda. 2007. Relationship between eyed-egg percentage and levels of cortisol and thyroid hormone in masu salmon *Oncorhynchus masou*. J Fish Biol, 70: 1045–1056.

Miranda, J.A., R.F. Oliveira, L.A. Canario et al. 2003. Neurochemical correlates of male polymorphism and alternative tactics in the Azorean rock-poll blenny *Parablennius parvicornis*. Gen Comp Endocrinol, 132: 183–189.

Mire, J.B. and L. Millett. 1994. Size of mother does not determine size of eggs or fry in the Owens pupfish, *Cyprinodon radiosus*. Copeia, 1994: 100–107.

Miura, C., T. Miura, M. Yamashita et al. 1996. Hormonal induction of all stages of spermatogenesis in germ-somatic cell coculture from immature Japanese eel testis. Develop Growth Differ, 38: 257–262.

Miura, C., R. Kuwahara and T. Miura. 2007a. Transfer of spermatogenesis-related cDNAs into eel testis germ somatic cell culture pellets by electroporation: Methods for analysis of gene function. Mol Reprod Dev, 74: 420–427.

Miura, C., T. Higashino and T. Miura. 2007b. A progestin and an estrogen regulate early stages of oogenesis in fish. Biol Reprod, 77: 822–828.

Miura, T. 1999. Spermatogenic cycle in fish. 4: 571–578. *In:* E. Knobil and J.D. Neil [eds.]. Encyclopaedia of Reproduction. Academic Press, San Diego, USA.

Miura, T. and C. Miura. 2001. Japanese eel: A model for analysis of spermatogenesis. Zool Sci, 18: 1055–1063.

Miura, T. and C.I. Miura. 2003. Molecular control mechanisms of fish spermatogenesis. Fish Physiol Biochem, 28: 181–186.

Miura, T., C. Miura, M. Ohta et al. 1991a. Estradiol-17β stimulates the renewal of spermatogonial stem cells in males. Biochem Biophys Res Commun, 144: 5504–5510.

Miura, T., K. Yamaguchi, H. Takahashi and Y. Nagahama. 1991b. Involvement of steroid hormones in gonadotropin-induced maturation in male Japanese eel (*Anguilla japonica*). Biomed Res, 12: 241–248.

Miura, T., C. Miura, K. Yamguchi and Y. Nagahama. 1995. Human recombinant activin induces proliferation of spermatogonia *in vito* in the Japanese eel (*Anguilla japonica*). Fish Sci, 61: 434–437.

Miura, T., S. Kawamura, C. Miura and K. Yamaguchi. 1997. Impaired spermatogenesis in the Japanese eel *Anguilla japonica:* Possibility for existence of factors that regulate entry of germ cells into meiosis. Dev Growth Differ, 39: 685–691.

Miura, T., C. Miura, T. Ohta et al. 1999. Estradiol 17β stimulates the renewal of spermatogonial stem cells in male. Fish Physiol Biochem, 28: 181–186.

Miura, T., C. Miura, Y. Konda and K. Yamaguchi. 2002. Spermatogenesis-preventing substance in Japanese eel. Development, 129: 2689–2697.

Miura, T., M. Higuchi, Y. Ozaki et al. 2006. Progestin is an essential factor for initiation of the meiosis in spermatogenetic cells of the eel. Proc Natl Acad Sci USA, 103: 7337–7338.

Miyamoto, K., Y. Hasegawa, T. Minegishi et al. 1982. Isolation and characterization of chicken hypothalamic luteinizing hormone releasing hormone. Biochem Biophys Res Commun, 107: 820–837.

Modesto, T. and A.V.M. Canario. 2003. Morphometric changes and sex steroid levels during annual reproductive cycle of the Lusitanean toadfish *Halobatranchus didactylus.* Gen Comp Endocrinol, 131: 220–231.

Modig, C., L. Westerlund and P.E. Olsson. 2007. Oocyte zona pellucida proteins. pp. 113–139. *In:* P.J. Babin, J. Cerda and E. Lubzens [eds.]. The Fish Oocyte: From Basic Studies to Biotechnological Applications. Springer, Dordrecht, The Netherlands.

Molloy, P.P., N.B. Godwin, I.M. Cole et al. 2007. Sperm competition and sex change: A comparative analysis across fishes. Evolution, 61: 640–652.

Moncaut, N., G. Somoza, D.M. Power and A.V. Canario. 2005. Five gonadotropin-releasing hormone receptors in a teleost fish: isolation, tissue distribution and phylogenetic relationships. J Mol Endocrinol, 34: 767–779.

Montero, M. and S. Dufour. 1996. Gonadotropin-releasing hormones (GnRH) in fishes: Evolutionary data on their structure, localization, regulation and function. Zool Stud, 35: 149–160.

Moore, M.C., D.K. Hews and R. Knapp. 1998. Hormonal control and evolution of alternative male phenotypes. Generalizations of models of sexual differentiation. Am Zool, 38: 133–151.

Morita, K. and Y. Takashima. 1998. Effect of female size on fecundity and egg size in white-spotted charr: comparison between sea-run and resident forms. J Fish Biol, 53: 1140–1142.

Morrey, C.E., Y. Nagahama and E.G. Grau. 2002. Terminal phase males stimulate ovarian function and inhibit sex change in the protogynous wrasse *Thalassoma duperrey.* Zool Sci, 19: 103–109.

Moser, H.G. 2007. Reproduction in viviparous South African clinid fish *Fuconimus mus.* S Afr J Mar Sci, 29: 423–436.

Moyer, J.T. and A. Nakazona. 1978. Protandrous hermaphroditism in six speices of anemonefish genus *Amphiprion* in Japan. Jap J Ichthyol, 24: 101–112.

Moyer, J.T. 1990. Social and reproductive behaviour of *Chaetodontoplus mesoleucus* (Pomacanthidae) at Bantayan Island, Philippines with notes on pomacanthid relationships. Jap J Ichthyol, 36: 459–462.

Muller, R. 1969. Die Einwirkung von sexual hormonalle die Geschlechts Differenzierung von *Hemihaplochromis multicolour* (Hilgendorf, Cichlidae). Zool Jahrb, Abt Allg Zool Physiol Tiere, 74: 519–562.

Muller-Belecke, A. and G. Horstgen-Schwark. 2007. A YY-male *Oreochromis niloticus* strain developed from an exceptional mitotic gynogenetic male and growth performance testing of genetically all-male progenies. Aquacult Res, 38: 773–775.

Munday, P.L., M.J. Caley and G.P. Jones. 1998. Bidirectional sex change in a coral-dwelling goby. Behav Ecol Sociobiol, 43: 371–377.

Munday, P.L., J.W. White and R.R. Warner. 2006. A social basis for the development of primary males in a sex changing fish. Proc R Soc Lond B, 273: 2845–2851.

Munehara, H., K. Takano and Y. Koya. 1989. Internal gametic association and external fertilization in the elkhorn sculpin *Alcichthys alcicornis.* Copeia, 1989: 675–678.

Munoz, M., M. Casadevall and S. Bonet. 2002. Testicular structure and semicystic spermatogenesis in a specialized ovuliparous species: *Scorpaena notata* (Pisces, Scorpaenidae). Acta Zool, 83: 213–219.

Murata, R., H. Karimata, M.A. Alam and M. Nakamura. 2010. Precocious sex change and spermatogenesis in the underyearling Malabar grouper *Epinephelus malabaricus* by androgen treatment. Aquacult Res, 41: 303–308.

Murua, H., L. Ibaibarriaga, P. Alvarez et al. 2010. The daily egg production method: A valid tool for application of European hake in the Bay of Biscay. Fish Res, 104: 100–110.

Mustonen, A.M., P. Nieminen and H. Hyvrinen. 2002. Leptin, ghrelin and energy metabolism of the spawning burbot (*Lota lota* L.). J Exp Zool, 293: 119–126.

Nagahama, Y. 1994. Endocrine regulation of gametogenesis in fish. Int J Dev Biol, 38: 217–229.

Nagahama, Y. 1997. 17 alpha, 20 beta-dihydroxy-4-e pregnen-3-one, a maturation inducing hormone in fish oocytes: Mechanisms of synthesis and action. Steroids, 62: 190–196.

Nagahama, Y. 2000. Gonadal steroid hormones: major regulators of gonadal set differentiation and gametogenesis in fish. pp. 211–222. *In:* B. Norberg, O.S. Kjesbu, G.L. Taranger et al. (eds.). Proc Sixth Internal Symp reprod Physiol Fish. University of Bergen, Norway.

Nagahama, Y. 2005. Molecular mechanisms of sex determination and gonadal sex differentiation in fish. Fish Physiol Biochem, 31: 105–109.

Nagahama, Y. and M. Yamashita. 2008. Regulation of oocyte maturation in fish. Dev Growth Diff, 215: 11–18.

Nagahama, Y., M. Yamashita and T. Tokumoto. 1994. Regulation of oocyte maturation in fish. Curr Topics Dev Biol, 30: 103–145.

Nagashima, Y., Y. Sakai, K. Karino and T. Kuwamura. 2000. Female-female spawning and sex change in a haremic coral–reef fish *Labroides dimidiatus.* Zool Sci, 17: 967–970.

Naish, K.A. and A.J. Ribbink. 1990. A preliminary investigation of sex change in *Pseudotropheus lombardoi* (Pisces: Cichlidae). Env Fish Biol, 28: 285–294.

Nakamura, M. 1975. Dosage-dependent changes in the effect of oral administration of methyltestosterone on gonadal sex differentiation in *Tilapia mossambica.* Bull Fac Fish, Hokkaido Uni, 26: 99–108.

Nakamura, M. 1978. Morphological and experimental studies on sex differentiation of the gonad in several teleost fishes. Ph.D. Thesis, Hokkaido University, Hokkaido, Japan.

Nakamura, M. 1981. Feminization of masu salmon *Oncorhynchus masou* by administration of oestradiol-17β. Bull Jap Soc Sci Fish, 47: 1529.

Nakamura, M. and Y. Nagahama. 1985. Steroid producing cells during ovarian differentiation in the tilapia *Sarotherodon niloticus.* Dev Growth Diff, 27: 701–708.

Nakamura, M. and Y. Nagahama. 1993. Ultrastructural study on the differentiation and development of steroid-producing cells during ovarian differentiation in the amago salmon, *Oncorhynchus rhodurus.* Aquaculture, 112: 237–251.

Nakamura, M., T. Hourigan, F. Yamaguchi et al. 1989. Histological and ultrastructural evidence for the role of gonadal steroid hormones in sex change in the protogynous wrasse *Thalassoma duperrey.* Env Biol Fish, 24: 117–136.

Nakamura, M., T. Kobayashi, X.T. Chang and Y. Nagahama. 1998. Gonadal sex differentiation in teleost fish. J Exp Zool, 281: 362–372.

Nakamura, M., R.H. Bhandari and M. Higa. 2003. The role estrogens play in sex differentiation and sex change of fish. Fish Physiol Biochem, 28: 113–117.

Nakamura, M., Y. Kobayashi, S. Miura et al. 2005. Sex change in coral reef fish. Fish Physiol Biochem, 31: 117–122.

Nam, Y.K., C.H. Noh and D.S. Kim. 1998. Effects of 17α-methyltestosterone treatments on sex reversal of mud loach *Misgurnus mizolepis.* Fish Sci, 64: 914–917.

Nam, Y.K., Y.S. Cho and D.S. Kim. 2000. Isogenic transgenic homozygous fish induced by artificial parthenogenesis. Transgenic Res, 9: 463–469.

NaNakorn, U., P. Sidthikraiwong, W. Tanchala-Nukit and T.R. Roberts. 1993a. Chromosome study of hybrid and gynogenetic offspring of artificial crosses between members of the catfish families Clariidae and Pangasiidae. Env Biol Fish, 37: 317–322.

NaNakorn, U., W. Rangsin and S. Witchasunscul. 1993b. Suitable conditions for induction of gynogenesis in the catfish *Clarias macrocephalus* using sperm of *Pangasius sutchi*. Aquaculture, 118: 53–68.

Nanda, I., M. Kondo, U. Hornung et al. 2002. A duplicated copy of *DMRT 1* in the sex-determining region of the Y chromosome of the medaka *Oryzias latipes*. Proc Natl Acad Sci USA, 99: 11778–11783.

Nanda, I., I. Schlupp, D.K. Lamatsch et al. 2007. Stable inheritance of host species derived microchromosomes in the gynogenetic fish *Poecilia formosa*. Proc Natl Acad Sci, USA, 99: 11778–11783.

Nash, R.D.M., G.M. Pilling, L.T. Kell et al. 2010. Investment in maturity-at-age and -length in northeast Atlantic cod stocks. Fish Res, 104: 89–99.

Nath, P. and B.I. Sundararaj. 1981. Isolation and identification of female-specific serum lipophosphoprotein (Vitellogenin) in the catfish *Heteropneustes fossilis* (Bloch). Gen Comp Endocrinol, 43: 184–190.

Nath, P., M. Bhakta, S. Maitra and S. Sarkar. 1997. Vitellogenin induces vitellogenesis in the catfish, *Clarias batrachus*. pp. 1475–1479. *In:* S. Kawashima and S. Kikuyama [eds.]. Advances in Comparative Endocrinology. Tome 2. Monduzzi Editore S.P.A. Bologna, Italy.

Nath, P., R. Sahu, S.K. Kabita and D. Bhattacharya. 2007. Vitellogenesis with special emphasis on Indian fishes. Fish Physiol Biochem, 33: 359–366.

Navarro-Martin, L., M. Blazquez and F. Piferrer. 2009. Masculinization of the European sea bass (*Dicentrarchus labrax*) by treatment with an androgen or aromatase inhibitor involves different gene expression and has distinct lasting effects on maturation. Gen Comp Endocrinol, 160: 3–11.

Nayyar, S.K. and B.I. Sundararaj. 1970. Seasonal reproductive activity in the teleosts and seminal vesicles of the catfish *Heteropneustes fossilis* (Bloch). J Morphol, 130: 207–266.

Neff, B.D. 2001. Genetic paternity analysis and breeding success in bluegill sunfish (*Lepomis macrochirus*). J Hered, 92: 111–119.

Neff, B.D. 2004. Increased performance of offspring sired by parasitic males in bluegill sunfish. Behav Ecol, 15: 327–331.

Neff, B.D., P. Fu and M.R. Gross. 2000. Microsatellite multiplexing fish. Trans Am Fish Soc, 129: 590–599.

Ngamvongchon, S., D.A. Lovejoy, W.H. Fischer et al. 1992. Primary structures of two forms of gonadotropin-releasing hormone, one distinct and one conserved from catfish brain. Mol Cell Neurosci, 3: 17–22.

Nimeitz, A., R. Kreutzfeldt, M. Schartl et al. 2002. Male mating behaviour of a molly *Poecilia latipuntctata*: a third host for the sperm-dependent Amazon molly *Poecilia formosa*. Acta Ethol, 5: 45–49.

Nirmala, A.R.C. and T.J. Pandian. 1983. Effect of steroid injection on food utilization in *Channa striatus*. Proc Ind Acad Sci Ani Sci, 92: 221–229.

Nobrega, R.H. and I. Quagio-Grassiotto. 2007. Morpho-functional changes in Leydig cells throughout the continuous spermatogenesis of the freshwhater teleost fish *Serrasalmus spilopleura* (Characiformes, Characidae): an ultrastructural and enzyme study. Cell Tissue Res, 329: 339–349.

Nobrega, R.H., S.R. Batlouni and L.R. Franca. 2009. An overview of functional and stereological evaluation of spermatogenesis and germ cell transplantation. Fish Physiol Biochem, 35: 197–206.

Nocillado, J., B. Levavi-Sivan, F. Carrick and A. Elizur. 2007. Temporal expression of G-protein-coupled receptor 54 (GPR54), gonadotropin-releasing hormones (GnRH), and dopamine receptor D2 (drd2) in pubertal grey mullet, *Mugil cephalus*. Gen Comp Endocrinol, 150: 278–287.

Noh, C.H., Y.K. Nam and D.S. Kim. 2002. Production of supermale (YY) and superfemale (YY) Nile tilapia (*Oreochromis niloticus*) by sex reversal and chromosome sex manipulation-III comparative study on male seed production traits of supermale and superfemale. J Aquacult, 13: 325–380.

Norris, D.O. 2007. Vertebrate Endocrinology. Academic Press, San Diego, USA.

Nunez, J. and F. Duponchelle. 2009. Towards a universal scale to assess sexual maturation and related life history traits in oviparous teleost fishes. Fish Physiol Biochem, 35: 167–180.

O'Dennel, L., P.G. Stahton, J.R. Bartles and D.M. Robertson. 2000. Sertoli cell ectoplasmic specialization in the seminiferous epithelium of the testosterone-suppressed adult rat. Biol Reprod, 63: 99–108.

O'Dennel, L., K.M. Robertson, M.E. Jones and E.R. Simpson. 2001. Estrogen and spermatogenesis. Endocrine Rev, 22: 289–318.

Ogawa, S., G. Akiyama, S. Kato et al. 2006. Immunoneutralization of gonadotropin-releasing hormone type III suppresses male reproductive behaviour of cichlids. Neurosci Lett, 403: 201–205.

Ogawa, S., M. Akiyoshi, M. Higuchi et al. 2008. 'Post-sex differntiational' sex reversal in the female common carp (*Cyprinus carpio*). Cybium, 32: 102–103.

Ogino, Y., H. Katoh and G. Yamada. 2004. Androgen dependent development of a modified anal fin, gonopodium, as a model to understand the mechanism of secondary sexual character expression in vertebrates. FEBS Lett, 575: 119–126.

Ogiwara, K., N. Takano, M. Shinohara et al. 2005. Gelatinase A and membrane-type matrix metalloproteinase 1 and 2 are responsible for follicle rupture during ovulation in the medaka. Proc Natl Acad Sci, USA, 102: 8442–8447.

Ohkubo, M., F. Aranish and A. Shimzu. 2010. Molecular cloning and brain distribution of three types of gonadotropin-releasing hormone from mummichog *Fundulus heteroclitus*. J Fish Biol, 76: 379–394.

Ohta, K., S. Yamaguchi, A. Yamaguchi et al. 2002. Biosynthesis of steroid metabolites for germinal vesicle break down *in vitro*. Comp Biochem Physiol, 133 B: 45–54.

Ohta, K., J.K. Sundaray, T. Okida et al. 2003. Bi-directional sex change and its steroidogenesis in the wrasse *Pseudolabrus sieboldi*. Fish Physiol Biochem, 28: 173–174.

Ohta, T., H. Miyake, C. Miura et al. 2007. Follicle stimulating hormone induces spermatogenesis mediated by androgen production in Japanese eel (*Anguilla japonica*). Biol Reprod, 77: 970–977.

Okada, Y.K. and H. Yamashita. 1994. Experimental investigation of the manifestation of secondary sexual characters in the fish using medaka *Oryzias latipes* as material. J Fac Sci Tokyo Univ, Tokyo Sect IV 6: 383–437.

Okubo, K., M. Amano, Y. Yoshiura et al. 2000. A novel form of gonadotropin-releasing hormone in the medaka *Oryza latipes*. Biochem Biophys Res Commun, 276: 298–303.

Okutsu, T., K. Suzuki, Y. Takeuchi et al. 2006. Testicular germ cells can colonize sexually undifferentiated embryonic gonad and produce functional egg in fish. Proc Natl Acad Sci USA, 103: 2725–2729.

Oldfield, R.G. 2005. Genetic, abiotic and social influences on sex differentiation in cichlid fishes and the evolution of sequential hermaphroditism. Fish Fish, 6: 93–110.

Oldfield, R.G. 2007. Behavioral interaction, body size and sex determination in the Midas cichlid *Amphilophus citrenellus*. J Fish Internatl, 2: 242–249.

Oliveiro, M. and J. Oliveiro. 1979. Effect of estradiol-17β on the cytology of the liver, gonads and pituitary and plasma electrolytes in the female freshwater eel. Cell Tissue Res, 199: 431–454.

Oliveira, R.F. 2006. Neuroendocrine mechanism of alternative reproductive tactics in fish. 24: 297–357. *In:* K. Sloman, S. Batshine and R. Wilson [eds.]. Fish Physiology, Behaviour and Physiology of Fish. Elsevier, Amsterdam, The Netherlands.

Oliveira, R.F. 2009. Social behavior in context: Hormonal modulation of behavioral plasticity and social competence. Integ Comp Biol, 49: 423–440.

Oliveira, R.F. and V.C. Almada. 1998. Androgenization of dominant males in cichlid fish: Androgens mediate the social modulation of sexually dimorphic traits. Ethology, 104: 841–858.

Oliveira, R.F. and D.M. Goncalves. 2008. Hormones and social behaviour of teleost fish. pp. 61–125. *In:* C. Magnhagen, V. Braithcuaite, E. Forsgren and B. Kapoor [eds.]. Fish Behaviour. Science Publishers, Enfield, NH, USA.

Oliveira, R.F., A.V.M. Canario and M.S. Grober. 2001a. Male sexual polymorphism, alternative reproductive tactics and androgens in combtooth blennies (Pisces: Blennidae). Horm Behav, 40: 266–275.

Oliveira, R.F., A.V.M. Canario, M.S. Grober and R.S. Santos. 2001b. Endocrine correlates of male polymorphism and alternative reproductive tactics in the Azorean peacock blenny *Parablennius sanguinolentus parvicornis.* Gen Comp Endocrinol, 121: 278–288.

Oliveira, R.F., E.J. Goncalves and R.S. Santos. 2001c. Gonadal investment of young males in two blennid fishes with alternative mating tactics. J Fish Biol, 59: 459–462.

Oliveira, R.F., L.A. Carnerio, D.M. Goncalves et al. 2001d. 11 Ketotestosterone inhibits the alternative mating tactic in sneaker males of the peacock blenny *Salaria pavo.* Brain Behav Evol, 58: 28–37.

Oliveira, R.F., L.A. Canerio, A.V.M. Canario and M.S. Grober. 2001e. Effects of androgens on social behavior and morphology of alternative reproductive males of the Azorean rock pool blenny. Horm Behav, 39: 157–166.

Oliveira, R.F., K. Hirschenhauser, L.A. Canario and A.V.M. Canario. 2002. Social modulation of androgen levels in male teleost fish. Comp Biochem Physiol, Part B, 132: 203–215.

Oliveira, R.F., K. Hirschenhauser, A.V.M. Canario and M. Taborsky. 2003. Androgen levels of reproductive competitors in a cooperatively breeding cichlid. J Fish Biol, 63: 1615–1620.

Oliveira, R.F., L.A. Carneiro and A.V.M. Carario. 2005a. No hormonal response in tied fights. Nature, 437: 207–208.

Oliveira, R.F., A.F.H. Ros and D.M. Goncalves. 2005b. Intra-sexual variation in male reproduction in teleost fish: a comparative approach. Horm Behav, 48: 430–439.

Omoto, N., M. Macbayashi, S. Adachi et al. 2005. The influence of oocyte maturational stage on hatching and triploidy rates in hybrid (bester) sturgeon (*Huso huso* x *Acipenser ruthenus*). Aquaculture, 245: 287–294.

Oppen-Bernstein, D.O., J.V. HeiVik and B.T. Walther. 1990. The major structural proteins of cod (*Gadus morhua*) egg shells and protein cross linking during teleost egg hardening. Dev Biol, 137: 258–265.

Orlando, E.F., Y. Katsu, S. Miyagawa et al. 2006. Cloning and differential expression of estrogen receptor and aromatase genes in the self-fertilizing hermaphrodite and male mangrove rivulus *Kryptolebias marmoratus.* J Mol Endocrinol, 37: 353–365.

Orlando, L., B. Afonso, G. Wassermann and R.T. De Oliveira. 2001. Sex reversal in Nile tilapia (*Oreochromis niloticus*) using a nonsteroidal aromatase inhibitor. J Exp Zool, 290: 177–181.

Ota, K. and M. Kohda. 2006. Descripton of alternative male reproductive tactics in a shell-brooding cichlid *Telematochromis vittatus* in Lake Tanganyika. J Ethol, 24: 9–15.

Ozaki, Y., M. Higuchi, C. Miura et al. 2006. Roles of 11 beta-hydroxysteroid dehydrogenase in spermatogenesis. Endocrinology, 147: 5139–5146.

Padoa, E. 1937. Differenziazione e inversion sessuale (feminizzazione) di avanotti ditrota (*Salmo irdeus*) trattati con ormone follicolare. Monit Zool Ital, 48: 195–203.

Pajuelo, J.C., J.M. Lorenzo, O.A. Bilbao and A.G. Ramos. 2006. Reproductive characteristic of the benthic coastal fish *Diplodus vulgaris* (Teleostei: Sparidae) in the Canarian Archipelago, Northwest Africa. J Appl Ichthyol, 22: 414–418.

Pall, M.K., I. Mayer and B. Borg. 2002. Androgen and behaviour in the male three-spined-stickleback *Gasterosteus aculeatus.* II castration and 11-ketotestosterone effects on courtship and parental care during nesting cycle. Horm Behav, 42: 337–344.

Pandian, T.J. 1993. Endocrine and chromosome manipulation techniques for the production of all male and all female population in food- and ornamental fishes. Proc Indian Acad Sci B, 59: 549–566.

Pandian, T.J. 1995. Hormonal regulation of sex in fish. Platinum Jubilee Lectures. 1995. Indian Science Congress Association, Calcutta, India. pp. 102–115.

Pandian, T.J. 2010. Sexuality in Fishes. Science Publishers, Enfield/CRC Press Boca Raton, FL., USA. p. 208.

Pandian, T.J. 2011. Sex Determination in Fish. Science Publishers, Enfield/CRC Press, Boca Raton,FL., USA. p. 282.

Pandian, T.J. 2012. Genetic Sex Differentiation in Fish. CRC Press, Roca Baton, FL., USA, p. 214.

Pandian. T.J. 2014. Environmental Sex Differentiation in Fish. CRC Press, Roca Baton, FL., USA. (in preparation)

Pandian, T.J. and J. Fluchter. 1968. Rate and efficiency of yolk utilization in the sole *Solea solea*. Helgolander wissens Meereunters, 18: 53–65.

Pandian, T.J. and K. Varadaraj. 1990. The development of monosex female *Oreochromis mossambicus* broodstock by integrating gynogenesis and endocrine sex reversal technique. J Exp Zool, 255: 88–96.

Pandian, T.J. and K. Varadaraj. 1991. Techniques to produce "cent percent" male tilapias. Naga ICLARM, 13: 3–5.

Pandian, T.J. and S.G. Sheela. 1995. Hormonal induction of sex reversal in fish. Aquaculture, 138: 1–22.

Pandian, T.J. and R. Koteeswaran. 1998. Ploidy induction and sex control in fish. Hydrobiologia, 384: 167–243.

Pandian, T.J. and R. Koteeswaran. 1999. Natural occurence of monoploid and polyploids in the Indian catfish *Heteropneustes fossilis*. Curr Sci, 76: 1134–1137.

Pandian, T.J. and S. Kirankumar. 2003. Recent advances in hormonal induction of sex reversal in fish. J Appl Aquacult, 13: 205–230.

Pandian, T.J., S.G. Sheela and S. Kavumpurath. 1994. Endocrine sex reversal in fishes: Masculinization evokes greater stress and mortality. Curr Sci, 66: 240–243.

Pang, Y. and W. Ge. 1999. Activin stimulation of zebrafish oocyte maturation *in vitro* and its potential role in mediating gonadotropin-induced oocyte maturation. Biol Reprod, 61: 987–992.

Pankhurst, N.W. 1998. Reproduction. pp. 1–26. *In:* K.D. Black and A.D. Pickering (eds.). *Biology of Farmed Fish*. CRC press, Boca Raton, FL, USA.

Pankhurst, N.W. 2008. Gonadal steroids: functions and patterns of change. pp. 67–111. *In:* M.J. Rocha, A. Arukwe and B.G. Kapoor [eds.]. Fish Reproduction. Science Publishers, Enfield, NH, USA.

Pankhurst, N.W. and M.J.R. Porter. 2003. Cold and dark or warm and light: variations on the theme of environmental control of reproduction. Fish Physiol Biochem, 28: 385–389.

Parenti, L.R. and H.J. Grier. 2004. Evolution and phylogeny of gonad morphology in bony fishes. Integr Comp Biol, 44: 333–348.

Parhar, I.S., T. Soga and Y. Sakuma. 2000. Thyroid hormone and estrogen regulate brain region-specific messenger ribonucleic acids encoding three gonadotropin-releasing hormone genes in sexually immature male fish *Oreochromis niloticus*. Endocrinology, 141: 1618–1626.

Parhar, I.S., H. Sato and Y. Sakuma. 2003. Ghrelin gene in cichlid fish is modulated by sex and development. Biochem Biophys Res Commun, 305: 169–175.

Parikh, V.N., T. Clement and R.D. Fernald. 2006. Physiological consequence of social descent: studies in *Astatotilapia burtoni*. J Endocrinol, 190: 183–190.

Park, E.H., H.H. Chang and W.N. Joo. 1994. Assessment of estuarine hermaphroditic fish *Rivulus marmoratus* as a useful euryhaline species for acute toxicity tests as shown as using cadmium. Can J Fish Aquat Sci, 51: 280–285.

Park, I.S., J.H. Kim, S.H. Cho and D.S. Kim. 2004. Sex differentiation and hormonal sex reversal in the bagrid catfish *Pseudobagrus fulvidraco* (Richardson). Aquaculture, 232: 183–193.
Pasmanik, M. and G. Callard. 1988. A high abundance of androgen receptor in goldfish brain: characteristics and seasonal changes. Endocrinology, 123: 1162–1171.
Pati, D., K. Balshaw, D.L. Grinwich et al. 1996. Experimental growth factor receptor binding and biological activity in the ovary of goldfish *Carrassius auratus.* Am J Physiol, 270: R1065–R1072.
Patil, J.G. and R.M. Gunasekera. 2008. Tissue and sexually dimorphic expression of ovarian and brian aromatase mRNA in the Japanese medaka (*Oryzias latipes*): Implications for their preferential roles in ovarian and neural differentiation and development. Gen Comp Endocrinol, 158: 131–137.
Patino, R. and C.V. Sullivan. 2002. Ovarian follicle growth, maturation and ovulation in teleost fish. Fish Physiol Biochem, 26: 57–70.
Patino, R., K.B. Davis, J.E. Schoore et al. 1996. Sex differentiation of channel catfish gonads: Normal development and effect of temperature. J Exp Zool, 276: 209–218.
Patino, R., G. Yoshizaki, P. Thomas and H. Kagawa. 2001. Gonadotropic control of ovarian follicle maturation: the two stage concept and its mechanisms. Comp Biochem Physiol B, 129: 427–439.
Patino, R., G. Yoshizaki, D. Bolamba and P. Thomas. 2003. Role of arachidonic acid and protein kinase C during maturation-inducing hormone-dependent meiotic resumption and ovulation in ovarian follicles of Atlantic croaker. Biol Reprod, 68: 516–523.
Payne, A.H. and D.B. Hales. 2004. Overview of steroidogenic enzymes in the pathway from cholesterol to active steroid hormone. Endocr Rev, 25: 947–970.
Pecio, A., F. Lahnsteiner and J. Rafinski. 2001. Ultrastructure of the epithelial cells in the aspermatogenic part of the testis in *Mimagoniatus barberi* (Teleostei: Characidae, Glanducaudinae) and the role of their secretions in spermatozeugmata formation. Ann Anat, 183: 427–435.
Petkam, P., R.L. Renauld, A.M.M.S. Freitas et al. 2002. *In vitro* metabolism of progesterone androgens and estrogens by rainbow trout embryos. Fish Physiol Biochem, 27: 117–128.
Pellegrini, E., K.J. Mouriec, I. Anglade et al. 2007. Identification of aromatase-positive radial glial cells as progenitor cells in the ventricular layer of the forebrain in zebrafish. J Comp Neurol, 501: 150–167.
Perrault, H.A.N., K. Semsar and J. Godwin. 2003. Fluoretine treatment decreases territorial aggression in a coral reef fish. Physiol Beh, 79: 719–724.
Perrot, V., E.B. Moisseva, Y. Gozes et al. 2000. Insulin-like growth of a hermaphroditic species, the gilthead seabream *Sparus aurata*: expression and cellular localization. Biol Reprod, 6: 229–241.
Perruzi, S., G. Rudolfson, R. Primicerio et al. 2009. Milt characteristics of diploid and triploid Atlantic cod (*Gadus morhua* L.). Aquacult Res, 40: 1160–1169.
Peter, R.E., L.W. Crim, H.J. Goos and J.W. Crim. 1978. Leisoning studies on the gravid female goldfish: neuroendocrine regulation of ovulation. Gen Comp Endocrinol, 35: 391–401.
Peter, R.E., J.P. Chang, C.S. Nahorniak et al. 1986. Injections of catecholamine and GnRH in regulation of gonadotropin secretion in teleost fish. Rec Prog Horm Res, 42: 513–548.
Peter, R.E., H.R. Lin and G. van de Kraak. 1988. Induced ovulation and spawning of culture of freshwater fish in China: advances in application of GnRH analogues and dopamine antagonists. Aquaculture, 74: 1–10.
Peters, H.M. 1975. Hermaphroditism in cichlid fishes. pp. 228–235. *In:* R. Reinboth [ed.]. Intersexuality in Animal Kingdom. Springer Verlag, Berlin, Germany.
Petrino, T.R., Y.W.P. Lin and R.A. Wallace. 1989. Steroidogenesis in *Fundulus heteroclitus.* 1. Production of 17 alpha-hydroxy, 20 beta-dihydroprogesterone, testosterone and 17 beta-estradiol by prematurational follicles *in vitro.* Gen Comp Endocrinol, 73: 147–153.

Pham, H.Q., E. Kjorsvik, A.T. Nguyen et al. 2010. Reproductive cycle in female Waigien sea perch (*Psammoperca waigiensis*) reared under different salinity levels and the effects of dopamine antagonist on steroid hormone levels. J Exp Mar Biol Ecol, 383: 137–145.

Phelps, R. and M. Okoko. 2011. A non-paradoxical dose response to 17α-methyltestosterone by Nile tilapia *Oreochromis niloticus* (L.): effects on the sex ratio, growth and gonadal development. Aquacult Res, 42: 549–558.

Phillipart, J.C. and P. Berrebi. 1990. Experimental hybridization of *Barbus barbus* and *Barbus meridionalis:* physiological, morphological and genetic aspects. Aquat Living Resour, 3: 325–332.

Phillipp, D.P. and M.R. Gross. 1994. Genetic evidence of cuckoldry in bluegill sunfish *Lepomis macrochirus*. Mol Ecol, 3: 563–569.

Piferrer, F. 2001. Endocrine sex control strategies for the feminization of teleost fish. Aquaculture, 197: 229–281.

Piferrer, F. and E.M. Donaldson. 1991. Dosage-dependent differences the effect of aromatizable and non-aromatizable androgens on the resulting phenotype of coho salmon (*Oncorhynchus kisutch*). Fish Physiol Biochem, 9: 145–150.

Piferrer, F. and E.M. Donaldson. 1992. The comparative effectiveness of the natural and synthetic estrogen for the direct feminization of Chinook salmon (*Oncorhynchus tshawytscha*). Aquaculture, 106: 183–193.

Piferrer, F. and M. Donaldson. 1994. Uptake and clearance of exogenous estradiol-17β and testosterone during the early development of coho salmon (*Oncorhynchus kisutch*), including eggs, alevins and fry. Fish Physiol Biochem, 13: 219–232.

Piferrer, F. and Y. Guiguen. 2008. Fish gonadogenesis, Part II Molecular biology and genomics of sex differentiation. Rev Fish Sci, 16 (SI): 35–55.

Piferrer, F., S. Zanury, M. Carillo et al. 1994a. Brief treatment with an aromatase inhibitor during sex differentiation causes chromosomally female salmon to develop as normal functional male. J Exp Zool, 270: 255–262.

Piferrer, F., M. Carillo, S. Zanury et al. 1994b. Induction of sterility in coho salmon (*Oncorhynchus kisutch*) by androgen immersion before first feeding. Aquaculture, 119: 409–423.

Pinto, L. and M. Gotz. 2007. Radial glial cell heterogeneity—the source of diverse progeny in the CNS. Prog Neurobiol, 83: 2–23.

Pires, MN. 2007. The evolution of placentas in Poeciliid fishes. Ph D Thesis, University of California Riverside, USA. pp. 228.

Pires, M.N., J. Arendt and R.D. Reznick. 2010. The evolution of placentals and superfoetation in the fish genus *Poecilia* (Cyprinodontiformes: Poecillidae subgenera *Micropoecilia* and *Acanthophacelus*). Biol J Linn Soc, 99: 784–796.

Plaut, I. 2002. Does pregnancy affect swimming performance of female mosquitofish, *Gambusia affinis*? Funct Ecol, 16: 290–295.

Polder, J.J.W. 1971. On gonads and reproductive behavior in the cichlid fish *Acquiden portelegrensis* (Hensel). Netherland J Zool, 21: 265–365.

Pollux, B.J.A. and D.N. Reznick. 2011. Matrotrophy limits a female's ability to adaptively adjust offspring size and fecundity in fluctuating environments. Funct Ecol, 25: 747–756.

Pollux, B.J.A., M.N. Pires, A.I. Banet and D.N. Reznick. 2009. Evolution of placentas in the fish family Poeciliidae: An empirical study of macroevolution. Annu Rev Ecol Evol Syst, 40: 271–289.

Ponce de León, J.L., R. Rodríguez, M. Acosta and M.C. Uribe. 2011. Egg size and its relationship with fecundity, newborn length and female size in Cuban poeciliid fishes (Teleostei: Cyprinodontiformes). Ecol Freshwater Fish, 20: 243–250.

Pongthana, N., P. Baoprasertkul, B. Tongmee and D. Penman. 1996. Hormonal feminization of snakeskin gourami (*Trichogaster pectoralis*). *In:* R.C. LeRoy [ed.]. World Aquaculture, 1996. Abstract 314, Bangkok, Thailand.

Popek, W., E. Luszczek-Trojnar, E. Drag-Kozak et al. 2005. Effect of the pineal gland and melatonin on dopamine release from perfused hypothalamus of mature female carp during spawning and winter regression. Acta Ichthyol et Piscat, 35: 65–71.

Popek, W., E. Luszczek-Trojnar, E. Drag-Kozak et al. 2006. Effect of melatonin on dopamine secretion in the hypothalamus of mature female common carp, *Cyprinus carpio* L. Acta Ichthyol et Piscat, 36: 135–141.

Powell, J.F., Y. Zohar, A. Elizur et al. 1994. Three forms of gonadotropin-releasing hormone characterized from brains of one species. Proc Natl Acad Sci USA, 91: 12081–12085.

Psenicka, M., V. Kaspar, M. Rodina et al. 2011. Comparative study on ultrastructure and motility parameters of spermatozoa of tetraploid and hexaploid Siberian sturgeon *Acipenser baerii.* J Appl Icthyol, 27: 683–686.

Quagio-Grassiotto, I. and C. Oliveira. 2008. Sperm ultrastructure and a new type of spermatogenesis in two species of Pimelodidae with a comparative review of sperm ultrastructure in Siluriformes. J Comp Zool, 247: 55–66.

Quinn, T.P., L.A. Vollestad, J. Peterson and V. Gallucci. 2004. Influences of freshwater and marine growth on the egg size-egg number trade off in coho and chinook salmon. Trans Am Fish Soc, 133: 55–65.

Raghuveer, K. and B. Senthilkumaran. 2010. Cloning and differential expression pattern of *vasa* in the developing and recrudescing gonads of catfish *Clarias gariepinus.* Comp Biochem Physiol, 157A: 79–85.

Raine, J.C. 2011. Thyroid hormones and reproduction in fishes. 1: 83–102. *In:* D.O. Norris and K.H. Lopez [eds.]. Hormones and Reproduction of Vertebrates: Fishes. Elsevier, Amsterdam, The Netherlands.

Ran, X., W. Li and H. Lin. 2004. Rat ghrelin stimulates GH mRNA expression in the pituitary of orange-spotted grouper *Epinephelus coioides.* Fish Physiol Biochem, 30: 95–102.

Rao, G.P.S., R.V.K. Sharma and G.Y. Keshappa. 1990. Elimination of testosterone in the fry of common carp *Cyprinus carpio (*Linm.). pp. 87–89. *In:* M.M. Joseph [ed.]. Proc Indian Fish Forum, Mangalore, India.

Rao, S.H.N. and G.P.S. Rao. 1983. Hormonal manipulation of sex in the common carp *Cyprinus carpio communis* (Linnaeus). Aquaculture, 35: 83–88.

Rasch, E.M. and J.S. Balsano. 1989. Trihybrids related to unisexual molly fish *Poecilia formosa.* Bulletin, 466: 252–267. *In:* R.M. Dawley and J. Bogard [eds.]. Evolution and Ecology of Unisexual Vertebrates. New York State Museum.

Rasch, E.M., P.J. Monaco and J.S. Balsano. 1982. Cytophotometric and autoradiographic evidence for functional apomixis in a gynogenetic fish *Poecilia formosa* and its related triploid unisexuals. Histochemistry, 73: 515–533.

Rasheeda, M.K., P. Sridevi and B. Senthilkumaran. 2010a. Cytochrome p450 aromatases: Impact on gonadal development, recrudescence and effect of hCG in the catfish *Clarias gariepinus.* Gen Comp Endocrinol, 167: 234–245.

Rasheeda, M.K., H. Kagawa, R. Kirubagaran et al. 2010b. Cloning, expression and enzyme activity of testicular 11β-hydoxysteroid dehydrogenase during seasonal cycle and after hCG induction in air breathing catfish *Clarias gariepinus.* J Steroid Biochem Mol Biol, 120: 1–10.

Rattaunayuvakorn, S., P. Mungkornkarn, A. Thongpan et al. 2006. Gonadal development and sex inversion in saddleback anemonefish *Amphiprion polymnus* Linnaeus (1758). Kasetsart J (Nat Sci), 40: 196–203.

Ravaglia, M.A., F.L. Lo Nostro, H.C. Maggese et al. 1997. Characterization of molecular variants of GnRH, induction of spermiation and sex reversal using salmon GnRH and domperidone in protogynous diandric fish *Synbramchus marmoratus* Bloch (Teleostei: Synbramchidae). Fish physiol Biochem, 16: 425–436.

Ravi, V. and B. Venkatesh. 2008. Rapidly evolving fish genomes and teleost diversity. Curr Opin Genet Dev, 18: 544–550.

Reavis, R.H. and M.S. Grober. 1999. An integrative approach to sex change: social, behavioural neurochemical changes in *Lythrypnus dalli* (Pisces). Acta Ethol, 2: 51–60.

Reinboth, R. and S. Brusle-Sigard. 1997. Histological and ultrastructural studies on the effects of hCG on sex reversion in the protogynous *Coris julis.* J Fish Biol, 51: 799–749.

Reinecke, M. 2010. Insulin-like growth factor and fish reproduction. Biol Reprod, 82: 656–661.
Reinecke, M., A. Schmid, R. Ermatlinger and D. Loffing-Cueni. 1997. Insulin-like growth factor 1 in the teleost *Oreochromis mossambicus* the tilapia: gene sequence, tissue expression and cellular localization. Endocrinology, 138: 3613–3619.
Reinecke, M., B.T. Bjomsson, W.W. Dickhoff et al. 2005. Growth hormone and insulin-like growth factors in fish: where we are and where to go. Gen Comp Endocrinol, 142: 20–24.
Remage-Healey, L. and A.H. Bass. 2007. Rapid hierarchial modulation of vocal patterning by steroid hormones. J Neurosci, 24: 5982–5900.
Reznick, D., H. Callahan and R. Llauredo. 1996. Maternal effects on offspring quality on poeciliid fishes. Am Zool, 36: 147–156.
Reznick, D., M. Mateos and M. Springer. 2002. Independent origins and rapid evolution of the placenta in the fish genus *Poeciliopsis*. Science, 298: 1018–1020.
Reznick, D., M. Bryant and D. Holmes. 2006. Evolution of senescnce and post- reproductive life span in guppies (*Poecilia reticulata*). PLoS Biol, 4: DOI:10. 1371/journal. Pbio.0010007.
Riberio, F., I.G. Cowx, P. Tago et al. 2003. Growth and reproductive traits of diploid and triploid forms of the *Squalius alburnoides,* cyprinid complex in the tributary of the Guadiana River, Portugal. Arch Hydrobiol, 156: 471–484.
Richter, C.J.J., E.H. Eding, H.J.T.H. Goos et al. 1987. The effect of pimozide/LHRHa and 17α-hydroxyprogesterone on plasma steroid levels and ovulation in the African catfish *Clarias gariepinus.* Aquaculture, 63: 157–168.
Riesch, R., M. Tobler, M. Plath and I. Schlupp. 2009. Offspring number in a livebearing fish (*Poecilia mexicana,* Poeciliidae): reduced fecundity and reduced plasticity in a population of cave mollies. Env Biol Fish, 84: 89–94.
Riesch, R., M. Plath, I. Schlupp and E. Marsh-Matthews. 2010. Matrotrophy in the cave molly: an unexpected provisioning strategy in an extreme environment. Evol Ecol, 24: 789–801.
Riesch, R., M. Plath and I. Schlupp. 2011a. Toxic hydrogen sulphide and dark caves: pronounced male life-history divergence among locally adapted *Poecilia mexicana* (Poeciliidae). J Evol Biol, 24: 596–606.
Riesch, R., I. Schlupp, R.B. Langerhans and M. Plath. 2011b. Shared and unique patterns of embryo development in extremophil poeciliids. PLoS ONE 6 (11: e27377. Doi: 10.1371/journal.pone.0027377.
Rinchard, J., K. Dabrowski, M.A. Garcia-Abiado and J. Ottobre. 1999. Uptake and depletion of plasma 17α methyltesterone during induction of masculinization in muskellunge *Esox masquinongy*: Effect of plasma steroids and sex reversal. Steroids, 64: 518–525.
Ripley, J.L. and C.M. Foran. 2009. Direct evidence for embryonic uptake of paternally-derived nutrients in two pipelines (Syngnathidae: *Syngnathus* sp). J Comp Physiol, 179B: 325–333.
Robalo, J.I., C. Sousa-Santos, A. Levy and A.C. Almada. 2006. Molecular insights on the taxonomic position of the paternal ancestors of the *Squalius alburnoides* hybridogenetic complex. Mol Phylogenet Evol, 39: 276–281.
Robertson, O.H. 1961. Prolongation of the life span of kokane salmon (*Oncorhynchus nerka kennerlyi*) by castration before beginning of gonad development. Proc Natl Acad Sci, USA, 47: 609–611.
Robertson, D.R. and R.R. Warner. 1978. Sexual patterns in the labroid fishes of the western Carribean. 2. The parrotfishes (Scaridae). Smithsonian Contributions to Zoology No 255.
Rocha, M.J. and M.A. Reis-Henriques. 2000. Steroid metabolism by ovarian follicles of the sea bass *Dicentrarchus labrax.* Comp Biochem Physiol, 125C: 85–91.
Rodriquez, S.J.S., S.I.P. Priesto and M.P.V. Minondo. 1993. Flow cytometric analyses of infectious pancreatic necrosis virus attachment to sperm. Dis. Aquat Org, 15: 153–156.

Rodriquez-Mari, A., Y.L. Yan, R.A. Brenuller et al. 2005. Characterization and expression pattern of zebrafish Anti-Mullerian hormone (*amh*) relative to *sox9a*, *sox9b* and *cyp19a1a* during gonadal development. Gene Expr Patterns, 5: 655–667.

Ros, A.F.H., A.V.M. Canario, E. Couto et al. 2003. Endocrine correlates of intra-specific variation in the mating system of the St Peter's fish (*Saratherodon galilaeus*). Horm Behav, 44: 365–373.

Ros, A.F.H., R. Bruintjes, R.S. Santos et al. 2004a. The role of androgens in the trade off between territorial and parental behavior in the Azorean rock pool blenny *Parablennius parvicornis*. Horm Behav, 46: 491–497.

Ros, A.F.H., K. Becker, A.V.M. Canario and R.F. Oliveira. 2004b. Androgen levels and energy metabolism in *Oreochromis mossambicus*. J Fish Biol, 65: 895–905.

Ros, A.F.H., K. Becker and R.F. Oliveira. 2006. Aggressive behavior and energy metabolism in a cichlid fish *Oreochromis mossambicus*. Physiol Behav, 89: 164–170.

Rosenblum, P.M., J. Pudney and I.P. Callard. 1987. Gonadal morphology, enzyme histochemistry and plasma steroid levels during the annual reproductive cycle of male and female brown bull head catfish *Ictalurus nebulosus* Lesueur. J Fish Biol, 31: 325–341.

Rosenstein, S. and G. Hulata. 1994. Sex reversal in the genus *Oreochromis*: optimization of feminization protocol. Aquacult Fish Manage, 25: 329–339.

Ross, R.M. 1978. Reproductive behaviour of anemonefish. *A. melanopus* on Guan, Pac Sci, 32: 100.

Ross, R.M. 1983. Annual, semi-lunar and diel reproductive rhythms in the Hawaiian labrid *Thalassoma duperrey*. Mar Biol, 72: 311–318.

Ross, R.M. 1990. The evolution of sex-change mechanisms in fishes. Env Biol Fish, 29: 81–93.

Ross, R.M., G.S. Losey and M. Diamond. 1983. Sex change in a coral-reef fish: Dependence of simulation and inhibition on relative size. Science, 221: 574–575.

Rothbard, S., B. Moav and Z. Yaron. 1987. Changes in steroid concentrations during sexual ontogenesis in tilapia. Aquaculture, 61: 59–74.

Rothbard, S., Y. Zohar, B. Zmora et al. 1990. Clearance of 17α ethynyltestosterone from muscle of sex reversed tilapia hybrids treated for growth enhancement with two doses of the androgen. Aquaculture, 89: 365–376.

Rougeot, C., A. Krim, S.N.M. Mandiki et al. 2007. Sex steroid dynamics during embryogenesis and sexual differentiation in Eurasian perch *Perca fluviatalis*. Theriogenology, 67: 1046–1052.

Rudolfsen, G., L. Figenschow, I. Folstad et al. 2006. Rapid adjustment of sperm characteristics in relation to social status. Proc R Soc Lond B, 273: 325–335.

Rurangwa, E., D.E. Kime, F. Ollevier and J.P. Nash. 2004. The measurement of sperm motility and factors affecting sperm quality in cultured fish. Aquaculture, 234: 1–28.

Ryan, M.J. and B.A. Causey. 1989. 'Alternative' mating behavior in the sword tail *Xiphophorus nigrensis* and *Xiphophorus pygmeas* (Pisces: Poeciliidae). Behav Evol Scociobiol, 24: 341–348.

Sadovy, Y. and D.Y. Shapiro. 1987. Criteria for the diagnosis of hermaphroditism in fishes. Copeia, 1987: 136–156.

Sadovy de Mitcheson, Y. and M. Liu. 2008. Functional hermaphroditism in teleosts. Fish Fish, 9: 1–43.

Sagebakken, G., I. Ahnesjo, K.B. Mobley et al. 2009. Brooding fathers, not siblings, take up nutrients from embryos. Proc R Soc Lond B, doi:10.1098/rspb. 2009.1767.

Saito, D., C. Morinaga, Y. Aoki et al. 2007. Proliferation of germ cells during gonadal sex differentiation in medaka. Insights from germ-cell depleted mutant zanzai. Dev Biol, 310: 280–290.

Saito, T., R. Goto-Kazeto, K. Arai et al. 2008. Xenogenesis in teleost fish through generation of germ cell chimeras by single primordial germ cell transplantation. Biol Reprod, 78: 159–166.

Sakai, Y. and M. Kohda. 1997. Harem structure of the protogynous angelfish *Centropyge ferrugatus* (Pomacaulthidae). Env Biol Fish, 49: 333–339.
Sakai, Y., K. Karino, T. Kuwamura and Y. Nagashima. 2003. Sexually dichromatic protogynous angelfish *Centropyge ferrugata* (Pomacanthidae) males can change back to females. Zool Sci, 20: 627–633.
Sakakura, S. and K. Tsukamoto. 1999. Ontogeny of aggressive behavior in schools of yellow tail *Seriola quinqueradiata*. Env Biol Fish, 56: 231–242.
Sakakura, Y. and D.L.G. Noakes. 2000. Age, growth, sexual development in the self-fertilizing hermaphroditic fish *Rivulus marmoratus*. Env Biol Fish, 59: 309–317.
Sampathkumar, R., S. Ijiri and J.M. Trant. 2000. Changes in the expression of genes encoding steroidogenic enzymes in the channel catfish (*Ictalurus punctatus*) ovary throughout a reproductive cycle. Biol Reprod, 63: 1676–1782.
Santangelo, N. and A.H. Bass. 2006. New insights into neuropeptide modulation of aggression: field studies of arginine vasotocin in a territorial tropical damselfish. Proc R Soc Lond B, 273: 3085–3092.
Santos, E.M., V.L. Workmann, G.C. Paull et al. 2007. Molecular basis of sex and reproductive status in breeding zebrafish. Physiol Genomics, 30: 111–122.
Santos, J.E., N. Bazzoli and G.B. Santos. 2001. Morphofunctional organization of the male reproductive system of the catfish *Ihering ichthyolabrosus* (Liitken, 1894). Tissue Cell, 33: 533–540.
Santos, R.S. 1985. Parentias e satellites: tacticas alternatives de acasalamento nos machos de *Blennius sanguinolentus* Pallas (Fisecs: Blenniidae). Arguipelago, Life & Earth Sci, 6: 119–146.
Sartar, K., M. Papadaki, S. Zanuy and C.C. Mylonas. 2006. Permanent sex inversion in 1-year old juveniles of the protogynous dusky grouper (*Epinephelus marginatus*) using controlled release of 17α-methyltestosterone implants. Aquaculture, 256: 443–456.
Satoh, N. 1974. An ultrastructural study of sex differentiation in the teleost *Oryzias latipes*. J Embryol Exp Morphol, 32: 192–215.
Satoh, N. and N. Egami. 1972. Sex differentiation of germ cells in the teleost *Oryzias latipes* during normal embryonic development. J Embryol Exp Morph, 28: 385–395.
Sattar, S.A., C. Jorgensen and O. Fixsen. 2008. Fisheries induced evolution of energy and sex allocation. Bull Mar Sci, 83: 235–250.
Sawaguchi, S., H. Kagawa, N. Ohkubo et al. 2006. Molecular characterization of three forms of vitellogenin and their yolk protein products during oocyte growth and maturation in red seabream (*Pagrus major*), a marine teleost spawning pelagic eggs. Mol Reprod Dev, 73: 719–736.
Sawatari, E., S. Shikina, T. Takeuchi and G. Yoshizaki. 2007. A novel transforming growth factor-beta super family member expressed in gonadal somatic cells enhances primordial germ cell and spermatogonial proliferation in rainbow trout (*Oncohynchus mykiss*). Dev Biol, 301: 266–275.
Sawyer, S.J., K.A. Gerstner and G.V. Callard. 2006. Real-time PCR analysis of cytochrome p450 aromatase expression in zebrafish: gene specific, tissue specific distribution of sex differentiation, development programming and estrogen regulation. Gen Comp Endocrinol, 147: 108–117.
Scaggiante, M., M.S. Grober, V. Lorenzi and M.B. Rasotto. 2004. Changes along male reproductive axis in response to social context in a gonochoric gobiid *Zosterisessor ophiocephalus* (Teleostei, Gobiidae), with alternative mating tactics. Horm Behav, 46: 607–617.
Schafhauser-Smith, D. and T.J. Benfey. 2002. The purification and development of a quantitative enzyme linked immunosorbent assay (ELISA) for the measurement of vitellogenin in diploid and triploid brook trout (*Salvelinus fontinalis*). Fish Physiol Biochem, 24: 287–298.
Schartl, M., I. Schlupp, A. Schartl et al. 1991. On the stability of dispensable constituents of the eukaryotic genome. Stability of coding sequences versus truly hypervariable sequences

in a clonal vertebrate the Amozan molly *Poecilia formosa.* Proc Natl Acad Sci, USA, 88: 8759–8763.
Schartl, M., I. Nanda, I. Schlupp et al. 1995a. Incorporation of subgenomic amounts of DNA as compensation for mutational load in a gynogenetic fish. Nature, 373: 68–71.
Schartl, M., B. Wilde, I. Schlupp and J. Parzefall. 1995b. Evolutionary origin of a parthenoform the Amazon molly *Poecilia formosa* on the basis of molecular geneology. Evolution, 49: 827–835.
Schmidt, P.J. and D.R. Idler. 1962. Steroid hormones in the plasma of salmon at various stages of maturation. Gen Comp Endocrinol, 2: 201–204.
Schmitz, M., S. Aroua, B. Vidal et al. 2005. Differential regulation of luteinizing hormone and follicle stimulating hormone expression during ovarian development under sexual steroid feedback in the European eel. Neuroendocrinology, 81: 107–119.
Scholz, S. and H.O. Gutzeit. 2000. 17-α ethylnylestradiol affects reproduction, sexual differentiation and aromatase gene expression of the medaka (*Oryzias latipes*). Aquatic Toxicol, 50: 363–373.
Scholz, S., S. Rosler, M. Schaffer et al. 2003. Hormonal induction and stability of monosex populations in the medaka (*Oryzias latipes*): expression of sex-specific marker genes. Biol Reprod, 69: 673–678.
Schrader, M. and J. Travis. 2009. Do embryos influence maternal investment? Evaluating maternal-fetal coadaptation and the potential for parent-offspring conflict in a placental fish. Evolution, 63: 2805–2815.
Schrader, M. and J. Travis. 2011. Variation in offspring size with birth order in placental fish: A role for asymmetric sibling competition? Evolution, Doi:10.1111/j.1558-5646.2011.01422.x.
Schreck, C.B. 1973. Hormonal treatment and sex manipulation in fishes. pp. 84–106. *In:* C.B. Schreck [ed.]. Control of Sex in Fishes. Virginia Polytechnic Institute and State University Sea Grant Program, Blackshury.
Schultz, R.J. 1989. Origins and relationships of unisexual poeciliids. pp. 69–87. *In:* G.K. Meffe and F.F. Snelson, Jr. [eds.]. Ecology and Evolution of Live-bearing Fishes (Poeciliidae). Prentice Hall Englewood Cliffs, New Jersey, USA.
Schultz, R.J. 1977. Evolution and ecology of unisexual fishes. Env Biol Fish, 10: 277–331.
Schulz, R.W. and T. Miura. 2002. Spermatogenesis and its endocrine regulation. Fish Physiol Biochem, 26: 43–56.
Schulz, R.W., W. van Dijk and J. Bogerd. 2003. Sertoli cell proliferation and FSH signaling in African catfish *Clarias gariepinus.* Fish Physiol Biochem, 28: 223–224.
Schulz, R.W., S. Menting, J. Bogerd et al. 2005. Sertoli cell proliferation in the adult testis-evidence from two fish species belonging to different orders. Biol Reprod, 73: 891–898.
Schulz, R.W., L.R. DeFranca, J.J. Lareyre et al. 2010. Spermatogenesis in fish. Gen Comp Endocrinol, 165: 390–411.
Scott, A.G., D.J. Penman, J.A. Beardmore and D.O.F. Skibinski. 1989. The YY supermale in *Oreochromis niloticus* (L.) and its potential in aquaculture. Aquaculture, 78: 237–251.
Scott, A.P. and T. Ellis. 2007. Measurement of fish steroids in water—a review. Gen Comp Endocrinol, 153: 392–400.
Scott, A.P., J.P. Sumpter and N. Stacey. 2010. The role of the maturation inducing steroid 17, 20β-dihydroxy-pregnen-4-en-3-one in male fishes: a review. J Fish Biol, 76: 183–224.
Scribner, K.T., K.S. Page and M.L. Bartron. 2001. Hybridization in freshwater fishes: a review of case studies and cytonuclear methods in biological inference. Rev Fish Biol Fish, 10: 293–323.
Sebert, M.-E., C. Legros, F.-A. Weltzien et al. 2008. Melatonin activates brain dopaminergic systems in the eel with an inhibitory impact on reproductive function. J Neuroendocrinol, 20: 917–929.
Segers, F.H.I.D., G. Berishvili and B. Taborsky. 2011. Egg size-dependent expression of growth hormone receptor accompanies compensatory growth in fish. Proc R Soc Lond B, doi:10.1098/rspb.2011.1104.

Semsar, K. and J. Godwin. 2003. Social influences on the arginine vasotocin system are independent of gonads in a sex-changing fish. J Neurosci, 23: 4386–4393.

Semsar, K. and J. Godwin. 2004. Multiple mechanisms of phenotypic development in the bluehead wrasse. Horm Behav, 45: 345–453.

Semsar, K., F.L.M. Kandal and J. Godwin. 2001. Manipulations of AVT system shift social status and related courtship and aggressive behavior in the bluehead wrasse. Horm Behav, 40: 21–31.

Semsar, K., H.A. Perreault and J. Godwin. 2004. Fluoxetine-treated male wrasse exhibit low AVT expression. Brian Res, 1029: 141–147.

Senthilkumaran, B. 2011. Recent advances in meiotic maturation and ovulation: comparing mammals and pisces. Front Biosci, 16: 1898–1914.

Senthilkumaran, B. and K.P. Joy. 1993. Annual cyclic and castration and crypoterone acetate induced changes in sialic acid content of seminal vesicle of the catfish *Heteropneustes fossilis* (Bloch). Fish Physiol Biochem, 10: 425–430.

Senthilkumaran, B. and K.P. Joy. 1995. Changes in hypothalamic catecholamines, dopamine-beta-hydroxylase, and phenylethanolamine-N-methyltransferase in the catfish *Heteropneustes fossilis* in relation to season, raised photoperiod and temperature, ovariectomy, and estradiol-17 beta replacement. Gen Comp Endocrinol, 97: 121–134.

Senthilkumaran, B., K. Okuzawa, T. Gen et al. 1999. Distribution and seasonal variations in levels of three native GnRHs in the brain and pituitary of perciform fish. J Neuroendocrinol, 11: 181–186.

Senthilkumaran, B., M. Yoshikuni and Y. Nagahama. 2004. A shift in steroidogenesis occurring in ovarian follicles prior to oocyte maturation. Mol Cell Endocrinol, 215: 11–18.

Shankar, D.S. and R.S. Kulkarni. 2000. Effects of cortisol on testis of freshwater fish *Notopterus notopterus* (Pallas). Indian J Exp Biol, 38: 1227–1280.

Shapiro, D.Y. and M.B. Rasotto. 1993. Sex differentiation and gonadal development in the diandric protogynous wrasse *Thalassoma bifasciatum* (Pisces, Labridae). J Zool Lond, 230: 231–245.

Shapiro, D.Y., A. Marconato and T. Yoshikawa. 1994. Sperm economy of a coral-reef fish. Ecology, 75: 1334–1344.

Shelbourn, J.E., W. Clarke, J.R. McBride et al. 1992. The use of 17α methyltestosterone and 3, 5, 3′ triido-L-thyronine for sterilizing and accelerating the growth of zero-age coho salmon (*Oncorhynchus kisutch*). Aquaculture, 103: 85–99.

Shelton, W.L. 1982. Reproduction of a reproductively limited grass carp for biological control of aquatic weeds-phase II. Water Resour Res Int Bull, 45.

Shelton, W.L. and G.L. Jensen. 1979. Production of reproductively limited grass carp for biological control of aquatic weeds. WRRI Bull (Auburn Univ), 39: 1–74.

Shen, X.Y., J.Z. Cui, Q.L. Gong et al. 2007. Cloning the full length coding sequence and expression analysis of Sox9b in guppy (*Poecilia reticulata*). Comp Physiol Biochem, 33: 195–202.

Sherwood, N., L. Eiden, M. Brownstein, J. Spiess et al. 1983. Characterization of a teleost gonadotropin-releasing hormone. Proc Natl Acad Sci USA, 80: 2794–2798.

Sherwood, N.M., S.A. Sower, D.R. Marshak et al. 1986. Primary structure of gonadotropin-releasing hormone from lamprey brain. J Biol Chem, 261: 4812–4819.

Shi, Z.-P., T.-J. Fan, R.-S. Cong et al. 2006. Purification and characterization of hatching enzyme from flounder *Paralichthys olivaceus*. Fish Physiol Biochem, 32: 35–42.

Shibata, N. and S. Hamaguchi. 1988. Evidence for the sexual bipotentiality of spermatogonia in the fish *Oryzias latipes*. J Exp Zool, 245: 71–77.

Shibuno,T., I. Chiba, H. Hashimoto et al. 1994. Reproductive behavior of the wrasse *Thalassoma lutescens* at Kuchierabu-jima. J Fac Appl Biol Sci, 33: 43–50.

Shimada, A. and H. Takeda. 2008. Production of a maternal zygotic medaka using hybrid sterility. Dev Growth Diff, 50: 421–426.

Shimizu, M. and H. Takahashi. 1980. Process of sex differentiation of the gonad and gonoducts of the three spinned stickleback *Gasteroesteus aculeatus* L. Bull Fac Fish Hokkaido Univ, 31: 137–148.

Shinomiya, A.I., N. Shibata, M. Sakaizumi and S. Hamaguchi. 2002. Sex reversal of genetic female (XX) induced by transplantation of XY somatic cells in the medaka *Oryzias latipes*. Int J Dev Biol, 46: 711–717.

Simpson, T.H. 1976. Endocrine aspects of salmonid culture. Proc R Soc Edinburgh, B 75: 241–252.

Singh, M.S. and K.P. Joy. 1998. A comparative study on histochemical distribution of some enzymes related to steroid and glucuornids synthesis in seminal vesicles and testis of the catfish *Clarias batrachus*. Zool Sci, 15: 955–961.

Singh, M.S. and K.P. Joy. 1999. Annual correlative changes in some biochemical contents of seminal vesicles and testis in the catfish *Clarias batrachus* (L.). Indian J Exp Biol, 39: 52–56.

Sismeros, J.A., P.W. Alderks, K. Leon and B. Sniffen. 2009. Morphometric changes associated with the reproductive cycle and behaviour of the intertidal-nesting male plainfin midshipman *Porichthys notatus*. J Fish Biol, 74: 18–36.

Skubic, E., M. Taborsky, J.M. McNamara and A.I. Houston. 2004. When to parasitize? A dynamic optimization model of reproductive strategies in a cooperative breeder. J Theor Biol, 227: 487–501.

Small, B.C. 2004. Effects of dietary cortisol administration on growth and reproductive success of channel catfish. J Fish Biol, 64: 589–596.

So, W.K., H.F. Kwok and W. Ge. 2005. Zebrafish gonadotropins and their receptors. II. Cloning and characterization of zebrafish follicle-stimulating hormone and luteinizing hormone subunits: their spatial-temporal expression patterns and receptor specificity. Biol Reprod, 72: 1382–1396.

Sokolowska-Mikolajczyk, M., M. Socha, P. Szczerbik and P. Epler. 2009. The effects of ghrelin on the *in vitro* spontaneous and sGnRH-A stimulated luteinizing hormone (LH) release from the pituitary cells of common carp (*Cyprinus carpio* L.). Comp Biochem Physiol A Mol Integr Physiol, 153: 386–390.

Solar, I.I., E.M. Donaldson and G.A. Hunter. 1984. Optimization of treatment regimes for controlled sex differentiation and sterilization in wild rainbow trout (*Salmo gairdneri* Richardson) by oral administration of 17α methyltestosterone. Aquaculture, 42: 129–139.

Sorensen, P.W., T.J. Hara, N.E. Stacey and F.W. Goetz. 1988. F prostoglandins function as potent olfactory stimulus that comprise the postovulatory female sex pheromone in goldfish. Biol Reprod, 39: 1039–1050.

Sorensen, P.W., A.P. Scott and R.L. Kihslinger. 2000. How common hormonal metabolites function as relatively specific pheromonal signals in goldfish. Proc Sixth Internatl Symp Reprod Physiol Fish, John Griegs AS Bergen, Norway. pp. 125–128.

Sorensen, P.W., M. Pinillos and A.P. Scott. 2005. Sexually mature male goldfish release large quantities of androstenedione into the water where it functions as a pheromone. Gen Comp Endocrinol, 140: 164–175.

Soto, C.G., J.F. Leatherland and D.L.G. Noakes. 1992. Gonadal histology in hermaphroditic fish *Rivulus marmoratus* (Pisces: Cyprinodontidae). Can J Zool, 70: 2338–2347.

Sousa-Santos, C., M.J. Collares-Pereira and V. Almada. 2007. Fertile triploid males—An uncommon case among hybrid vertebrates. J Exp Zool, 307A: 220–227.

Sower, S.A., Y.C. Chiang, S. Lovas and J.M. Conlon. 1993. Primary structure and biological activity of a third gonadotropin-releasing hormone from lamprey brain. Endocrinology, 132: 1125–1131.

Spadella, M.A., C. Oliveira and I. Quagio-Grassiotto. 2007. Comparative analyses of spermiogenesis and sperm ultrastructure in Callichthyidae (Teleostei: Ostariophysi: Siluriformes). Neotropical Ichthyol, 5: 337–350.

Sprando, R.L. and R.D. Russel. 1987. Germ cell—somatic cell relationships: a comparative study of intracellular functions during spermatogenesis in selected non-mammalian vertebrates. Scanning Micros, 1: 1249–1255.

Sprando, R.L., R.C. Heidinger and R.D. Russel. 1988. Spermiogenesis in the bluegill (*Lepomis macrochirus*): a study of cytoplasmic events including cell volume changes and cytoplasmic elimination. J Morphol, 198: 165–167.

Sreenivasulu, G. and B. Senthilkumaran. 2009a. New evidences for the involvement of 20β-hydroxysteroid dehydrogenase in final oocyte maturation of air-breathing catfish. Gen Comp Endocrinol, 163: 259–269.

Sreenivasulu, G. and B. Senthilkumaran. 2009b. A role for cytochrome p450 17α-hydroxylase/c17-20 lyase during shift in steroidogenesis occurring in ovarian follicles prior to oocyte maturation. J Steroid Biochem Mol Biol, 115: 77–85.

Sreenivasulu, G., B. Senthilkumaran, P. Sridevi et al. 2012. Expression and immunolocalization of 20β-hydroxysteroid dehydrogenase during testicular cycle and after hCG induction *in vivo* in the catfish *Clarias gariepinus.* Gen Comp Endocrinol, 175: 48–54.

Stacey, N. 1976. Effects of indomethacin and prostaglandins on spawning behavior of female goldfish. Prostaglandins, 12: 113–126.

Stacey, N. 2011. Hormonally derived sex pheromones in fishes. 1: 169–192. *In:* D.O. Norris and K.H. Lopez [eds.]. Hormones and Reproduction in Vertebrates: Fishes. Elsevier, Amsterdam, The Netherlands.

Stacey, N.E. and R.E. Peter. 1979. Central action of prostaglandins in spawning behaviour of female goldfish. Physiol Behav, 22: 1191–1996.

Stacey, N.E. and P.W. Sorensen. 2002. Fish hormonal pheromones. 2: 375–435. *In:* D.W. Pfaff, A.P. Arnold, A.M. Etgen, S.E. Fahrbach and R.T. Rubin [eds.]. Fish Physiology: Behaviour and Physiology of Fish. Academic Press, New York, USA.

Stacey, N.E. and P.W. Sorensen. 2006. Reproductive pheromones. 24: 359–412. *In:* K.A. Soloman, R.W. Wilson and S. Bahlshine [eds.]. Fish Physiology: Behaviour and Physiology of Fish. Elsevier, San Diego, USA.

Stacey, N.E. and P.W. Sorensen. 2009. Fish hormonal pheromones. 1: 639–681. *In:* D.W. Pfaff, A.P. Arnold, A.M. Etgen, S.E. Fahrbach and R.T. Rubin [eds.]. Hormones, Brain and Behavior. Elsevier, San Diego, USA.

Stauffer, Jr, J.R. and R.A. Ruffing. 2008. Behaviorally induced sex reversal of *Matriaclima* cf. *livingstoni.* Copeia, 2008: 618–620.

Steinke, D., S. Hoegg, H. Brinkmann and A. Meyer. 2006. Three rounds (1R/2R/3R) of genome duplications and the evolution of the glycolytic pathway in vertebrates. BMC Biol, 4: 16.

St. Mary, C.M. 1998. Characteristic gonad structure in the gobiid genus *Lythrypnus* and comparison with other hermaphrodite gobies. Copeia, 1998: 720–724.

Stocco, D.M. 2001. StAR protein and the regulation of steroid hormone biosynthesis. Annu Rev Physiol, 63: 193–213.

Stocco, D.M., X. Wang, Y. Jo et al. 2005. Multiple signaling pathway regulating steroidogenesis and steroidogenic acute regulatory protein expression: more complicated than we thought. Mol Endocrinol, 19: 2647–2659.

Stratholt, M.L., E.M. Donaldson and N.R. Liley. 1997. Stress induced elevation of plasma cortisol in adult female coho salmon (*Oncorhynchus kisutch*) is reflected in egg cortisol content, but does not appear to affect early development. Aquaculture, 158: 141–153.

Strum, A., N. Bury, L. Dengreville et al. 2005. 11-deoxycorticosterone is a potent agonist of the rainbow trout (*Oncorhynchus mykiss*) mineral corticoid receptor. Endocrinology, 146: 47–55.

Strussmann, C.A. and R. Patino. 1999. Sex determination, environmental role. pp. 402–409. *In:* E. Knobil and J.D. Neil [eds.]. Excyclopaedia of Reproduction. Academic Press, San Diego, USA.

Strussmann, C.A. and M. Nakamura. 2002. Morphology, endocrinology and environmental modulation of gonadal sex differentiation in teleost fishes. Fish Physiol Biochem, 26: 13–29.

Strussmann, C.A., J.C. Calsina Cota, G. Phonlor et al. 1996. Temperature effects on sex differentiation of two South American athernids *Odontethes argentinensis* and *Patagonia hatchery.* Env Biol Fish, 47: 143–154.

Sudhakumari, C.C., B. Senthilkumaran, K. Raghuveer et al. 2010. Dimorphic expression of tryptophan hydroxylase in the brain of XX and XY Nile tilapia during early development. Gen Comp Endocrinol, 166: 320–329.

Sunobe, T., M. Nakamura, Y. Kobayashi et al. 2005. Aromatase immunoreactivity and the role of enzyme in the steroid pathways of including sex change in the hermaphrodite gobiid fish *Trimma okinawae.* Comp Biochem Physiol A. Mol Integr Physiol, 141: 54–59.

Suzuki, A., M. Tanaka and N. Shibata. 2004. Expression of aromatase mRNA and effects of aromatase inhibitor during ovarian development in the medaka, *Oryzias latipes.* J Exp Zool, 301A: 266–273.

Swanson, P., K. Suzuki, H. Kawauchi and W.W. Dickhoff. 1991. Isolation and characterization of two coho salmon gonadotropins, GtH I and GtH II. Biol Reprod, 44: 29–38.

Swanson, P., J.T. Dickey and B. Campbell. 2003. Biochemistry and physiology of fish gonadotropins. Fish Physiol Biochem, 28: 53–59.

Swapna, I. and B. Senthilkumaran. 2009. Influence of ethynylestradiol and methyltestosterone on the hypothalmo-hypophyseal gonadal axis of adult air-breathing catfish. *Clarias gariepinus.* Aquatic Toxicol, 95: 222–229.

Swapna, I., M. Rajasekhar, A. Supriya et al. 2006. Thiurea-induced thyroid hormone depletion impairs testicular recrudescence in the air-breathing catfish *Clarias gariepinus.* Comp Biochem Physiol, 144A: 1–10.

Taborsky, B. 2006. Mothers determine offspring size in response to own juvenile growth conditions. Biol Lett, 2: 225–228.

Taborsky, B., E. Skubic and R. Bruintjes. 2007. Mothers adjust egg size to helper number in a cooperatively breeding cichlid. Behav Ecol, 18: 652–657.

Taborsky, M. 1994. Sneakers, satellites and helpers. Parasitic and cooperative behavior in fish reproduction. Adv Study Behav, 23: 1–100.

Taborsky, M. 1997. Bourgeois and parasitic tactics: do we need collective functional forms for alternative reproductive behaviours? Behav Ecol Sociobiol, 41: 361–362.

Taborsky, M. 1998. Sperm competition in fish bourgeois males and parasitic spawning. Trends Ecol Evol, 13: 222–227.

Taborsky, M. 2001. The evolution of bourgeois parasitic and cooperative reproductive behaviours in fishes. J Hered, 92: 100–110.

Takagi, M., K. Sakai and N. Taniguchi. 2008. Direct evidence of multiple paternities in natural population of viviparous Japanese surfperch by allelic markers of microsatellite DNA loci. Fish Sci, 74: 976–982.

Takamoto, G., S. Seki, Y. Nagashima et al. 2003. Protogynous sex change in the haremic triggerfish *Sufflamen chrysopterus* (Tetradontiformes). Ichthyol Res, 50: 281–283.

Takahashi, H. 1975. Functional feminization of genetic male of the guppy *Poecilia reticulata* treated with estrogen after birth. Bull Fac Fish Hokkaido Uni, 26: 223–234.

Takeuchi, Y., G. Yoshizaki, T. Kobayashi and T. Takeuchi. 2004. Surrogate-broodstock produces salmonids. Nature, 430: 629–630.

Tamate, T. and K. Maekawa. 2000. Intrapopulation variation in reproductive traits of female masu salmon *Oncorhynchus masou.* Oikos, 90: 209–218.

Tan-Fermin, J.D. 1992. Withdrawal of exogenous 17 alpha methyltestosterone causes reversal of sex-inversed male grouper *Epinephelus suillus* (Bloch and Schneider). Philippine Sci, 29: 33–39.

Tan-Fermin, J.D., L.M.B. Garcia A.R. Castillo, Jr. 1994. Induction of sex inversion in juvenile grouper *Epinephelus suillus* (Valenciennes) by injections of 17α-methyltestosterone. Jap J Ichthyol, 40: 413–420.

Tanaka, H., K. Hirose, K.-Y. Nogami et al. 1990. Sexual maturation and sex reversal in red spotted grouper *Epinephelus akaara*. Bull Natl Res Inst Aquacult, 17: 1–15.

Tanaka, H., Y. Tsuchihashi and Y. Kuromiya. 1999. Induction of sex reversal in the seven band grouper *Epinephelus septemfasciatus*. Proc Sixth Internatl Symp Reprod Physiol Fish. Bergen, Norway. p. 423.

Tanaka, M., D. Saito, C. Morinaga and H. Kurokawa. 2008. Cross talk between germ cells and gonadal somatic cells is critical for sex differentiation of the gonads in the teleost fish (*Oryzias latipes*). Dev Growth Differ, 50: 273–278.

Tao, Y.-X., H.R. Lin, G. Van der Kraak and R.E. Peter. 1993. Hormonal induction of precocious sex reversal in the rice field eel *Monopterus albus*. Aquaculture, 118: 131–140.

Targonska, K., D. Kucharczyk, R. Kujawa et al. 2010. Controlled reproduction of asp, *Aspius aspius* (L.) using luteinizing hormone releasing hormone (LHRH) analogues with dopamine inhibitors. Aquaculture, 306: 407–410.

Tatarenkov, A., S.M.Q. Lina, D.S. Taylor et al. 2009. Long term retention of self-fertilization in the fish clade. Proc Natl Acad Sci, USA, 106: 14456–14459.

Tena-Sempere, M. 2008. Ghrelin as a pleiotrophic modulator of gonadal function and reproduction. Nat Clin Pract Endocrinol Metab, 4: 666–674.

Teugels, G.G. 1996. Taxonomy, phylogeny and biogeography of catfishes (Ostariophysi: Siluroidei): an overview. Aquat Living Resour, 9: 9–34.

Thangaraj, M. and A.P. Lipton. 2009. Relationship between parental size and reproductive output in *Hippocampus kuda* Bleeker 1852. (personal communication).

Thibault, R.E. and R.J. Schultz. 1978. Reproductive adaptiveness among viviparous fishes (Cyprinodontiformes, Poeciliidae). Evolution, 32: 320–333.

Tinguard-Sequeira, A., F. Chauvigne, M. Fabra et al. 2008. Structural and functional divergence of two fish aquaporin-1 water channels following teleost-specific gene duplication. BMC Evol Biol, 8: 259.

Tiwary, B.K., R. Kirubagaran and A.K. Ray. 1998. Testosterone-mediated complete feminization with precocious ovarian maturation in catfish *Heteropneustes fossilis* (Bloch). Aquacult Res, 29: 333–336.

Tong, S.-K., H.-J. Hsu and B.-C. Chung. 2010. Zebrafish monosex population reveals female dominance in sex determination and earliest events of gonad differentiation. Dev Biol, 344: 849–859.

Trant, J.M., P. Thomas and C.H. Shackleton. 1986. Identification of 17 alpha 20 beta 21 trihydroxy-4-pregnen-3-one as the major ovarian steroid produced by the teleost *Micropogonias undulatus* during final oocyte maturation. Steroids, 47: 89–99.

Trant, J.M., S. Gavasso, J. Ackers et al. 2001. Developmental expression of cytochrome p450 aromatase gene (*cyp19a* and *cyp19b*) in zebrafish fry (*Danio rerio*). J Exp Zool, 290: 475–783.

Trexler, J.C. 1997. Resource availability and plasticity in offspring provisioning: embryo nourishment in sailfin mollies. Ecology, 78: 1370–1381.

Trexler, J.C. and D.L. DeAngelis. 2003. Resource allocation in offspring provisioning: An evaluation of the conditions favoring the evolution of matrotrophy. Am Nat, 162: 574–585.

Tsai, P.S. 2006. Gonadotropin-releasing hormone in invertebrates: structure, function, and evolution. Gen Comp Endocrinol, 148: 48–53.

Turner, C.L. 1947. The rate of morphogenesis and regeneration of the gonopodium in normal and castrated males of *Gambusia affinis*. J Exp Zool, 106: 125–143.

Twan, W.H., J.S. Hwang, Y.H. Lee et al. 2006. The presence and ancestral role of gonadotropin-releasing hormone in the reproduction of scleractinian coral, *Euphyllia ancora*. Endocrinology, 147: 397–406.

Tyler, C.R. and J.P. Sumpter. 1996. Oocyte growth and development in teleosts. Rev Fish Biol Fish, 6: 287–318.

Tyler, C., J. Sumpter, H. Kawuchi and P. Swanson. 1991. Involvement of gonadotropins I and II in the uptake of vitellogenin into vitellogenic oocytes of rainbow trout *Oncorhynchus mykiss*. Gen Comp Endocrinol, 84: 291–299.

Tyler, C.R., T.G. Pottinger, K. Coward et al. 1997. Salmonid follicle-stimulating hormone (GtHI) mediates vitellogenic development of oocytes in the rainbow trout *Oncorhynchus mykiss*. Biol Reprod, 57: 1238–1244.

Uchida, D., M. Yamashita, T. Kitano and T. Iguchi. 2002. Oocyte apoptosis during the transition from ovary-like tissue to testes during sex differentiation of juvenile zebrafish. J Exp Biol, 205: 711–718.

Uglem, I., G. Rosenqvist and H. Wassalavik. 2000. Phenotypic variation between dimorphic males in corkwing wrasse (*Symphodus melops* L.). J Fish Biol, 57: 1–14.

Uglem, I., I. Mayer and G. Rosenqvist. 2002. Variation in plasma steroids and reproductive traits in dimorphic males of corkwing wrasse (*Symphodus melops* L.). Horm Behav, 41: 396–404.

Umino, T., K. Arai, K. Maeda et al. 1997. Natural clones detected by multilocus DNA fingerprinting in gynogenetic triploid ginbuna *Carassius auratus langsdorfi*: in Kurose River, Hiroshima. Fish Sci, 63: 147–148.

Underwood, J.L., R.S.I.I. Hestand and B.J. Thompson. 1986. Gonadal regeneration in grass carp following bilateral gonadectomy. Prog Fish Cult, 48: 54–56.

Unniappan, S. 2010. Ghrelin: An emerging player in the regulation of reproduction in non-mammalian vertebrates. Gen Comp Endocrinol, 167: 340–343.

Unniappan, S. and R.E. Peter. 2004. *In vitro* and *in vivo* effects of ghrelin on luteinizing hormone and growth hormone release in goldfish. Am J Physiol Regul Integr Comp Physiol, 286: R1093–R1101.

Urbatzka, R., M.J. Rocha and E. Rocha. 2011. Regulation of ovarian development and function in teleosts.1: 65–82. *In:* D.O. Norris and K.H. Lopez [eds.]. Hormones and Reproduction of Vertebrates: Fishes. Elsevier, Amsterdam, The Netherlands.

Uribe, M.C. and H.J. Grier 2011. Oogenesis of microlecithal oocytes in the viviparous teleost *Heterandria formosa*. J Morphol. 272: 241–57. doi: 10.1002/jmor.10912.

van Aerle, R., P. Kille, A. Lange and C.R. Tyler. 2008. Evidence for the existence of a functional Kiss/Kiss 1 receptor pathway in fish. Peptides, 29: 57–64.

van den Hurk, R., J.G.D. Lambert and J. Peute. 1982. Steroidogenesis in the gonads of rainbow trout fry (*Salmo gairdneri*) before and after the onset of gonadal sex differentiation. Reprod Nutr Dev, 22: 413–425.

van den Hurk, R., C.J.J. Richter and J.J. Dommerholt. 1989. Effects of 17α methyltestosterone and 11β-hydroxy androstenodione on gonad differentiation in the African catfish *Clarias gariepinus*. Aquaculture, 83: 179–191.

Varadaraj, K. and T.J. Pandian. 1987. Masculinization of *Oreochromis mossambicus* by administration of 17α methyl-5-androsten-3β-17β- diol through rearing water. Curr Sci, 56: 412–413.

Varadaraj, K. and T.J. Pandian. 1989. First report on production of supermale tilapia by integrating endocrine sex reversal with gynogenetic technic. Curr Sci, 58: 434–441.

Varadaraj, K. and T.J. Pandian. 1991. Effect of solubulizing 17α-ethylnyltestosterone in three different solvents on sex reversal of Mozambique tilapia. Prog Fish Cult, 53: 67–71.

Varadaraj, K., S. Sindhukumari and T.J. Pandian. 1994. Comparison of conditions for hormonal sex reversal of Mozambique tilapia. Prog Fish Cult, 56: 81–83.

Vasil'yev, V.I., E.D. Vasil'eva and A.G. Osinov. 1990. The problem of reticulate species formation of vertebrates of the diploid-triploid-tetraploid complex in the genus *Cobitis* (Gobitidae). 4. Tetraploid form. Veprosy Ikiol, 30: 908–919.

Vaz Serrano, J., I. Folsted, G. Rudolfsen and L. Figenschow. 2006. Do the fastest sperm within an ejaculate swim faster in subordinate than dominant males of Arctic charr? Can J Zool, 84: 1019–1024.

Verma, D.K., P. Routray, C. Dash et al. 2009. Physical and biochemical characteristics of semen and ultrastructure of spermatozoa in six carp species. Turkish J Fish Aquat Sci, 9: 67–76.

Vermeirssen, E.L.M., R.J. Shields, C. Mazorra-de-Quero and A.P. Scott. 2000. Gonadotropin-releasing hormone agonist raises plasma concentration of progestogen and enhances milt quality in male Atlantic halibut (*Hippoglossius hippoglossus*). Fish Physiol Biochem, 22: 77–87.

Vilela, D.A.R., S.G.B. Silva, M.T.D. Peixoto et al. 2003. Spermatogenesis in teleost: insights from Nile tilapia (*Oreochromis niloticus*) model. Fish Physiol Biochem, 28: 187–150.

Villeneuve, D.L., P. Larkin, I. Knoebl et al. 2007. A graphical systems model to fractionate hypothesis-driven ecotoxicogenesis research on the teleosts brain-pituitary-gonadal axis. Environ Sci Technol, 41: 321–330.

Vinas, J. and F. Piferrer. 2008. Stage-specific gene expression during fish spermatogenesis in teleost: insights from the Nile tilapia (*Oreochromis niloticus*) model. Biol Reprod, 79: 738–747.

Vischer, H.F. and J. Bogerd. 2003. Cloning and functional characterization of a testicular TSH receptor cDNA from the African catfish (*Clarias gariepinus*). J Mol Endocrinol, 30: 227–238.

Vladic, T. and T. Jarvi. 2001. Sperm quality in alternative reproductive tactics of Atlantic salmon: the importance of loaded raffle. Proc R Soc Lond B, 268: 2375–2381.

Vladic, T., L.A. Forsberg and T. Jarvi. 2010. Sperm competition between alternative reproductive tactics of the Atlantic salmon *in vitro*. Aquaculture, 302: 265–269.

Volf, J.N., M. Kondo and M. Schartl. 2003. Medaka dmY/dmrt1aY is not the universal primary sex determining gene in fish. Trends Genet, 19: 196–199.

Volkoff, H. and R.E. Peter. 1999. Actions of two forms of gonadotropin releasing hormone and a GnRH antagonist on spawning behaviour of the goldfish *Carassius auratus*. Gen Comp Endocrinol, 116: 347–355.

Vrijenhoek, R.C. and R.J. Schultz. 1974. Evolution of a trihybrid unisexual fish (*Poeciliopsis*, Poeciliidae). Evolution, 28: 305–319.

Wallace, B.M.N. and H. Wallace. 2003. Synaptonemal complex karyotype of zebrafish. Heredity, 90: 136–140.

Wang, D.S., T. Kobayashi, Y. Zhou et al. 2007. Foxl2 up-regulates aromatase gene transcripts in a female specific manner by binding to the promoter as well as interacting with Ad4 binding protein/steroidogenic factor 1. Mol Endocrinol, 21: 712–725.

Wang, H.P., Z. Gao, B. Beres et al. 2008. Effects of estradiol 17β on survival growth performance, sex reversal and gonadal structure of bluegill sunfish *Lepomis macrochirus*. Aquaculture, 285: 216–223.

Wang, X.G. and L. Orban. 2007. Anti-Mullerian hormone and 11β-hydroxylase show reciprocal expression to that of aromatase in the transforming gonad of zebrafish males. Dev Dyn, 236: 1329–1338.

Wang, X.G., R. Bartfai, I. Sleptosova-Feidrich and L. Orban. 2007. The timing and extend of 'juvenile ovary' phase are highly variable during zebrafish testis differentiation. J Fish Biol, 70A: 33–44.

Wang, Y., L. Zhou, B. Yao et al. 2004. Differential expression of thyroid-stimulating hormone beta sub unit in gonads during sex reversal orange-spotted and red-spotted groupers. Mol Cell Endocrinol, 220: 77–88.

Warner, R.R. 1978. The Evolution of Hermaphroditism and Unisexuality in Aquatic and Terrestrial Vertebrates. *In:* E.S. Reese and F.J. Lighter [eds.]. Contrasts in Behaviour. John Wiley & Sons, New York, USA.

Warner, R.R. and D.R. Robertson. 1978. Sexual patterns in the labroid fishes of the western Carribean. 1. The wrasses (Labridae). Smithsonian Contributions to Zoology, No 254.

Warner, R.R. and S.E. Swearer. 1991. Social control of sex change in the bluehead wrasse *Thalassoma bifasciatum* (Pisces: Labridae), Biol Bull, 181: 199–204.

Weaver, I.C.G., N. Cervoni, F.A. Champagne et al. 2004. Epigenetic programming by maternal behaviour. Nat Neurosci, 7: 847–854.

Wei, Q., F. Ke and J. Zhang. 1997. Biology of fisheries and conservation of sturgeons and paddlefish in China. Env Biol Fish, 48: 241–255.

Weltzien, F.-A., C. Pasqualini, M.E. Serbert et al. 2006. Androgen-dependent stimulation of brain dopaminergic systems in the female European eel (*Anguilla anguilla*). Endocrinology, 147: 2964–2973.

Wendelaar Bonga, S.E. 1997. The stress response in fish. Physiol Rev, 77: 591–625.

White, S.A., T. Nquyan and R.D. Fernald. 2002. Social regulation of gonadotropin releasing hormone. J Exp Biol, 205: 2567–2581.

Wilson, A.B., I. Ahnesjo, A.C.J. Vincent and A. Meyer. 2003. The dynamics of male brooding, mating patterns, and sex roles in pipefishes and seahorses (Family, Syngnathidae). Evolution, 57: 1374–1386.

Winberg, S., G.E. Nilson and K.H. Olsen. 1992. Changes in serotonergic activity during hierarchial behavior in Arctic charr (*Salvelinus alpinus* L.) are socially induced. J Comp Physiol A, Sec Neural Behav Physiol, 170: 93–99.

Winberg, S., C.G. Carter, J.D. McCarthy et al. 1993. Feeding rank and brain serotonergic activity in rainbow *Oncorhynchus mykiss*. J Exp Biol, 179: 197–211.

Winberg, S., Y. Winberg and R.D. Fernald. 1997. Effect of social rank on brain monoaminergic activity in a cichlid fish. Brain Behav Evol, 49: 230–236.

Winemiller, K.O. and K.A. Rose. 1992. Patterns of life-history diversification in North American fishes: implications for population regulation. Can J Fish Aquat Sci, 49: 2196–2218.

Wingfield, J.C. 1984a. Environmental and endocrine control of reproduction in the song sparrow *Melospiza melodia*. I. Temporal organization of the breeding cycle. Gen Comp Endocrinol, 56: 406–416.

Wingfield, J.C. 1984b. Environmental and endocrine control of reproduction in the song sparrow *Melospiza melodia*. II. Agonistic interactions as environmental information stimulating secretion of testosterone. Gen Comp Endocrinol, 56: 417–424.

Wingfield, J.C., R.E. Hegner, A.M. Dufty and G.F. Ball. 1990. The 'challenge hypothesis': Theoretical implication for patterns of testosterone secretion, mating systems and breeding strategies. Am Nat, 136: 829–846.

Wong, T.T., S. Ijiri and Y. Zohar. 2006. Molecular biology of ovarian aromatase in sex reversal: Complementary DNA and 5′ flanking region isolation and differential expression of ovarian aromatase in the gilthead sea bream (*Sparus aurata*). Biol Reprod, 74: 857–864.

Wong, T.T., T. Saito, J. Crodian et al. 2010. Zebrafish germline chimeras produced by transplantation of ovarian germ cells into sterile host larvae. Biol Reprod, 84: 1190–1197.

Wourms, J. 2005. Functional morphology, development, and evolution of trophotaenia. pp. 525–569. *In:* H.J. Grier and M.C. Uribe [eds.]. Viviparous Fishes. New Life Publications, Homestead, Florida, USA.

Wu, G.C., S. Tomy and C.F. Chang. 2008a. The nrob1 and nr504 during gonadal development and sex change in protandrous black porgy fish. Biol Reprod, 78: 200–210.

Wu, G.C., S. Tomy, M. Nakamura and C.F. Chang. 2008b. Dual roles of *cyp19a1a* in gonadal sex differentiation and sex change in protandrous black porgy *Acanthopagrus schlegeli*. Biol Reprod, 78: 200–210.

Wu, G.C., S. Tomy, M.F. Lee et al. 2010. Sex differentiation and sex change in the protandrous black porgy *Acanthopagrus schlegeli*. Gen Comp Endocrinol, 167: 417–421.

Wu, Q., Y. Ye and D. Xinnhong. 2003. Two unisexual artificial polyploid clones constructed by genome addition of common carp (*Cyprinus carpio*) and crucian carp (*Carassius auratus*). Sci China 46C: 595–604.

Wuertz, S., J. Gessner, F. Kirschbaum and W. Kloas. 2007. Expression of IGF-1 receptor in male and female starlet, *Acipenser ruthenus*: evidence for an important role in gonad maturation. Comp Biochem Physiol, A147: 223–230.

Yamaha, E., M. Kazama-Wakabayashi, S. Otani et al. 2001. Germ line chimera by lower-part blastoderm transplantation between diploid goldfish and triploid crucian carp. Genetica, 111: 227–236.

Yamaha, E., M. Murakami, K. Hada et al. 2003. Recovery of fertility in male hybrids of a cross between goldfish and common carp by transplantation of PGC (primordial germ cell) containing graft. Genetica, 119: 121–131.

Yamamoto, T. 1953. Artificially induced sex-reversal in genotypic males of the medaka (*Oryzias latipes*). J Exp Zool, 123: 571–594.

Yamamoto, T. 1955. Progeny of artificially induced sex reversal of male genotype (XY) in the medaka (*Oryzias latipes*) with special reference to YY male. Genetics, 40: 406–416.

Yamamoto, T. 1958. Artificial induction of functional sex reversal in genotypic females of the medaka (*Oryzias latipes*). J Exp Zool, 137: 227–263.

Yamamoto, T. 1969. Sex differentiation. 3: 117–175. *In:* W.S. Hoar and D.J. Randall [eds.]. Fish Physiology. Academic Press, New York, USA.

Yamamoto, T. 1975. A YY male goldfish from mating estrone-induced XY female and normal male. J Hered, 66: 2–4.

Yamamoto, Y. and G. Yoshizaki. 2008. Heterologous gap junctions between granulose cells and oocytes in ayu (*Plecoglossus ativelis*): formation and role during luteinizing hormone-dependent acquisition of oocyte maturational competence. J Reprod Dev, 54: 1–5.

Yamazaki, F. 1976. Application of methylestroterone on the skin and the gonads of salmonids. Gen Comp Endocrinol, Suppl, 3: 741–750.

Yamazaki, F. 1983. Sex control and manipulation in fish. Aquaculture, 33: 329–354.

Yao, B., L. Zhou, Y. Wang et al. 2007. Differential expression and dynamic changes of Sox3 during gametogenesis and sex reversal in protogynous hermaphroditic fish. J Exp Zool, 307A: 207–219.

Yao, H.H. 2005. The pathway to femaleness: current knowledge on embryonic development of the ovary. Mol Cell Endocirnol, 230: 87–93.

Yeh, S.L., C.M. Kuo, Y.Y. Ting and C.F. Chang. 2003a. Androgens stimulate sex change in protogynous grouper *Epinephelus coioides*: spawning performance in sex changed males. Comp Biochem Physiol, 135C: 375–382.

Yeh, S.L., Q.C. Dai, Y.T. Chu et al. 2003b. Induced sex change, spawning and larviculture of potato grouper *Epinephelus tukula*. Aquaculture, 228: 371–381.

Yeoh, C.G., C.B. Schreck, G.W. Feist and M.S. Fitzpatrick. 1996a. Endogenous steroid metabolism is indicated by fluctuations of endogenous steroid and steroid glucoronide levels in early development of the steelhead trout (*Oncorhynchus mykiss*). Gen Comp Endocrinol, 103: 107–114.

Yeoh, C.G., C.B. Schreck, M.S. Fitzpatrick and G.W. Feist. 1996b. *In vitro* steroid metabolism in embryonic and newly hatched steelhead trout (*Oncorhynchus mykiss*). Gen Comp Endocrinol, 102: 197–209.

Yoo, M.S., H.M. Kang, H.S. Choi et al. 2000. Molecular cloning, distribution and pharmacological characterization of a novel gonadotropin-releasing hormone ([Trp8] GnRH) in frog brain. Mol Cell Endocrinol, 164: 197–204.

Yoshizaki, G., M. Ichigawa, M. Hayashi et al. 2010. Sexual plasticity of ovarian germ cells in rainbow trout. Development, 137: 1227–1230.

Young, G., D.M. Lokman, M. Kusakabe et al. 2005. Gonadal steroidogenesis in teleost fish. pp. 155–223. *In:* P. Melamed and N. Sherwood [eds.]. Hormones and their Receptors in Fish Reproduction. World Scientific Press, Hackensack, NJ, USA.

Yufera, M., S. Halm and S. Beltran. 2012. Transcriptomic characterization of the larval stage in gilthead seabream (*Sparus aurata*) by 454 pyrosequencing. Mar biotechnol, 14: 423–435.

Zakes, K.D. and Z. Zakes. 1999. The effect of 11β-hydroxy androstenedione on pikeperch *Stizostediona luciperca* (L.). Aquacult Res, 30: 731–735.

Zarkower, D. 2006. Somatic sex determination. *In:* Worm Book. The *C. elegans* Research Community. Worm Book, doi/10.1895/wormbook. 184.1.

Zhang, D., H. Xiong, J.A. Mennigen et al. 2009. Defining global neuroendocrine gene expression patterns associated with reproductive seasonality in fish. PLoS ONE, 4: e5816.

Zhang, F., T. Oshiro and F. Takashima. 1992a. Fertility of triploid backcross progeny (Gengoroubuna *Carassius auratus cuvieri* ♀ x carp *Cyprinus carpio* ♂) F_1 x carp or genogoroubuna ♂. Jap J Ichthyol, 39: 229–233.

Zhang, F., T. Oshiro and F. Takashima. 1992b. Chromosome synopsis and recombination during meiotic division of gynogenetic triploid ginbuna *Carassius auratus langsdorfi*. Jap J Ichthyol, 39: 151–159.

Zhang, Q., K. Arai and S. Yamashita. 1998. Cytogenetic mechanisms for triploid and haploid egg formation in the triploid loach *Misgurnus anguillicaudatus*. J Exp Zool, 281: 608–619.

Zhang, Q., K. Hanada and K. Arai. 2002. Aberrant meiosis in diploid and triploid progeny of gynogenetic diploids produced from eggs of natural tetraploid loach *Misgurnus anguillicaudatus*. Folia Zool, 51: 165–176.

Zhang, W.-L., L.-Y. Zhou, B. Senthilkumaran et al. 2010. Molecular cloning of two isoforms of 11β-hydroxylase and their expressions in the Nile tilapia *Oreochromis niloticus*. Gen Comp Endocrinol, 165: 34–41.

Zhang, Z., Y. Wang, S. Wang et al. 2011. Transcriptome analysis of female and male *Xiphophorus maculatus* Jp 163A. PLoS ONE, 6: e18379. doi: 10.1371/journal. pone. 0018379.

Zhao, J., P. Mak, A. Tchoudakova et al. 2001. Different catalytic properties and inhibitor responses of the goldfish brain and ovary aromatase isozymes. Gen Comp Endocrinol, 123: 180–191.

Zhou, L.Y., D.S. Wang, Y. Shibata et al. 2007. Characterization expression and transcriptional regulation of P450c17-I and -II in the medaka, *Oryzias latipes*. Biochem Biophys Res Commun, 362: 619–625.

Zimmerer, E.J. and K.D. Kallmann. 1989. The genetic basis for alternative reproductive tactics in the pygmy swordtail, *Xiphophorus nigrensis*. Evolution, 43: 1298–1307.

Zohar, Y. and C.C. Mylonas. 2001. Endocrine manipulations of spawning in cultured fish: from hormones to genes. Aquaculture, 197: 99–136.

Zohar, Y., J.A. Munoz-Cueto, A. Elizur and O. Kah. 2010. Neuroendocrinology of reproduction in teleost fish. Gen Comp Endocrinol, 165: 438–455.

Author Index

C

D

P

Q

R

Species Index

O

P

R

Subject Index

Color Plate Section

Chapter 2

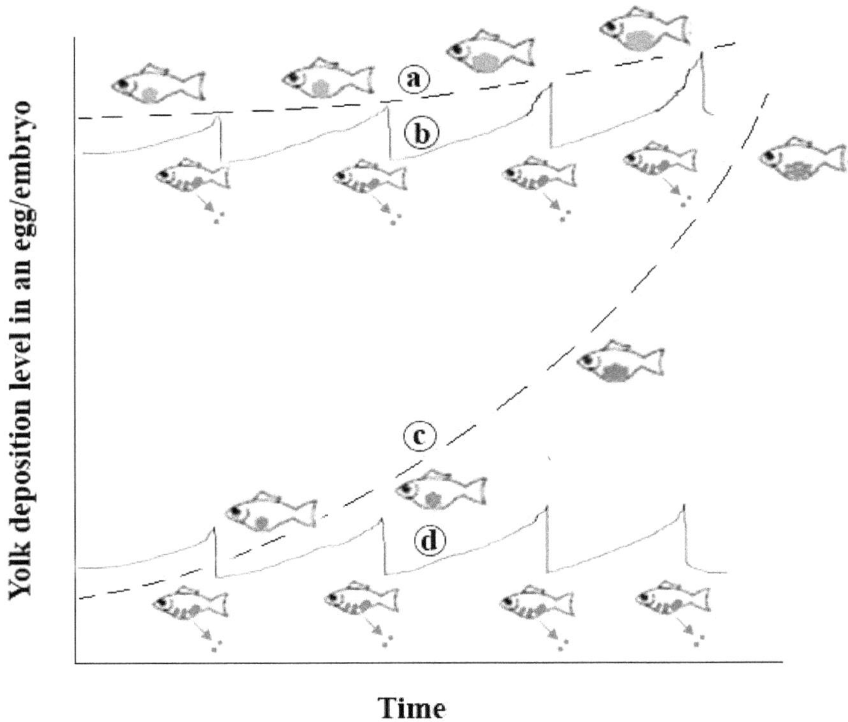

Fig. 9. Suggested effects of superfertation in oviparous/viviparous lecithotrophic and viviparous matrotrophic fishes. Note the relationship in the levels of reproductive allocation for yolk deposition. Also note water uptake (- - -) in non-superfetatious matrotrophics producing micro/mesolecithal eggs. Incidentally, the hydration, i.e., water uptake process necessitates in lecithotrophic oviparous pelagic spawners to repeatedly spawn during spawning season in an analogous similarity to matrotrophy with superfetation. (a) Lecithotrophy without superfetation, (b) Lecithotrophy with superfetation/hydration, (c) Matrotrophy without superfetation, (d) Matrotrophy with superfetation (compiled from Pollux et al., 2009 and others).

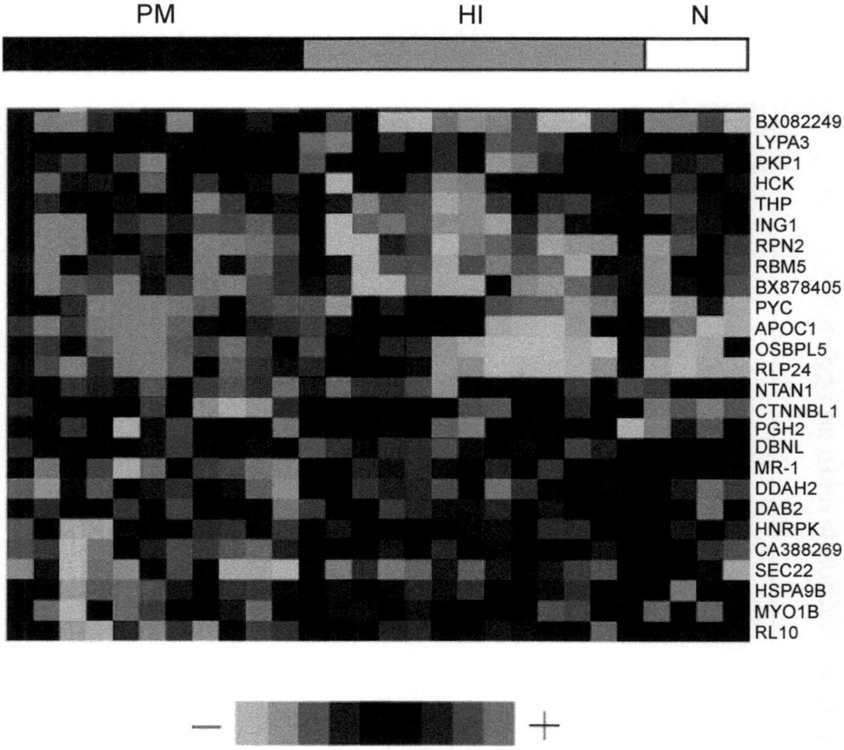

Fig. 12. Mean linkage clustering analysis of the 26 differentially abundant genes in eggs collected after photoperiod-manipulated ovulation (PM), hormonally-induced ovulation (HI) and natural ovulation (N). Each row represents a gene and each column represents an egg RNA sample. For each gene, the expression level within the sample set is indicated using a color intensity scale. Red and green are used for over and under abundance respectively, while black is used for median abundance (from Bonnet et al., 2007).

Chapter 5—Fig. 30 (Facing page). Simplified illustration of regulatory neurological pathways of aggression common to mammals and fishes (highlighted in red color). 5-HIAA = 5-hydroindoleacetic acid, 5-HT = 5-hydroxytryptamine (serotonin), 5-HTP = 5-hydroxytryptophan, AANAT = arylalkylamine N-acetyltransferase, ACTH = adrenocorticotropic hormone, ALD-DH = acetaldehyde dehydrogenase, AR = androgen receptor, AVT = arginine vasotocin, COMT = catechol-O-methyl transferase, CRH = corticotropin releasing hormone, DAO = diamine oxidase, DAT = dopamine transporter, DOPA = dihydroxyphenylalanine, ER = estrogen receptor, FSH = follicle-stimulating hormone, GH = growth hormone, GHRH = growth hormone releasing hormone, GnRH = gonadotropin releasing hormone, GR = glucocorticoid receptor, HIOMT = hydroxyindole-O-methyltransferase, HMT = histamine-N-methyltransferase, HNS = hypothalamo-neurohypophysial system, HPI = hypothalamo-pituitary-interrenal, HPG = hypothalamo pituitary-gonadal, IL1β = interleukin 1β, LH = luteinizing hormone, MAO = monoamine oxidase, MR = mineralocorticoid receptor, NOS = nitric oxide synthase, NPY = neuropeptide Y, SERT, 5-HT transporter, TSH = thyrotropin-stimulating hormone. Plus (+) and negative (–) symbols indicate the proposed action of a gene/neurotransmitter/pathway on aggression (i.e., stimulatory or inhibitory, respectively) (from Filby et al., 2010).

Chapter 5

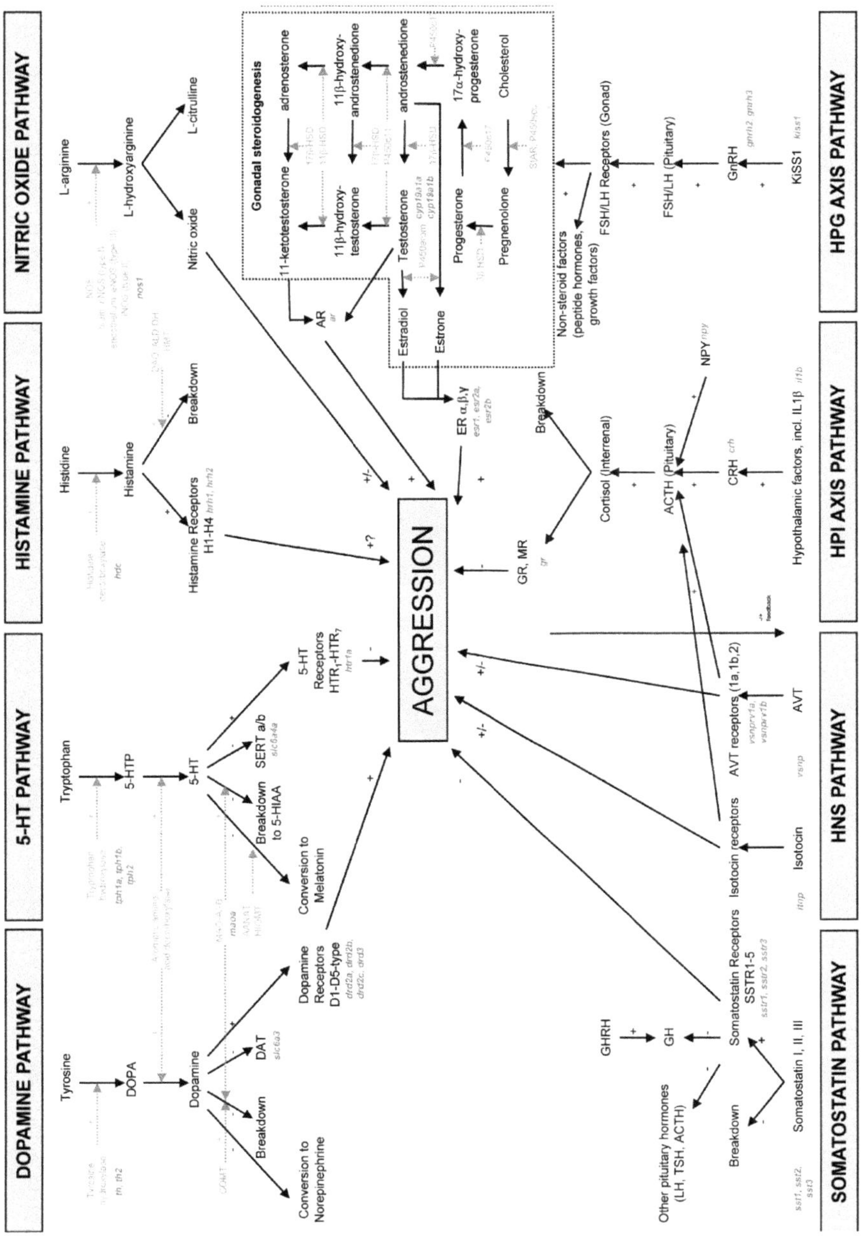

DOPAMINE PATHWAY
5-HT PATHWAY
HISTAMINE PATHWAY
NITRIC OXIDE PATHWAY
SOMATOSTATIN PATHWAY
HNS PATHWAY
HPI AXIS PATHWAY
HPG AXIS PATHWAY
AGGRESSION
Tyrosine
DOPA
Dopamine
Breakdown
DAT
Conversion to Norepinephrine
Dopamine Receptors D1-D5-type
Tryptophan
5-HTP
5-HT
SERT a/b
Breakdown to 5-HIAA
Conversion to Melatonin
5-HT Receptors
Histidine
Histamine
Histamine Receptors H1-H4
L-arginine
L-hydroxyarginine
L-citrulline
Nitric oxide
Gonadal steroidogenesis
11-ketotestosterone
adrenosterone
11β-hydroxy-testosterone
11β-hydroxy-androstenedione
Testosterone
androstenedione
Progesterone
17α-hydroxy-progesterone
Pregnenolone
Cholesterol
AR
Estradiol
Estrone
ER α,β,γ
Non-steroid factors (peptide hormones, growth factors)
FSH/LH Receptors (Gonad)
FSH/LH (Pituitary)
GnRH
KiSS1
NPY
Cortisol (Interrenal)
ACTH (Pituitary)
CRH
Hypothalamic factors, incl. IL1β
GR, MR
AVT receptors (1a,1b,2)
AVT
Isotocin receptors
Isotocin
Somatostatin Receptors SSTR1-5
Somatostatin I, II, III
GHRH
GH
Other pituitary hormones (LH, TSH, ACTH)

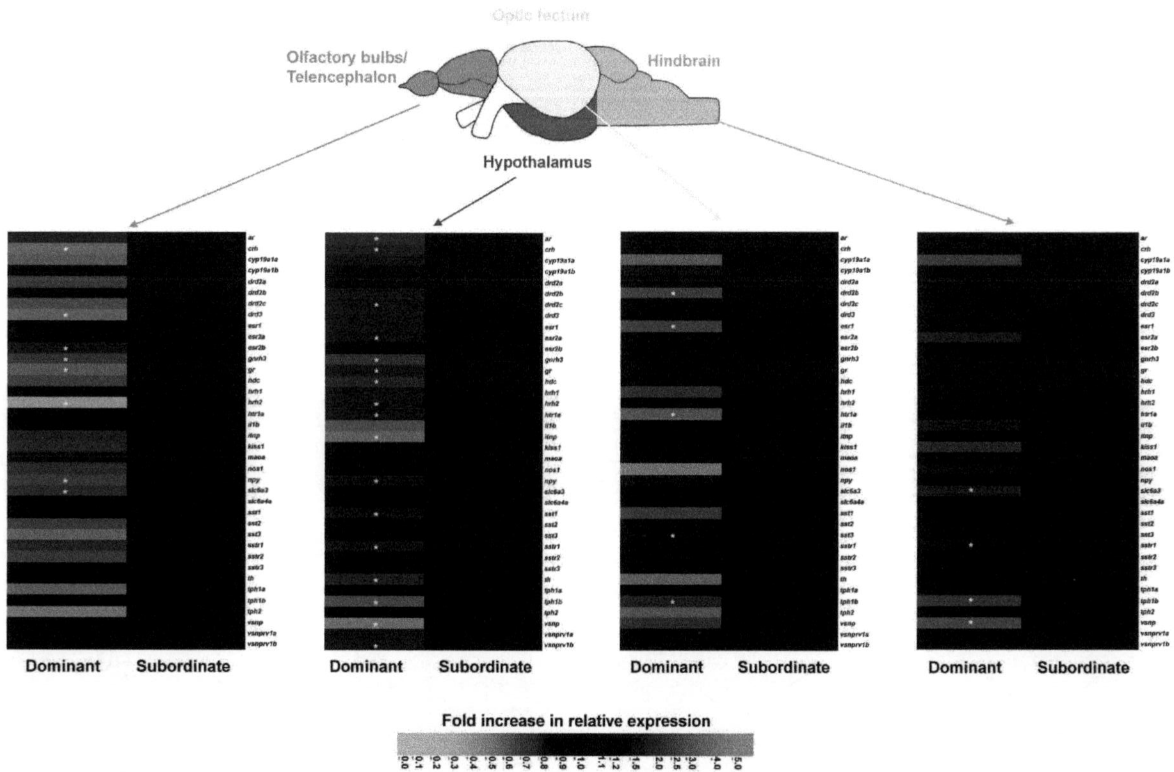

Fig. 31. Regions of the brain associated with differences in aggressiveness in *Danio rerio*. Heat maps show the consistently expressed genes in four regions of the brain in dominant and subordinate males. Genes that are differentially expressed, are denoted with an asterix (from Filby et al., 2010).

Chapter 6

Fig. 40. Left upper: *Amphiprion sebae* in its host *Stichodactyla haddoni*, Right upper: *Premnas biaculeatus* in its host *S. haddoni*, Left lower: *Amphiprion percula* in its host *S. haddoni*, Right lower: *Amphiprion ocellaris* in its artificial sea anemone (curtsey, TT Ajithkumar and T Balasubramanian).

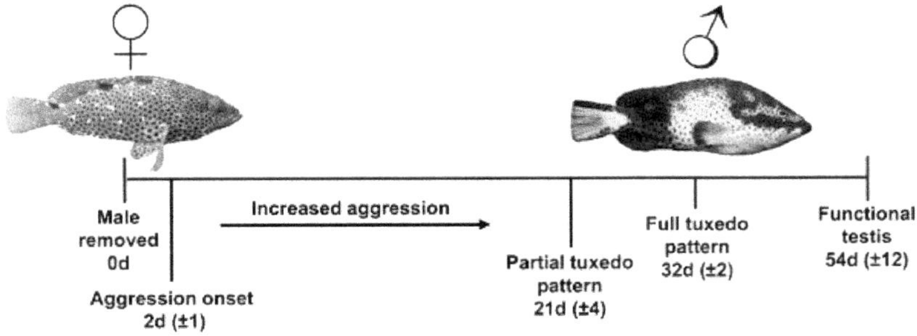

Fig. 41. Chronological sequence of sex change in rock hind *Epinephelus adscensionis* (from Kline et al., 2011).

For Product Safety Concerns and Information please contact our EU
representative GPSR@taylorandfrancis.com
Taylor & Francis Verlag GmbH, Kaufingerstraße 24, 80331 München, Germany

www.ingramcontent.com/pod-product-compliance
Ingram Content Group UK Ltd.
Pitfield, Milton Keynes, MK11 3LW, UK
UKHW022218190926
475702UK00003B/418
* 9 7 8 1 4 6 6 5 7 5 6 0 8 *